Modelling Survival Data
in Medical Research

TEXTS IN STATISTICAL SCIENCE

Editors:

Dr Chris Chatfield, Reader in Statistics, School of Mathematical Sciences, University of Bath, UK

Professor Jim V. Zidek, Department of Statistics, University of British Columbia, Canada

OTHER TITLES IN THE SERIES INCLUDE

Practical Statistics for Medical Research
D. G. Altman

The Analysis of Time Series
C. Chatfield

Problem Solving – A statisticians guide
C. Chatfield

Statistics for Technology
C. Chatfield

Introduction to Multivariate Analysis
C. Chatfield and A. J. Collins

Statistical Analysis of Reliability Data
M.J. Crowder, A.C. Kimber, T.J. Sweeting and R.L. Smith

An Introduction to Generalized Linear Models
Second edition
A. J. Dobson

Multivariate Statistics – A practical approach
B. Flury and H. Riedwyl

Readings in Decision Analysis
S. French

Multivariate Analysis of Variance and Repeated Measures
D. J. Hand and C. C. Taylor

Applied Nonparametric Statistical Methods
Second edition
P. Sprent

Probability: Methods and measurement
A. O'Hagan

Decision Analysis: A Bayesian approach
J. Q. Smith

Applied Statistics: A handbook of BMDP analyses
E. J. Snell

Elementary Applications of Probability Theory
H. C. Tuckwell

Statistical Process Control
Theory and practice
Third edition
G. B. Wetherill and D. W. Brown

Modelling Binary Data
D. Collett

Statistical Methods in Agriculture and Experimental Biology
Second edition
R. Mead, R. N. Curnow and A. M. Hasted

Statistical Methods for SPC and TQM
D. Bissell

The Theory of Linear Models
B. Jorgensen

Statistical Theory
Fourth edition
B. Lindgren

Large Sample Methods in Statistics
P. Sen and J. Singer

*Full information on the complete range of Chapman & Hall statistics books
is available from the publishers.*

Modelling Survival Data in Medical Research

D. Collett

Senior Lecturer in Applied Statistics, University of Reading, UK

CHAPMAN & HALL

London · Glasgow · Weinheim · New York · Tokyo · Melbourne · Madras

Published by Chapman & Hall, 2—6 Boundary Row, London SE1 8HN

Chapman & Hall, 2—6 Boundary Row, London SE1 8HN, UK

Blackie Academic & Professional, Wester Cleddens Road, Bishopbriggs, Glasgow G64 2NZ, UK

Chapman & Hall GmbH, Pappelallee 3, 69469 Weinheim, Germany

Chapman & Hall USA, 115 Fifth Avenue, New York, NY 10003, USA

Chapman & Hall Japan, ITP-Japan, Kyowa Building, 3F, 2-2-1 Hirakawacho, Chiyoda-ku, Tokyo 102, Japan

Chapman & Hall Australia, 102 Dodds Street, South Melbourne, Victoria 3205, Australia

Chapman & Hall India, R. Seshadri, 32 Second Main Road, CIT East, Madras 600 035, India

First edition 1994
Reprinted 1995, 1996

© 1994 D. Collett

Printed in Great Britain by Clays Ltd, St Ives Plc, Bungay, Suffolk

ISBN   0 412 44880 7 (HB)   0 412 44890 4 (PB)

A catalogue record for this book is available from the British Library

∞ Printed on acid-free text paper, manufactured in accordance with ANSI/NISO Z39.48-1992 and ANSI/NISO Z39.48-1984 (Permanence of Paper).

To my mother

and

the memory of my father

Contents

Preface

In the course of medical research, data on the time to the occurrence of a particular event, such as the death of a patient, are frequently encountered. Such data are generically referred to as survival data. However, the event of interest need not necessarily be death, but could, for example, be the end of a period spent in remission from a disease, relief from symptoms, or the recurrence of a particular condition. Although there are a number of books devoted to the analysis of survival data, this book is designed to meet the need for an intermediate text which emphasises the application of the methodology to survival data arising from medical studies, which shows how widely-available computer software can be used in survival analysis, and which will appeal to statisticians engaged in medical research.

This book is based on a course on the analysis of survival data from clinical trials which has been given annually by the Statistical Services Centre of the Department of Applied Statistics, University of Reading, since 1986. Although it is written primarily for those working as statisticians in the pharmaceutical industry and in medical research institutes, much of the text should be accessible to numerate scientists and clinicians working alongside statisticians on the analysis of their own data sets. This book could also be used as a text to accompany undergraduate and postgraduate courses on survival analysis in universities and other institutes of higher education.

Many illustrative examples have been included in the text. In addition, sufficient methodological development is given to enable the reader to understand the assumptions on which particular techniques are based, and to help in adapting the methodology to deal with non-standard problems. A number of data sets are based on fewer observations than would normally be encountered in medical research programmes. This enables certain methods of analysis to be illustrated more easily, and means that tabular presentations of results are not too unwieldy. Naturally, the methods described in this book can be applied without modification to larger data sets.

The book begins with an introduction to survival analysis, and a description of four studies in which survival data were obtained. These data sets, and others besides, are then used to illustrate the techniques for analysing survival data presented in subsequent chapters. In Chapter 2, some methods for summarizing survival data are introduced, and non-parametric methods for comparing the survival times of patients in two or more treatment groups are described.

A modelling approach to the analysis of survival data, based on the Cox proportional hazards model, is presented in Chapter 3. Models which assume a Weibull distribution for survival times are developed in Chapter 4, and Chapter 5 gives a comprehensive account of diagnostics which can be used to check the adequacy of both the Cox and Weibull proportional hazards models. Some other parametric models for survival data, including the accelerated failure time model and the proportional odds model, are described in Chapter 6. This is followed by a chapter which shows how variables whose values change over time can be incorporated in models for survival data.

When the survival times of patients are not known exactly, methods used to analyse interval-censored data may be appropriate, and these are described in Chapter 8. The important issue of sample size requirements for a survival study is considered in Chapter 9, and this is followed by a chapter which contains a brief discussion of some additional topics in the analysis of survival data.

In order to implement many of the techniques for analysing survival data, appropriate computer software is needed. Accordingly, the final chapter of the book contains details on the use of some widely-available statistical packages for survival analysis, particularly SAS, BMDP and SPSS. In this chapter, the facilities for survival analysis in these three packages are summarised, and illustrated using a particular data set.

Bibliographic notes and suggestions for further reading are given at the end of each chapter, but so as not to interrupt the flow, references in the text itself will be kept to a minimum. Some sections contain more mathematical details than others, and these have been denoted with an asterisk. These sections can be omitted without loss of continuity.

In writing this book, I have assumed that the reader has a basic knowledge of statistical methods, and has some familiarity with topics such as linear regression analysis and analysis of variance. Matrix algebra is used to give an expression for the standard error of a percentile of the Weibull proportional hazards model in Chapter 4 and Appendix B, and to express some diagnostics for model checking in Chapter 5. However, an understanding of matrices is not an essential requirement.

I have received help from a number of colleagues while writing this book. Mike Patefield and Anne Whitehead provided constructive comments on initial drafts of some of the chapters, and Marilyn Collins provided valuable assistance in writing the SAS macros described in Chapter 11. I would also like to thank Doug Altman for his many comments and suggestions on the first six chapters of the book. I owe a particular debt of gratitude to John Whitehead for many helpful discussions and for his comments on several draft chapters. In addition, the chapter on interval-censored survival data, and sections in some of the other chapters, are

based heavily on material prepared by John for courses that he has given on survival analysis. However, I take full responsibility for any errors in the text. Finally,
I would again like to thank my wife Janet for her support and encouragement over
the period that this book was written.

D. Collett
Reading, UK.

1

Survival analysis

Survival analysis is the phrase used to describe the analysis of data that correspond to the time from a well-defined *time origin* until the occurrence of some particular event or *end-point*. In medical research, the time origin will often correspond to the recruitment of an individual into an experimental study, such as a clinical trial to compare two or more treatments. This in turn may coincide with the diagnosis of a particular condition, the commencement of a treatment regime, or the occurrence of some adverse event. If the end-point is the death of a patient, the resulting data are literally survival times. However, data of a similar form can be obtained when the end-point is not fatal, such as the relief of pain, or the recurrence of symptoms. The methods for analysing survival data that are presented in this book are not restricted to survival times in their literal form, but apply equally to data that refer to the time until other end-points. The methodology can also be applied to data from other application areas, such as the survival times of animals in an experimental study, the time taken by an individual to complete a task in a psychological experiment, the storage times of seeds being kept in a seed bank, or the lifetimes of industrial or electronic components. The focus of this book is on the application of survival analysis to data arising from medical research, and for this reason much of the general discussion will be phrased in terms of the survival time of an individual patient from entry to a study until death.

1.1 Special features of survival data

We must first consider the reasons why survival data are not amenable to standard statistical procedures used in data analysis. The first is that survival data are generally not symmetrically distributed. Typically, a histogram constructed from the survival times of a group of similar individuals will tend to be *positively skewed*, that is, the histogram will have a longer 'tail' to the right of the interval that contains the largest number of observations. As a consequence, it will not be reasonable to assume that data of this type have a normal distribution. This difficulty could be resolved by first transforming the data to give a more symmetric distribution, for example by taking logarithms. However, a more satisfactory approach is to adopt an alternative distributional model for the original data.

A second feature of survival data which renders standard methods inappropriate is that survival times are frequently *censored*. The survival time of an individual is said to be censored when the end-point of interest has not been observed for that individual. This may be because the data from a study are to be analysed at a point in time when some individuals are still alive. Alternatively, the survival status of an individual at the time of the analysis might not be known because that individual has been *lost to follow-up*. As an example, suppose that after being recruited to a clinical trial, a patient moves to another part of the country, or to a different country, and can no longer be traced. The only information available on the survival experience of that patient is the last date on which he or she was known to be alive. This date may well be the last time that the patient reported to a clinic for a regular check-up.

An actual survival time can also be regarded as censored when death is from a cause that is known to be unrelated to the treatment. In many cases, it is difficult to be sure that the death is not related to a particular treatment that the patient is undergoing. For example, consider a patient in a clinical trial to compare alternative therapies for prostatic cancer who experiences a fatal road traffic accident. The accident could have resulted from an attack of dizziness, which might be a side-effect of the treatment to which that patient has been assigned. If so, the death is not unrelated to the treatment. In circumstances such as these, the survival time until death from all causes, or the time to death from causes other than the primary condition for which the patient is being treated, might also be subjected to a survival analysis.

In each of these situations, a patient who entered a study at time t_0 dies at time $t_0 + t$. However, t is unknown, either because the individual is still alive or because he or she has been lost to follow-up. If the individual was last known to be alive at time $t_0 + c$, the time c is called a censored survival time. This censoring occurs after the individual has been entered into a study, that is, to the right of the last known survival time, and is therefore known as *right censoring*. The right-censored survival time is then less than the actual, but unknown, survival time.

Another form of censoring is *left censoring*, which is encountered when the actual survival time of an individual is less than that observed. To illustrate this form of censoring, consider a study in which interest centres on the time to recurrence of a particular cancer following surgical removal of the primary tumour. Three months after their operation, the patients are examined to determine if the cancer has recurred. At this time, some of the patients may be found to have a recurrence. For such patients, the actual time to recurrence is less than three months, and the recurrence times of these patients is left-censored. Left censoring occurs far less commonly than right censoring, and so the emphasis of this book will be on the analysis of right-censored survival data.

Yet another type of censoring is *interval censoring*. Here, individuals are known

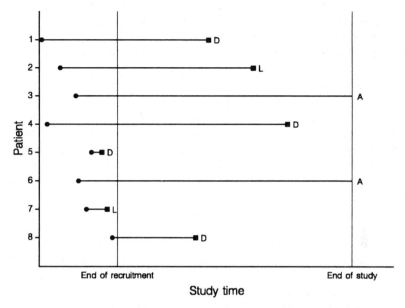

Figure 1.1 Study time for eight patients in a survival study.

to have experienced a failure within an interval of time. Consider again the example concerning the time to recurrence of a tumour used in the above discussion of left censoring. If a patient is observed to be free of the disease at three months, but is found to have had a recurrence when examined six months after surgery, the actual recurrence time of that patient is known to be between three months and six months. The observed recurrence time is then said to be interval-censored. We will return to interval censoring later in Chapter 8.

1.1.1 Patient time and study time

In a typical study, patients are not all recruited at exactly the same time, but accrue over a period of months or even years. After recruitment, patients are followed up until they die, or until a point in calendar time which marks the end of the study, when the data are analysed. Although the actual survival times will be observed for a number of patients, after recruitment some patients may be lost to follow-up, while others will still be alive at the end of the study. The calendar time period in which an individual is in the study is known as the *study time*. The study time for eight individuals in a clinical trial is illustrated diagrammatically in Figure 1.1, in which the time of entry to the study is represented by a '•'. Individuals 1, 4, 5 and 8 die (D) during the course of the study, individuals 2 and 7 are lost to follow-up (L), and individuals 3 and 6 are still alive (A) at the end of the observation period.

Figure 1.2   Patient time for eight patients in a survival study.

As far as each patient is concerned, the trial begins at some time t_0. The corresponding survival times for the eight individuals depicted in Figure 1.1 are shown in order in Figure 1.2. The period of time that a patient spends in the study, measured from that patient's time origin, is often referred to as *patient time*. The period of time from the time origin to the death of a patient (D) is then the survival time, and this is recorded for individuals 1, 4, 5 and 8. The survival times of the remaining individuals are right-censored (C).

In practice, the actual data recorded will be the date on which each individual enters the study, and the date on which each individual dies or was last known to be alive. The survival time in days, weeks or months, whichever is the most appropriate, can then be calculated. A number of computer software packages for survival analysis have facilities for performing this calculation from input data in the form of dates.

An important assumption which will be made in the analysis of censored survival data is that the actual survival time of an individual, t, is independent of any mechanism which causes that individual's survival time to be censored at time c, where $c < t$. This means that if we consider a group of individuals, all of whom have the same values of relevant prognostic variables, an individual whose survival time is censored at time c must be representative of all other individuals in that group who have survived to that time. A patient whose survival time is censored will be representative of those at risk at the censoring time if the censoring process operates randomly. Similarly, when survival data are to be analysed at a

Table 1.1 Time in weeks to discontinuation of the use of an IUD.

10	13*	18*	19	23*	30	36	38*	54*
56*	59	75	93	97	104*	107	107*	107*

predetermined point in calendar time, or at a fixed interval of time after the time origin for each patient, the prognosis for individuals who are still alive can be taken to be independent of the censoring, so long as the time of analysis is specified before the data are examined. However, this assumption cannot be made if, for example, the survival time of an individual is censored through treatment being withdrawn as a result of a deterioration in the physical condition of a patient. This type of censoring is known as *informative censoring*. Great care should be taken to ensure that any censoring is non-informative, for otherwise the methods presented in this book for the analysis of survival data are no longer valid.

1.2 Some examples

In this section, the essential features of survival data are illustrated through a number of examples. Data from these examples will then be used to illustrate some of the statistical techniques presented in subsequent chapters.

Example 1.1 Time to discontinuation of the use of an IUD
In trials involving contraceptives, prevention of pregnancy is an obvious criterion for acceptability. However, modern contraceptives have very low failure rates, and so the occurrence of bleeding disturbances, such as amenorrhoea (the prolonged absence of bleeding), irregular or prolonged bleeding, become important in the evaluation of a particular method of contraception. To promote research into methods for analysing menstrual bleeding data from women in contraceptive trials, the World Health Organization have made available data from clinical trials involving a number of different types of contraceptive (WHO, 1987). Part of this data set relates to the time from which a woman commences use of a particular method until discontinuation, with the discontinuation reason being recorded when known. The data in Table 1.1 refer to the number of weeks from the commencement of use of a particular type of intrauterine device (IUD), known as the Multiload 250, until discontinuation because of menstrual bleeding problems. Data are given for 18 women, all of whom were aged between 18 and 35 years and who had experienced two previous pregnancies. Discontinuation times that are censored are labelled with an asterisk.

In this example, the time origin corresponds to the first day in which a woman uses the IUD, and the end-point is discontinuation because of bleeding problems. Some women in the study ceased using the IUD because of the desire for pregnancy, or because they had no further need for a contraceptive, while others were simply lost to follow-up. These reasons account for the censored discontinuation

times of 13, 18, 23, 38, 54 and 56 weeks. The study protocol called for the menstrual bleeding experience of each woman to be documented for a period of two
years from the time origin. For practical reasons, each woman could not be examined exactly two years after recruitment to determine if they were still using the
IUD, and this is why there are three discontinuation times greater than 104 weeks
that are right-censored.

One objective in an analysis of these data would be to summarize the distribution of discontinuation times. We might then wish to estimate the median time to
discontinuation of the IUD, or the probability that a woman will stop using the
device after a given period of time. Indeed, a graph of this estimated probability,
as a function of time, will provide a useful summary of the observed data.

Example 1.2 Prognosis for women with breast cancer
Breast cancer is one of the most common forms of cancer occurring in women
living in the Western world. However, the biological behaviour of the tumour is
unpredictable, and there is at present no reliable method for determining whether
or not a tumour is likely to have metastasized, or spread, to other organs in the
body. Around 80% of women presenting with primary breast cancer are likely
to have tumours which have already metastasized to other sites. If these patients
could be identified, adjunctive treatment could be focused on them, whilst the
remaining 20% could be reassured that their disease is surgically curable. The
most accurate assessment of the prognosis of a patient is based on whether or
not there is lymph node involvement. However, as a result of the trend towards
more conservative forms of breast surgery, this indication will often be unknown.
This has prompted research into alternative procedures for predicting the survival
prospects of breast cancer patients.

The aim of an investigation carried out at the Middlesex Hospital, and documented in Leathem and Brooks (1987), was to evaluate a histochemical marker
which discriminates between primary breast cancer that has metastasized and that
which has not. The marker under study was a lectin from the albumin gland of
the Roman snail, *Helix pomatia*, known as *Helix pomatia* agglutinin, or HPA. The
marker binds to those breast cancer cells associated with metastasis to local lymph
nodes, and the HPA stained cells can be identified by microscopic examination.
In order to investigate whether HPA staining can be used to predict the survival
experience of women who present with breast cancer, a retrospective study was
carried out, based on the records of women who had received surgical treatment for
breast cancer. Sections of the tumours of these women were treated with HPA and
each tumour was subsequently classified as being positively or negatively stained,
positive staining corresponding to a tumour with the potential for metastasis. The
study was concluded in July 1987, when the survival times of those women who
had died of breast cancer were calculated. For those women whose survival status
in July 1987 was unknown, the time from surgery to the date on which they were
last known to be alive is regarded as a censored survival time. The survival times
of women who had died from causes other than breast cancer are also regarded as

Table 1.2 Survival times of women with tumours which were negatively or positively stained with HPA.

Negative staining	Positive staining	
23	5	68
47	8	71
69	10	76*
70*	13	105*
71*	18	107*
100*	24	109*
101*	26	113
148	26	116*
181	31	118
198*	35	143
208*	40	154*
212*	41	162*
224*	48	188*
	50	212*
	59	217*
	61	225*

right-censored. The data given in Table 1.2 refer to the survival times in months of women who had received a simple or radical mastectomy to treat a tumour of Grade II, III or IV, between January 1969 and December 1971. In the table, the survival times of each woman are classified according to whether their tumour was positively or negatively stained. Censored survival times are labelled with an asterisk.

In the analysis of the data from this study, we will be particularly interested in whether or not there is a difference in the survival experience of the two groups of women. If there were evidence that those women with negative HPA staining tended to live longer after surgery than those with a positive staining, we would conclude that the prognosis for a breast cancer patient was dependent on the result of the staining procedure.

Example 1.3 Survival of multiple myeloma patients
Multiple myeloma is a malignant disease characterized by the accumulation of abnormal plasma cells, a type of white blood cell, in the bone marrow. The proliferation of the abnormal plasma cells within the bone causes pain and the destruction of bone tissue. Patients with multiple myeloma also experience anaemia, haemorrhages, recurrent infections and weakness. Unless treated, the condition is invariably fatal. The aim of a study carried out at the Medical Centre of the University of West Virginia, USA, was to examine the association between the values of certain *explanatory variables* or *covariates* and the survival time of patients. In

the study, the primary response variable was the time, in months, from diagnosis until death from multiple myeloma.

The data in Table 1.3, which were obtained from Krall, Uthoff and Harley (1975), relate to 48 patients, all of whom were aged between 50 and 80 years. Some of these patients had not died by the time that the study was completed, and so these individuals contribute right-censored survival times. The coding of the survival status of an individual in the table is such that zero denotes a censored observation and unity death from multiple myeloma. At the time of diagnosis, the values of a number of explanatory variables were recorded for each patient. These included the age of the patient in years, their sex (1 = male, 2 = female), the levels of blood urea nitrogen (BUN), serum calcium (CA) and haemoglobin (HB), the percentage of plasma cells in the bone marrow (PC) and an indicator variable (BJ) which denotes whether or not Bence–Jones protein was present in the urine (0 = absent, 1 = present).

The main aim of an analysis of these data would be to investigate the effect of the risk factors BUN, CA, HB, PC and BJ on the survival time of the multiple myeloma patients. The effects of these risk factors may be modified by the age or sex of a patient, and so the extent to which the relationship between survival and the important risk factors is consistent for each sex and for each of a number of age-groups will also need to be studied.

Example 1.4 Comparison of two treatments for prostatic cancer
A randomized controlled clinical trial to compare treatments for prostatic cancer was begun in 1967 by the Veteran's Administration Cooperative Urological Research Group. The trial was double blind and two of the treatments used in the study were a placebo and 1.0 mg of diethylstilbestrol (DES). The treatments were administered daily by mouth. The time origin of the study is the date on which a patient was randomized to a treatment, and the end-point is the death of the patient from prostatic cancer.

The full data set is given in Andrews and Herzberg (1985), but the data used in this example are from patients presenting with Stage III cancer, that is, patients for whom there was evidence of a local extension of the tumour beyond the prostatic capsule, but without elevated serum prostatic acid phosphatase. Furthermore, the patients were those who had no history of cardiovascular disease, had a normal ECG result at trial entry, and who were not confined to bed during the daytime. In addition to recording the survival time of each patient in the study, information was recorded on a number of other prognostic factors. These included the age of the patient at trial entry, their serum haemoglobin level in gm/100 ml, the size of their primary tumour in cm^2, and the value of a combined index of tumour stage and grade. This index is known as the Gleason index; the more advanced the tumour, the greater the value of the index.

Table 1.4 gives the data recorded for 38 patients, where the survival times are given in months. The survival times of patients who died from other causes, or who were lost during the follow-up process are regarded as censored. A variable

Table 1.3 Survival times of patients in a study on multiple myeloma.

Patient number	Survival time	Status	Age	Sex	BUN	CA	HB	PC	BJ
1	13	1	66	1	25	10	14.6	18	1
2	52	0	66	1	13	11	12.0	100	0
3	6	1	53	2	15	13	11.4	33	1
4	40	1	69	1	10	10	10.2	30	1
5	10	1	65	1	20	10	13.2	66	0
6	7	0	57	2	12	8	9.9	45	0
7	66	1	52	1	21	10	12.8	11	1
8	10	0	60	1	41	9	14.0	70	1
9	10	1	70	1	37	12	7.5	47	0
10	14	1	70	1	40	11	10.6	27	0
11	16	1	68	1	39	10	11.2	41	0
12	4	1	50	2	172	9	10.1	46	1
13	65	1	59	1	28	9	6.6	66	0
14	5	1	60	1	13	10	9.7	25	0
15	11	0	66	2	25	9	8.8	23	0
16	10	1	51	2	12	9	9.6	80	0
17	15	0	55	1	14	9	13.0	8	0
18	5	1	67	2	26	8	10.4	49	0
19	76	0	60	1	12	12	14.0	9	0
20	56	0	66	1	18	11	12.5	90	0
21	88	1	63	1	21	9	14.0	42	1
22	24	1	67	1	10	10	12.4	44	0
23	51	1	60	2	10	10	10.1	45	1
24	4	1	74	1	48	9	6.5	54	0
25	40	0	72	1	57	9	12.8	28	1
26	8	1	55	1	53	12	8.2	55	0
27	18	1	51	1	12	15	14.4	100	0
28	5	1	70	2	130	8	10.2	23	0
29	16	1	53	1	17	9	10.0	28	0
30	50	1	74	1	37	13	7.7	11	1
31	40	1	70	2	14	9	5.0	22	0
32	1	1	67	1	165	10	9.4	90	0
33	36	1	63	1	40	9	11.0	16	1
34	5	1	77	1	23	8	9.0	29	0
35	10	1	61	1	13	10	14.0	19	0
36	91	1	58	2	27	11	11.0	26	1
37	18	0	69	2	21	10	10.8	33	0
38	1	1	57	1	20	9	5.1	100	1
39	18	0	59	2	21	10	13.0	100	0
40	6	1	61	2	11	10	5.1	100	0
41	1	1	75	1	56	12	11.3	18	0
42	23	1	56	2	20	9	14.6	3	0
43	15	1	62	2	21	10	8.8	5	0
44	18	1	60	2	18	9	7.5	85	1
45	12	0	71	2	46	9	4.9	62	0
46	12	1	60	2	6	10	5.5	25	0
47	17	1	65	2	28	8	7.5	8	0
48	3	0	59	1	90	10	10.2	6	1

associated with the status of an individual at the end of the study takes the value unity if the patient has died from prostatic cancer, and zero if the survival time is right-censored. The variable associated with the treatment group takes the value 2 when an individual is treated with DES and unity if an individual is on the placebo treatment.

The main aim of this study is to determine the extent of any evidence that patients treated with DES survive longer than those treated with the placebo. Since the data on which this example is based are from a randomized trial, one might expect that the distributions of the prognostic factors, that is the age of patient, serum haemoglobin level, size of tumour and Gleason index, will be similar over the patients in each of the two treatment groups. However, it would not be wise to rely on this assumption. For example, it could turn out that patients in the placebo group had larger tumours on average than those in the group treated with DES. If patients with large tumours have a poorer prognosis than those with small tumours, the size of the treatment effect would be over-estimated unless proper account was taken of size of tumour in the analysis. Consequently, it will first be necessary to determine if any of the covariates are related to survival time. If so, the effect of these variables will need to be allowed for when comparing the survival experiences of the patients in the two treatment groups.

1.3 Survivor function and hazard function

In summarizing survival data, there are two functions of central interest, namely the *survivor function* and the *hazard function*. These functions are therefore defined in this first chapter.

The actual survival time of an individual, t, can be regarded as the value of a variable T, which can take any non-negative value. The different values that T can take have a *probability distribution*, and we call T the *random variable* associated with the survival time. Now suppose that the random variable T has a probability distribution with underlying *probability density function* $f(t)$. The *distribution function* of T is then given by

$$F(t) = P(T < t) = \int_0^t f(u)\, du,$$

and represents the probability that the survival time is less than some value t.

The survivor function, $S(t)$, is defined to be the probability that the survival time is greater than or equal to t, and so

$$S(t) = P(T \geq t) = 1 - F(t). \tag{1.1}$$

The survivor function can therefore be used to represent the probability that an individual survives from the time origin to some time beyond t.

The *hazard function* is the probability that an individual dies at time t, conditional on he or she having survived to that time. The hazard function therefore represents the instantaneous death rate for an individual surviving to time t. To

Table 1.4 Survival times of prostatic cancer patients in a clinical trial to compare two treatments.

Patient number	Treat-ment	Survival time	Status	Age	Serum Haem.	Size of tumour	Gleason index
1	1	65	0	67	13.4	34	8
2	2	61	0	60	14.6	4	10
3	2	60	0	77	15.6	3	8
4	1	58	0	64	16.2	6	9
5	2	51	0	65	14.1	21	9
6	1	51	0	61	13.5	8	8
7	1	14	1	73	12.4	18	11
8	1	43	0	60	13.6	7	9
9	2	16	0	73	13.8	8	9
10	1	52	0	73	11.7	5	9
11	1	59	0	77	12.0	7	10
12	2	55	0	74	14.3	7	10
13	2	68	0	71	14.5	19.	9
14	2	51	0	65	14.4	10	9
15	1	2	0	76	10.7	8	9
16	1	67	0	70	14.7	7	9
17	2	66	0	70	16.0	8	9
18	2	66	0	70	14.5	15	11
19	2	28	0	75	13.7	19	10
20	2	50	1	68	12.0	20	11
21	1	69	1	60	16.1	26	9
22	1	67	0	71	15.6	8	8
23	2	65	0	51	11.8	2	6
24	1	24	0	71	13.7	10	9
25	2	45	0	72	11.0	4	8
26	2	64	0	74	14.2	4	6
27	1	61	0	75	13.7	10	12
28	1	26	1	72	15.3	37	11
29	1	42	1	57	13.9	24	12
30	2	57	0	72	14.6	8	10
31	2	70	0	72	13.8	3	9
32	2	5	0	74	15.1	3	9
33	2	54	0	51	15.8	7	8
34	1	36	1	72	16.4	4	9
35	2	70	0	71	13.6	2	10
36	2	67	0	73	13.8	7	8
37	1	23	0	68	12.5	2	8
38	1	62	0	63	13.2	3	8

obtain a more formal definition of the hazard function, consider the probability that the random variable associated with an individual's survival time, T, lies between t and $t + \delta t$, conditional on T being greater than or equal to t, written $P(t \leq T < t + \delta t \mid T \geq t)$. The hazard function $h(t)$ is then the limiting value of this probability divided by the time interval δt, as δt tends to zero, so that

$$h(t) = \lim_{\delta t \to 0} \left\{ \frac{P(t \leq T < t + \delta t \mid T \geq t)}{\delta t} \right\}. \qquad (1.2)$$

From this definition, we can obtain some useful relationships between the survivor and hazard functions. According to a standard result from probability theory, the probability of an event A, conditional on the occurrence of an event B, is given by $P(A \mid B) = P(AB)/P(B)$, where $P(AB)$ is the probability of the joint occurrence of A and B. Using this result, the conditional probability in the definition of the hazard function in equation (1.2) is

$$\frac{P(t \leq T < t + \delta t)}{P(T \geq t)},$$

which is equal to

$$\frac{F(t + \delta t) - F(t)}{S(t)},$$

where $F(t)$ is the distribution function of T. Then,

$$h(t) = \lim_{\delta t \to 0} \left\{ \frac{F(t + \delta t) - F(t)}{\delta t} \right\} \frac{1}{S(t)}.$$

Now,

$$\lim_{\delta t \to 0} \left\{ \frac{F(t + \delta t) - F(t)}{\delta t} \right\}$$

is the definition of the derivative of $F(t)$ with respect to t, which is $f(t)$, and so

$$h(t) = \frac{f(t)}{S(t)}. \qquad (1.3)$$

It then follows that

$$h(t) = -\frac{d}{dt} \{ \log S(t) \}, \qquad (1.4)$$

and so

$$S(t) = \exp \{ -H(t) \}, \qquad (1.5)$$

where

$$H(t) = \int_0^t h(u) \, du. \qquad (1.6)$$

The function $H(t)$ features widely in survival analysis and is called the *integrated* or *cumulative hazard*. From equation (1.5), the cumulative hazard can be obtained from the survivor function, since

$$H(t) = -\log S(t). \qquad (1.7)$$

In the analysis of survival data, the survivor function and hazard function are estimated from the observed survival times. Methods of estimation which do not require the form of the probability density function of T to be specified are described in Chapters 2 and 3, while methods based on the assumption of a particular survival time distribution are presented in Chapter 4.

1.4 Further reading

An introduction to the techniques used in the analysis of survival data is included in general books on statistics in medical research, such as those of Armitage and Berry (1987) and Altman (1991).

A number of texts on the analysis of survival data appeared during the 1970s, including the books by Mann, Shafer and Singpurwalla (1974), and Barlow and Proschan (1975), which are concerned more with reliability testing, and Gross and Clark (1975) who consider applications in the biomedical sciences. However, these have now been overtaken by methodological and computational advances in the subject.

More recent texts which describe the application of survival analysis to industrial life-testing or actuarial science include those of Elandt-Johnson and Johnson (1980), Lawless (1982) and Nelson (1982). An up-to-date account of the analysis of reliability data is given by Crowder et al. (1991). Kalbfleisch and Prentice (1980) and Lee (1992) illustrate the methodology of survival analysis using biological and medical data. The books by Kalbfleisch and Prentice (1980) and Lawless (1980) provide comprehensive accounts of the subject. Both are written for the postgraduate statistician or research worker, and are usually regarded as reference books rather than introductory texts. The mathematical text of Miller (1981) is not as comprehensive as either Kalbfleisch and Prentice (1980) or Lawless (1980), and dwells mainly on non-parametric methods and the Cox regression model. A concise account of survival analysis is given in the research monograph of Cox and Oakes (1984). See also the chapter on survival analysis in Hinkley, Reid and Snell (1991).

Although medical applications have prompted much recent methodological development in survival analysis, the only book to be written primarily for medical statisticians is the text by Harris and Albert (1991). This book is less mathematical than its competitors, but like Miller (1981) it is limited to non-parametric methods and the Cox regression model. Applications of survival analysis in the analysis of epidemiological data are described by Breslow and Day (1987). The papers of Peto et al. (1976, 1977) give a non-mathematical introduction to the design and analysis of clinical trials in which the response variable is a survival time.

2

Some non-parametric procedures

An initial step in the analysis of a set of survival data is to present numerical or graphical summaries of the survival times for individuals in a particular group. Such summaries may be of interest in their own right, or as a precursor to a more detailed analysis of the data. Survival data are conveniently summarized through estimates of the survivor function and hazard function. Methods for estimating these functions from a single sample of survival data are described in Sections 2.1 and 2.2. These methods are said to be *non-parametric* or *distribution-free*, since they do not require specific assumptions to be made about the underlying distribution of the survival times.

Once the estimated survivor function has been found, the median and other percentiles of the distribution of survival times can be estimated graphically, as shown in Section 2.3. Estimates of the location and dispersion of a set of data, derived on the basis of assumptions about the probability distribution from which the data have been obtained, will be considered later in Chapter 4.

When the survival times of two groups of patients are being compared, an informal comparison of the survival experience of each group of individuals can be made using the estimated survivor functions. However, there are more formal procedures which enable two groups of survival data to be compared. Two non-parametric procedures for comparing two or more groups of survival times, namely the *log-rank test* and the *Wilcoxon test*, are described in Section 2.4.

2.1 Estimating the survivor function

Suppose first that we have a single sample of survival times, where none of the observations are censored. The survivor function $S(t)$, defined in equation (1.1), is the probability that an individual survives for a time greater than or equal to t. This function can be estimated by the *empirical survivor function*, given by

$$\tilde{S}(t) = \frac{\text{Number of individuals with survival times} \geq t}{\text{Number of individuals in the data set}}. \qquad (2.1)$$

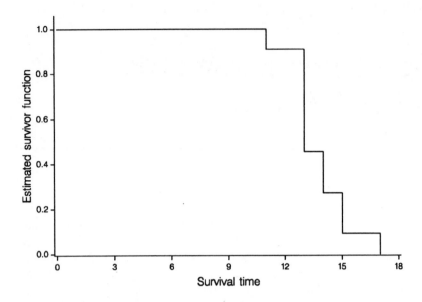

Figure 2.1   Estimated survivor function for the data from Example 2.1.

Equivalently, $\tilde{S}(t) = 1 - \tilde{F}(t)$, where $\tilde{F}(t)$ is the *empirical distribution function*, that is, the ratio of the total number of individuals alive at time t to the total number of individuals in the study. Notice that the empirical survivor function is equal to unity for values of t before the first death time, and zero after the final death time.

The estimated survivor function $\tilde{S}(t)$ is assumed to be constant between two adjacent death times, and so a plot of $\tilde{S}(t)$ against t is a step-function. The function decreases immediately after each observed survival time.

Example 2.1 Pulmonary metastasis

One complication in the management of patients with a malignant bone tumour, or osteosarcoma, is that the tumour often spreads to the lungs. This pulmonary metastasis is life-threatening. In a study concerned with the treatment of pulmonary metastasis arising from osteosarcoma, Burdette and Gehan (1970) give the following survival times, in months, of eleven male patients.

$$11 \quad 13 \quad 13 \quad 13 \quad 13 \quad 13 \quad 14 \quad 14 \quad 15 \quad 15 \quad 17$$

Using equation (2.1), the estimated values of the survivor function at times 11, 13, 14, 15 and 17 months are 1.000, 0.909, 0.455, 0.273, and 0.091. The estimated value of the survivor function is unity from the time origin until 11 weeks, and zero after 17 weeks. A graph of the estimated survivor function is given in Figure 2.1.

The method of estimating the survivor function illustrated in the above example cannot be used when there are censored observations. The reason for this is that the method does not allow information provided by an individual whose survival time is censored before time t to be used in computing the estimated survivor function at t. Non-parametric methods for estimating $S(t)$, which can be used in the presence of censored survival times, are described in the following sections.

2.1.1 Life-table estimate of the survivor function

The *life-table estimate* of the survivor function, which is also known as the *actuarial estimate*, is obtained by first dividing the period of observation into a series of time intervals. These intervals need not necessarily be of equal length, although they usually are. Suppose that the j'th of m such intervals, $j = 1, 2, \ldots, m$, extends from time t'_j to t'_{j+1}, and let d_j and c_j denote the number of deaths and the number of censored survival times, respectively, in this time interval. Also let n_j be the number of individuals who are alive, and therefore at risk of death, at the start of the j'th interval. We now make the assumption that the censoring process is such that the censored survival times occur uniformly throughout the j'th interval, so that the average number of individuals who are at risk during this interval is

$$n'_j = n_j - c_j/2. \tag{2.2}$$

This assumption is sometimes known as the *actuarial assumption*.

In the j'th interval, the probability of death can be estimated by d_j/n'_j, so that the corresponding survival probability is $(n'_j - d_j)/n'_j$. Now consider the probability that an individual survives beyond time t'_k, $k = 1, 2, \ldots, m$, that is, until some time after the start of the k'th interval. This will be the product of the probabilities that an individual survives beyond the start of the k'th interval and through each of the $k - 1$ preceding intervals, and so the life-table estimate of the survivor function is given by

$$S^*(t) = \prod_{j=1}^{k} \left(\frac{n'_j - d_j}{n'_j} \right), \tag{2.3}$$

for $t'_k \leq t < t'_{k+1}$, $k = 1, 2, \ldots, m$. The estimated probability of surviving until the start of the first interval, t'_1, is of course unity, while the estimated probability of surviving beyond t'_{m+1} is zero. A graphical estimate of the survivor function will then be a step-function with constant values of the function in each time interval.

Example 2.2 Survival of multiple myeloma patients

To illustrate the computation of the life-table estimate, consider the data on the survival times of the 48 multiple myeloma patients given in Table 1.3. In this illustration, the information collected on other explanatory variables for each individual will be ignored.

The survival times are first grouped to give the number of patients who die, d_j, and the number who are censored, c_j, in each of the first five years of the

Table 2.1 Life-table estimate of the survivor function for the data from Example 1.3.

Interval	Time period	d_j	c_j	n_j	n'_j	$(n'_j - d_j)/n'_j$	$S^*(t)$
1	0–	16	4	48	46.0	0.6522	0.6522
2	12–	10	4	28	26.0	0.6154	0.4013
3	24–	1	0	14	14.0	0.9286	0.3727
4	36–	3	1	13	12.5	0.7600	0.2832
5	48–	2	2	9	8.0	0.7500	0.2124
6	60–	4	1	5	4.5	0.1111	0.0236

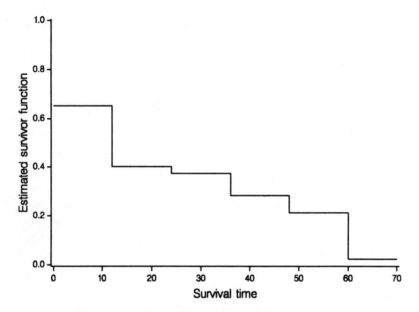

Figure 2.2   Life-table estimate of the survivor function.

study, and in the subsequent three-year period. The number at risk of death at the start of each of these intervals, n_j, is then computed, together with the adjusted number at risk, n'_j. Finally, the probability of survival through each interval is estimated, from which the estimated survivor function is obtained using equation (2.3). The calculations are shown in Table 2.1, in which the time period is given in months, and the interval which begins at time t'_k and ends just before time t'_{k+1}, for $k = 1, 2, \ldots, m$, is denoted t'_k-. A graph of the life-table estimate of the survivor function is shown in Figure 2.2.

The form of the estimated survivor function obtained using this method is sen-

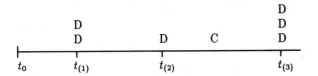

Figure 2.3 Construction of intervals used in the derivation of the Kaplan–Meier estimate.

sitive to the choice of the intervals used in its construction, just as the shape of a histogram depends on the choice of the class intervals. On the other hand, the life-table estimate is particularly well suited to situations in which the actual death times are unknown, and the only available information is the number of deaths and the number of censored observations which occur in a series of consecutive time intervals. In practice, such interval-censored survival data occur quite frequently.

When the actual survival times are known, the life-table estimate can still be used, as in Example 2.2, but the grouping of the survival times does result in some loss of information. This is particularly so when the number of patients is small, less than about thirty, say. A method of estimating the survivor function from un-grouped survival data is given in the next section.

2.1.2 Kaplan–Meier estimate of the survivor function

To determine the Kaplan–Meier estimate of the survivor function from a sample of censored survival data, a series of time intervals is formed, as for the life-table estimate. However, each of these intervals is constructed to be such that one death time is contained in the interval, and this death time is taken to occur at the start of the interval.

For example, suppose that $t_{(1)}$, $t_{(2)}$ and $t_{(3)}$ are three observed survival times arranged in rank order so that $t_{(1)} < t_{(2)} < t_{(3)}$, and that c is a censored survival time which falls between $t_{(2)}$ and $t_{(3)}$. The constructed intervals then begin at times $t_{(1)}$, $t_{(2)}$ and $t_{(3)}$, and each interval includes one death time, although there could be more than one individual who dies at any particular death time. Notice that no interval begins at the censored time of c. The situation is illustrated diagrammatically in Figure 2.3, in which D represents a death and C a censored survival time. Notice that two individuals die at $t_{(1)}$, one dies at $t_{(2)}$, and three die at $t_{(3)}$.

The time origin is denoted by t_0, and so there is an initial period commencing at t_0 which ends just before $t_{(1)}$, the time of the first death. This means that the interval from t_0 to $t_{(1)}$ will not include a death time. The first constructed interval extends from $t_{(1)}$ to just before $t_{(2)}$, and since the second death time is at $t_{(2)}$, this interval includes the single death time at $t_{(1)}$. The second interval begins at time

$t_{(2)}$ and ends just before $t_{(3)}$, and includes the death time at $t_{(2)}$ and the censored time c. There is also a third interval beginning at $t_{(3)}$ which contains the longest survival time, $t_{(3)}$.

In general, suppose that there are n individuals with observed survival times $t_1, t_2, \ldots, t_n$. Some of these observations may be right-censored, and there may also be more than one individual with the same observed survival time. We therefore suppose that there are r death times amongst the individuals, where $r \leq n$. After arranging these death times in ascending order, the j'th is denoted $t_{(j)}$, for $j = 1, 2, \ldots, r$, and so the r ordered death times are $t_{(1)} < t_{(2)} < \cdots < t_{(r)}$. The number of individuals who are alive just before time $t_{(j)}$, including those who are about to die at this time, will be denoted n_j, for $j = 1, 2, \ldots, r$, and d_j will denote the number who die at this time. The time interval from $t_{(j)} - \delta$ to $t_{(j)}$, where δ is an infinitesimal time interval, then includes one death time. Since there are n_j individuals who are alive just before $t_{(j)}$ and d_j deaths at $t_{(j)}$, the probability that an individual dies during the interval from $t_{(j)} - \delta$ to $t_{(j)}$ is estimated by d_j/n_j. The corresponding estimated probability of survival through that interval is then $(n_j - d_j)/n_j$.

It sometimes happens that there are censored survival times which occur at the same time as one or more deaths, so that a death time and a censored survival time occur simultaneously. In this event, the censored survival time is taken to occur immediately after the death time when computing the values of the n_j.

From the manner in which the time intervals are constructed, the interval from $t_{(j)}$ to $t_{(j+1)} - \delta$, the time immediately before the next death time, contains no deaths. The probability of surviving from $t_{(j)}$ to $t_{(j+1)} - \delta$ is therefore unity, and the joint probability of surviving from $t_{(j)} - \delta$ to $t_{(j)}$ and from $t_{(j)}$ to $t_{(j+1)} - \delta$ can be estimated by $(n_j - d_j)/n_j$. In the limit, as δ tends to zero, $(n_j - d_j)/n_j$ becomes an estimate of the probability of surviving from $t_{(j)}$ to $t_{(j+1)}$.

We now make the assumption that the deaths of the individuals in the sample occur independently of one another. Then, the estimated survivor function at any time in the k'th constructed time interval from $t_{(k)}$ to $t_{(k+1)}$, $k = 1, 2, \ldots, r$, where $t_{(r+1)}$ is defined to be ∞, will be the estimated probability of surviving beyond $t_{(k)}$. This is actually the probability of surviving through the interval from $t_{(k)}$ to $t_{(k+1)}$, and all preceding intervals. This is the Kaplan–Meier estimate of the survivor function, which is given by

$$\hat{S}(t) = \prod_{j=1}^{k} \left(\frac{n_j - d_j}{n_j} \right), \qquad (2.4)$$

for $t_{(k)} \leq t < t_{(k+1)}$, $k = 1, 2, \ldots r$, with $\hat{S}(t) = 1$ for $t < t_{(1)}$ and where $t_{(r+1)}$ is taken to be ∞. Strictly speaking, if the largest observation is a censored survival time, t^*, say, $\hat{S}(t)$ is undefined for $t > t^*$. On the other hand, if the largest observed survival time, $t_{(r)}$, is an uncensored observation, $n_r = d_r$, and so $\hat{S}(t)$ is zero for $t \geq t_{(r)}$. A plot of the Kaplan–Meier estimate of the survivor function is a step-function, in which the estimated survival probabilities are constant between

Table 2.2 Kaplan–Meier estimate of the survivor function for the data from Example 1.1.

Time interval	n_j	d_j	$(n_j - d_j)/n_j$	$\hat{S}(t)$
0–	18	0	1.0000	1.0000
10–	18	1	0.9444	0.9444
19–	15	1	0.9333	0.8815
30–	13	1	0.9231	0.8137
36–	12	1	0.9167	0.7459
59–	8	1	0.8750	0.6526
75–	7	1	0.8571	0.5594
93–	6	1	0.8333	0.4662
97–	5	1	0.8000	0.3729
107	3	1	0.6667	0.2486

adjacent death times and decrease at each death time.

The Kaplan–Meier estimate is also known as the *product-limit estimate* of the survivor function. The reason for this will become clear in Section 3.8.2 when a further estimate of the survivor function, known as *Altshuler's estimate* is introduced.

Note that if there are no censored survival times in the data set, $n_j - d_j = n_{j+1}$, $j = 1, 2, \ldots, k$, in equation (2.4), and on expanding the product we get

$$\hat{S}(t) = \frac{n_2}{n_1} \times \frac{n_3}{n_2} \times \cdots \times \frac{n_{k+1}}{n_k}. \tag{2.5}$$

This reduces to n_{k+1}/n_1, for $k = 1, 2, \ldots r - 1$, with $\hat{S}(t) = 1$ for $t < t_{(1)}$ and $\hat{S}(t) = 0$ for $t \geq t_{(r)}$. Now, n_1 is the number of individuals at risk just before the first death time, which is the number of individuals in the sample, and n_{k+1} is the number of individuals with survival times greater than or equal to $t_{(k+1)}$. Consequently, in the absence of censoring, $\hat{S}(t)$ is simply the empirical survivor function defined in equation (2.1).

Example 2.3 Time to discontinuation of the use of an IUD
Data from 18 women on the time to discontinuation of the use of an IUD were given in Table 1.1. For these data, the survivor function, $S(t)$, represents the probability that a woman discontinues the use of the contraceptive device after any time t. The Kaplan–Meier estimate of the survivor function is readily obtained using equation (2.4), and the required calculations are set out in Table 2.2. The estimated survivor function, $\hat{S}(t)$, is plotted in Figure 2.4. Note that since the largest discontinuation time of 107 days is censored, $\hat{S}(t)$ is not defined beyond $t = 107$.

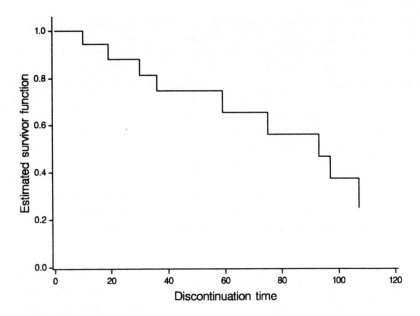

Figure 2.4 Kaplan–Meier estimate of the survivor function for the data from Example 1.1.

2.1.3 * Standard error of the Kaplan–Meier estimate

Because the Kaplan–Meier estimate is the most important and widely used estimate of the survivor function, the standard error of $\hat{S}(t)$ will be derived in this section. The details of this derivation can be omitted on a first reading.

The Kaplan–Meier estimate of the survivor function for any value of t in the interval from $t_{(k)}$ to $t_{(k+1)}$ can be written as

$$\hat{S}(t) = \prod_{j=1}^{k} \hat{p}_j,$$

for $k = 1, 2, \ldots, r$, where $\hat{p}_j = (n_j - d_j)/n_j$ is the estimated probability that an individual survives through the time interval which begins at $t_{(j)}, j = 1, 2, \ldots, r$. Taking logarithms,

$$\log \hat{S}(t) = \sum_{j=1}^{k} \log \hat{p}_j,$$

and so the variance of $\log \hat{S}(t)$ is given by

$$\text{var} \left\{ \log \hat{S}(t) \right\} = \sum_{j=1}^{k} \text{var} \left\{ \log \hat{p}_j \right\}. \tag{2.6}$$

Now, the number of individuals who survive through the interval beginning at $t_{(j)}$ can be assumed to have a binomial distribution with parameters n_j and p_j, where p_j is the true probability of survival through that interval. The observed number who survive is $n_j - d_j$, and using the result that the variance of a binomial random variable with parameters n, p is $np(1 - p)$, the variance of $n_j - d_j$ is given by

$$\operatorname{var}(n_j - d_j) = n_j p_j (1 - p_j).$$

Since $\hat{p}_j = (n_j - d_j)/n_j$, the variance of $\hat{p}_j$ is $\operatorname{var}(n_j - d_j)/n_j^2$, that is, $p_j(1 - p_j)/n_j$. The variance of $\hat{p}_j$ may then be estimated by

$$\hat{p}_j(1 - \hat{p}_j)/n_j. \tag{2.7}$$

In order to obtain the variance of $\log \hat{p}_j$, we make use of a general result for the approximate variance of a function of a random variable. According to this result, the variance of a function $g(X)$ of the random variable X is given by

$$\operatorname{var}\{g(X)\} \approx \left\{\frac{\mathrm{d}g(X)}{\mathrm{d}X}\right\}^2 \operatorname{var}(X). \tag{2.8}$$

This is known as the *Taylor series approximation* to the variance of a function of a random variable. Using equation (2.8), the approximate variance of $\log \hat{p}_j$ is $\operatorname{var}(\hat{p}_j)/\hat{p}_j^2$, and using expression (2.7), the approximate estimated variance of $\log \hat{p}_j$ is $(1 - \hat{p}_j)/(n_j \hat{p}_j)$, which on substitution for $\hat{p}_j$, reduces to

$$\frac{d_j}{n_j(n_j - d_j)}. \tag{2.9}$$

From equation (2.6),

$$\operatorname{var}\left\{\log \hat{S}(t)\right\} \approx \sum_{j=1}^{k} \frac{d_j}{n_j(n_j - d_j)}, \tag{2.10}$$

and a further application of the result in equation (2.8) gives

$$\operatorname{var}\left\{\log \hat{S}(t)\right\} \approx \frac{1}{\left[\hat{S}(t)\right]^2} \operatorname{var}\left\{\hat{S}(t)\right\},$$

so that

$$\operatorname{var}\left\{\hat{S}(t)\right\} \approx \left[\hat{S}(t)\right]^2 \sum_{j=1}^{k} \frac{d_j}{n_j(n_j - d_j)}. \tag{2.11}$$

Finally, the standard error of the Kaplan–Meier estimate of the survivor function, defined to be the square root of the estimated variance of the estimate, is given by

$$\text{s.e.}\left\{\hat{S}(t)\right\} \approx [\hat{S}(t)]\left\{\sum_{j=1}^{k} \frac{d_j}{n_j(n_j - d_j)}\right\}^{\frac{1}{2}}, \tag{2.12}$$

for $t_{(k)} \le t < t_{(k+1)}$. This result is known as *Greenwood's formula*.

If there are no censored survival times, $n_j - d_j = n_{j+1}$, and expression (2.9) becomes $(n_j - n_{j+1})/n_j n_{j+1}$. Now,

$$\sum_{j=1}^{k} \frac{n_j - n_{j+1}}{n_j n_{j+1}} = \sum_{j=1}^{k} \left(\frac{1}{n_{j+1}} - \frac{1}{n_j} \right) = \frac{n_1 - n_{k+1}}{n_1 n_{k+1}},$$

which can be written as

$$\frac{1 - \hat{S}(t)}{n_1 \hat{S}(t)},$$

since $\hat{S}(t) = n_{k+1}/n_1$ for $t_{(k)} \leq t < t_{(k+1)}$, $k = 1, 2, \ldots r - 1$, in the absence of censoring. Hence, from equation (2.11), the estimated variance of $\hat{S}(t)$ is $\hat{S}(t)[1 - \hat{S}(t)]/n_1$. This is an estimate of the variance of the empirical survivor function, $\tilde{S}(t)$, on the assumption that the number of individuals at risk at time t has a binomial distribution with parameters $n_1, S(t)$.

2.1.4 * Confidence intervals for values of the survivor function

Once the standard error of $\hat{S}(t)$ has been calculated, a *confidence interval* for the corresponding value of $\hat{S}(t)$ can be found. A confidence interval is an interval estimate of the survivor function, and is the interval which is such that there is a prescribed probability that the value of the true survivor function is included within it.

A confidence interval for the true value of the survivor function at time t is obtained by assuming that the estimated value of the survivor function at t is normally distributed with mean $S(t)$ and estimated variance given by equation (2.11). The interval is computed from *percentage points* of the standard normal distribution. Thus, if Z is a random variable that has a standard normal distribution, the upper (one-sided) $\alpha/2$-point, or the two-sided α-point, of this distribution is that value $z_{\alpha/2}$ which is such that $P(Z > z_{\alpha/2}) = \alpha/2$. This probability is the area under the standard normal curve to the right of $z_{\alpha/2}$, as illustrated in Figure 2.5. For example, the two-sided 5% and 1% points of the standard normal distribution, $z_{0.025}$ and $z_{0.005}$, are 1.96 and 2.58, respectively.

A $100(1-\alpha)\%$ confidence interval for $S(t)$, for a given value of t, is the interval from $\hat{S}(t) - z_{\alpha/2}$ s.e. $\{\hat{S}(t)\}$ to $\hat{S}(t) + z_{\alpha/2}$ s.e. $\{\hat{S}(t)\}$, where s.e. $\{\hat{S}(t)\}$ is found from equation (2.12). These intervals for $S(t)$ can be superimposed on a graph of the estimated survivor function, as shown in Example 2.4.

One difficulty with this procedure arises from the fact that the confidence intervals are symmetric. When the estimated survivor function is close to zero or unity, symmetric intervals are inappropriate, since they can lead to confidence limits for the survivor function that lie outside the interval (0,1). A pragmatic solution to this problem is to replace any limit that is greater than unity by 1.0, and any limit that is less than zero by 0.0.

An alternative procedure is to transform $\hat{S}(t)$ to a value in the range $(-\infty, \infty)$, and obtain a confidence interval for the transformed value. The resulting confi-

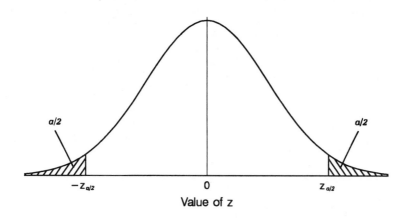

Figure 2.5 Upper and lower $\alpha/2$-points of the standard normal distribution.

dence limits are then back-transformed to give a confidence interval for $S(t)$ itself. Possible transformations are the logistic transformation, $\log[S(t)/\{1-S(t)\}]$, and the complementary log-log transformation, $\log\{-\log S(t)\}$. Note that from equation (1.7), the latter quantity is the logarithm of the cumulative hazard function. In either case, the standard error of the transformed value of $\hat{S}(t)$ can be found using the approximation in equation (2.8).

For example, the variance of $\log\{-\log \hat{S}(t)\}$ is obtained from the expression for var $\{\log \hat{S}(t)\}$ in equation (2.10). Using the general result in equation (2.8),

$$\text{var}\{\log(-X)\} \approx \frac{1}{X^2} \text{var}(X),$$

and setting $X = \log \hat{S}(t)$ gives

$$\text{var}\left[\log\{-\log \hat{S}(t)\}\right] \approx \frac{1}{[\log \hat{S}(t)]^2} \sum_{j=1}^{k} \frac{d_j}{n_j(n_j - d_j)}.$$

The standard error of $\log\{-\log \hat{S}(t)\}$ is the square root of this quantity. This leads to $100(1 - \alpha)\%$ limits of the form

$$\hat{S}(t)^{\exp[\pm z_{\alpha/2}\text{s.e.}\{\log[-\log \hat{S}(t)]\}]},$$

where $z_{\alpha/2}$ is the upper $\alpha/2$-point of the standard normal distribution.

A further problem is that in the tails of the distribution of the survival times, that

Table 2.3 Standard error of $\hat{S}(t)$ and confidence intervals for $S(t)$ for the data from Example 1.1.

Time interval	$\hat{S}(t)$	s.e. $\{\hat{S}(t)\}$	95% confidence interval
0–	1.0000	0.0000	
10–	0.9444	0.0540	(0.839, 1.000)
19–	0.8815	0.0790	(0.727, 1.000)
30–	0.8137	0.0978	(0.622, 1.000)
36–	0.7459	0.1107	(0.529, 0.963)
59–	0.6526	0.1303	(0.397, 0.908)
75–	0.5594	0.1412	(0.283, 0.836)
93–	0.4662	0.1452	(0.182, 0.751)
97–	0.3729	0.1430	(0.093, 0.653)
107	0.2486	0.1392	(0.000, 0.522)

is, when $\hat{S}(t)$ is close to zero or unity, the variance of $\hat{S}(t)$ obtained using Greenwood's formula can underestimate the actual variance. In these circumstances, an alternative expression for the standard error of $\hat{S}(t)$ may be used. Peto *et al.* (1977) propose that the standard error of $\hat{S}(t)$ should be obtained from the equation

$$\text{s.e.} \{\hat{S}(t)\} = \frac{\hat{S}(t)\sqrt{\{1 - \hat{S}(t)\}}}{\sqrt{(n_k)}},$$

for $t_{(k)} \leq t < t_{(k+1)}$, $k = 1, 2, \ldots r$, where $\hat{S}(t)$ is the Kaplan–Meier estimate of $S(t)$ and n_k is the number of individuals at risk at $t_{(k)}$, the start of the k'th constructed time interval.

This expression for the standard error of $\hat{S}(t)$ is conservative, in the sense that the standard errors obtained will tend to be larger than they ought to be. For this reason, the Greenwood estimate is recommended for general use.

Example 2.4 Time to discontinuation of the use of an IUD
The standard error of the estimated survivor function, and 95% confidence limits for the corresponding true value of the function, for the data from Example 1.1 on the times to discontinuation of use of an IUD, are given in Table 2.3. In this table, confidence limits outside the range (0,1) have been replaced by zero or unity.

From this table we see that in general the standard error of the estimated survivor function increases with the discontinuation time. The reason for this is that estimates of the survivor function at later times are based on fewer individuals. A graph of the estimated survivor function, with the 95% confidence limits shown as dashed lines, is given in Figure 2.6.

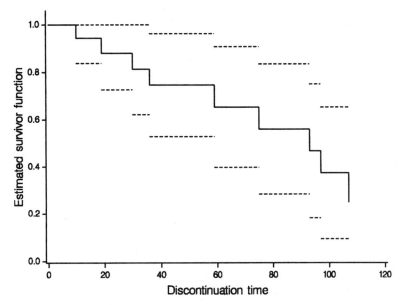

Figure 2.6   Estimated survivor function and 95% confidence limits for $S(t)$.

2.2 Estimating the hazard function

A single sample of survival data may also be summarized through the hazard function, which shows the dependence of the instantaneous risk of death on time. There are a number of ways of estimating this function, two of which are described in this section.

2.2.1 Life-table estimate of the hazard function

Suppose that the observed survival times have been grouped into a series of m intervals, as in the construction of the life-table estimate of the survivor function. An appropriate estimate of the average hazard of death per unit time over each interval is the observed number of deaths in that interval divided by the average time survived in that interval. This latter quantity is the average number of persons at risk in the interval, multiplied by the length of the interval. Let the number of deaths in the j'th time interval be d_j, $j = 1, 2, \ldots, m$, and suppose that n'_j is the average number of individuals at risk of death in that interval, where n'_j is given by equation (2.2). Assuming that the death rate is constant during the j'th interval, the average time survived in that interval is $(n'_j - d_j/2)\tau_j$, where τ_j is the length of the j'th time interval. The life-table estimate of the hazard function in the j'th

Table 2.4 Life-table estimate of the hazard function for the data from Example 1.3.

Time period	τ_j	d_j	n'_j	$h^*(t)$
0–	12	16	46.0	0.0351
12–	12	10	26.0	0.0397
24–	12	1	14.0	0.0062
36–	12	3	12.5	0.0227
48–	12	2	8.0	0.0238
60–	36	4	4.5	0.0444

time interval is then given by

$$h^*(t) = \frac{d_j}{(n'_j - d_j/2)\tau_j},$$

for $t'_j \leq t < t'_{j+1}, j = 1, 2, \ldots, m$, so that $h^*(t)$ is a step-function.

The asymptotic standard error of this estimate has been shown by Gehan (1969) to be given by

$$\text{s.e.}\left\{h^*(t)\right\} = \frac{h^*(t)\sqrt{\{1 - [h^*(t)\tau_j/2]^2\}}}{\sqrt{(d_j)}}.$$

Confidence intervals for the corresponding true hazard over each of the m time intervals can be obtained in the manner described in Section 2.1.4.

Example 2.5 Survival of multiple myeloma patients

The life-table estimate of the survivor function for the data from Example 1.3 on the survival times of 48 multiple myeloma patients was given in Table 2.1. Using the same time intervals as were used in Example 2.2, calculations leading to the life-table estimate of the hazard function are given in Table 2.4.

The estimated hazard function is plotted as a step-function in Figure 2.7. The general pattern is for the hazard to remain roughly constant over the first two years from diagnosis, after which time it declines and then increases gradually. However, some caution is needed in interpreting this estimate, as there are few deaths two years after diagnosis.

2.2.2 Kaplan–Meier type estimate

A natural way of estimating the hazard function for ungrouped survival data is to take the ratio of the number of deaths at a given death time to the number of individuals at risk at that time. If the hazard function is assumed to be constant between successive death times, the hazard per unit time can be found by further dividing by the time interval. Thus if there are d_j deaths at the j'th death time,

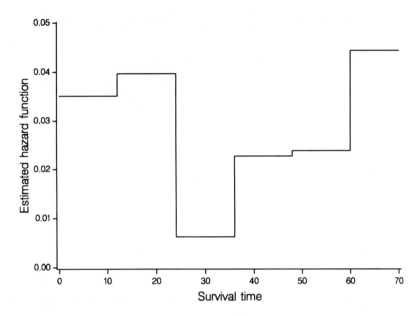

Figure 2.7 Life-table estimate of the hazard function for the data from Example 1.3.

$t_{(j)}, j = 1, 2, \ldots, r$, and n_j at risk at time $t_{(j)}$, the hazard function in the interval from $t_{(j)}$ to $t_{(j+1)}$ can be estimated by

$$\hat{h}(t) = \frac{d_j}{n_j \tau_j}, \tag{2.13}$$

for $t_{(j)} \le t < t_{(j+1)}$, where here $\tau_j = t_{(j+1)} - t_{(j)}$. Notice that it is not possible to use equation (2.13) to estimate the hazard in the interval which begins at the final death time, since this interval is open-ended. The estimate in equation (2.13) is referred to as a Kaplan–Meier type estimate because the estimated survivor function derived from it is the Kaplan–Meier estimate.

To show this, note that since $\hat{h}(t)$, $t_{(j)} \le t < t_{(j+1)}$, is an estimate of the risk of death per unit time in the j'th interval, the probability of death in that interval is $\hat{h}(t)\tau_j$, that is, d_j/n_j. Hence an estimate of the corresponding survival probability in that interval is $1 - (d_j/n_j)$, and the estimated survivor function is as given by equation (2.4).

Example 2.6 Time to discontinuation of the use of an IUD
Consider again the data on the time to discontinuation of the use of an IUD for 18 women, given in Example 1.1. The Kaplan–Meier estimate of the survivor function for these data was given in Table 2.2, and Table 2.5 gives the corresponding Kaplan–Meier type estimate of the hazard function computed from equa-

Table 2.5 Kaplan–Meier type estimate
of the hazard function for the data from
Example 1.1.

Time interval	τ_j	n_j	d_j	$\hat{h}(t)$
0–	10	18	0	0.0000
10–	9	18	1	0.0062
19–	11	15	1	0.0061
30–	6	13	1	0.0128
36–	23	12	1	0.0036
59–	16	8	1	0.0078
75–	18	7	1	0.0079
93–	4	6	1	0.0417
97–	10	5	1	0.0200

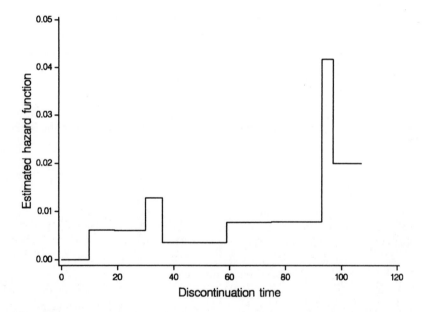

Figure 2.8 Kaplan–Meier type estimate of the hazard function for the data from
Example 1.1.

tion (2.13).

Figure 2.8 shows a plot of the estimated hazard function. From this figure, there
is some evidence that the longer the IUD is used, the greater is the risk of discontinuation, but the picture is not very clear.

2.2.3 *Estimating the cumulative hazard function*

We will see later that the cumulative hazard function is important in the identification of models for survival data. Accordingly, methods which can be used to estimate this function will be described here.

The cumulative hazard at time t, $H(t)$, was defined in equation (1.6) to be the integral of the hazard function, but is more conveniently found using equation (1.7). According to this result, $H(t) = -\log S(t)$, and so if $\hat{S}(t)$ is the Kaplan–Meier estimate of the survivor function, $\hat{H}(t) = -\log \hat{S}(t)$ is an appropriate estimate of the cumulative hazard to time t.

Now, using equation (2.4),

$$\hat{H}(t) = -\sum_{j=1}^{k} \log \left(\frac{n_j - d_j}{n_j} \right).$$

Moreover, the series expansion of $\log(1 - x)$ is $-x - x^2/2 + \cdots$, and so

$$\log \left(\frac{n_j - d_j}{n_j} \right) = \log \left(1 - \frac{d_j}{n_j} \right) \approx -\frac{d_j}{n_j},$$

on neglecting terms involving $(d_j/n_j)^2$ and higher-order terms. Consequently,

$$\hat{H}(t) \approx \sum_{j=1}^{k} \frac{d_j}{n_j},$$

which is the cumulative sum of the estimated probabilities of death from the first to the k'th time interval, $k = 1, 2, \ldots, r$. This quantity therefore has immediate intuitive appeal as an estimator of the cumulative hazard.

2.3 Estimating the median and percentiles of survival times

Since the distribution of survival times tends to be positively skew, the median is the preferred summary measure of the location of the distribution. Once the survivor function has been estimated, it is straightforward to obtain an estimate of the *median survival time*. This is the time beyond which 50% of the individuals in the population under study are expected to survive, and is given by that value $t(50)$ which is such that $S\{t(50)\} = 0.5$.

Because the non-parametric estimates of $S(t)$ are step-functions, it will not usually be possible to realize an estimated survival time that makes the survivor function exactly equal to 0.5. Instead, the estimated median survival time, $\hat{t}(50)$, is defined to be the smallest observed survival time for which the value of the estimated survivor function is less than 0.5.

In mathematical terms,

$$\hat{t}(50) = \min\{t_i \mid \hat{S}(t_i) \leq 0.5\},$$

where t_i is the observed survival time for the i'th individual, $i = 1, 2, \ldots, n$. Since the estimated survivor function only changes at a death time, this is equivalent to the definition

$$\hat{t}(50) = \min\{t_{(j)} \mid \hat{S}(t_{(j)}) \le 0.5\},$$

where $t_{(j)}$ is the j'th ordered death time, $j = 1, 2, \ldots, r$.

In the situation where there are no censored survival times, the estimated median survival time will be the smallest time beyond which 50% of the individuals in the sample survive.

Example 2.7 Time to discontinuation of the use of an IUD
The Kaplan–Meier estimate of the survivor function for the data from Example 1.1 on the time to discontinuation of the use of an IUD was given in Table 2.2. The estimated survivor function, $\hat{S}(t)$, for these data was shown in Figure 2.4. From the estimated survivor function, the smallest discontinuation time beyond which the estimated probability of discontinuation is less than 0.5 is 93 weeks. This is therefore the estimated median time to discontinuation of the IUD for this group of women.

It sometimes happens that the estimated survivor function is greater than 0.5 for all values of t. In such cases, the median survival time cannot be estimated. It would then be natural to summarize the data in terms of the estimated survival probabilities at particular time points.

A similar procedure to that described above can be used to estimate other *percentiles* of the distribution of survival times. The p'th percentile of the distribution of survival times is defined to be the value $t(p)$ which is such that $F\{t(p)\} = p/100$. In terms of the survivor function, $t(p)$ is such that $S\{t(p)\} = 1 - (p/100)$, so that for example the 10th and 90th percentiles are given by

$$S\{t(10)\} = 0.9, \qquad S\{t(90)\} = 0.1,$$

respectively. Using the estimated survivor function, the estimated p'th percentile is the smallest observed survival time, $\hat{t}(p)$, for which $\hat{S}\{\hat{t}(p)\} < 1 - (p/100)$.

Estimates of the dispersion of a sample of survival data are not widely used, but should such an estimate be required, the *semi-interquartile range* (SIQR) can be calculated. This is defined to be half the difference between the 75th and 25th percentiles of the distribution of survival times. Hence,

$$\text{SIQR} = \frac{1}{2}\{t(75) - t(25)\},$$

where $t(25)$ and $t(75)$ are the 25th and 75th percentiles of the survival time distribution. These two percentiles are also known as the *first* and *third quartiles*, respectively. The corresponding sample-based estimate of the SIQR is $\{\hat{t}(75) - \hat{t}(25)\}/2$. Like the variance, the larger the value of the SIQR, the more dispersed is the survival time distribution.

Example 2.8 Time to discontinuation of the use of an IUD
From the Kaplan–Meier estimate of the survivor function for the data from Ex-

ample 1.1, given in Table 2.2, the 25th and 75th percentiles of the distribution of discontinuation times are 36 and 107 weeks, respectively. Hence, the SIQR of the distribution is estimated to be 35.5 weeks.

2.4 * Confidence intervals for the median and percentiles

Approximate confidence intervals for the median and other percentiles of a distribution of survival times can be found once the variance of the estimated percentile has been obtained. An expression for the approximate variance of a percentile can be derived from a direct application of the general result for the variance of a function of a random variable in equation (2.8). Using this result,

$$\text{var}\,[\hat{S}\{t(p)\}] = \left(\frac{\mathrm{d}\hat{S}\{t(p)\}}{\mathrm{d}t(p)}\right)^2 \text{var}\,\{t(p)\}, \tag{2.14}$$

where $t(p)$ is the p'th percentile of the distribution and $\hat{S}\{t(p)\}$ is the Kaplan–Meier estimate of the survivor function at $t(p)$. Now,

$$\frac{\mathrm{d}\hat{S}\{t(p)\}}{\mathrm{d}t(p)} = -\hat{f}\{t(p)\},$$

an estimate of the probability density function of the survival times at $t(p)$, and on rearranging equation (2.14), we get

$$\text{var}\,\{t(p)\} = \left(\frac{1}{\hat{f}\{t(p)\}}\right)^2 \text{var}\,[\hat{S}\{t(p)\}].$$

The standard error of $\hat{t}(p)$, the estimated p'th percentile, is therefore given by

$$\text{s.e.}\,\{\hat{t}(p)\} = \frac{1}{\hat{f}\{\hat{t}(p)\}} \text{s.e.}\,[\hat{S}\{\hat{t}(p)\}]. \tag{2.15}$$

The standard error of $\hat{S}\{\hat{t}(p)\}$ is found using Greenwood's formula for the standard error of the Kaplan–Meier estimate of the survivor function, given in equation (2.12), while an estimate of the probability density function at $\hat{t}(p)$ is

$$\hat{f}\{\hat{t}(p)\} = \frac{\hat{S}\{\hat{u}(p)\} - \hat{S}\{\hat{l}(p)\}}{\hat{l}(p) - \hat{u}(p)},$$

where

$$\hat{u}(p) = \max\{t_{(j)} \mid \hat{S}(t_{(j)}) \geq 1 - \frac{p}{100} + \epsilon\},$$

and

$$\hat{l}(p) = \min\{t_{(j)} \mid \hat{S}(t_{(j)}) \leq 1 - \frac{p}{100} - \epsilon\},$$

for $j = 1, 2, \ldots, r$, and small values of ϵ. In many cases, taking $\epsilon = 0.05$ will be satisfactory, but a larger value of ϵ will be needed if $\hat{u}(p)$ and $\hat{l}(p)$ turn out to be equal. In particular, from equation (2.15), the standard error of the median survival time is given by

$$\text{s.e.} \{\hat{t}(50)\} = \frac{1}{\hat{f}\{\hat{t}(50)\}} \text{ s.e.} [\hat{S}\{\hat{t}(50)\}], \qquad (2.16)$$

where $\hat{f}\{\hat{t}(50)\}$ can be found from

$$\hat{f}\{\hat{t}(50)\} = \frac{\hat{S}\{\hat{u}(50)\} - \hat{S}\{\hat{l}(50)\}}{\hat{l}(50) - \hat{u}(50)}. \qquad (2.17)$$

In this expression, $\hat{u}(50)$ is the largest survival time for which the Kaplan–Meier estimate of the survivor function exceeds 0.55, and $\hat{l}(50)$ is the smallest survival time for which the survivor function is less than or equal to 0.45.

Once the standard error of the estimate p'th percentile has been found, a $100(1 - \alpha)\%$ confidence interval for $t(p)$ has limits of

$$\hat{t}(p) \pm z_{\alpha/2} \text{ s.e.} \{\hat{t}(p)\},$$

where $z_{\alpha/2}$ is the upper (one-sided) $\alpha/2$-point of the standard normal distribution.

This interval estimate is only approximate, in the sense that the probability that the interval includes the true percentile will not be exactly $1 - \alpha$. A number of methods have been proposed for constructing confidence interval for the median with superior properties, although these alternatives are more difficult to compute than the interval estimate derived in this section. The discussion on computer software for survival analysis in Chapter 11 includes information on the methods used by packages when computing standard errors and confidence intervals for percentiles.

Example 2.9 Time to discontinuation of the use of an IUD
The data on the discontinuation times for users of an IUD given in Example 1.1 is now used to illustrate the calculation of a confidence interval for the median discontinuation time. From Example 2.7, the estimated median discontinuation time for this group of women is given by $\hat{t}(50) = 93$ weeks. Also, from Table 2.3, the standard error of the Kaplan–Meier estimate of the survivor function at this time is given by s.e. $[\hat{S}\{\hat{t}(50)\}] = 0.1452$.

To obtain the standard error of $\hat{t}(50)$ using equation (2.16), we need an estimate of the density function at the estimated median discontinuation time. This is obtained from equation (2.17). The quantities $\hat{u}(50)$ and $\hat{l}(50)$ needed in this equation are such that

$$\hat{u}(50) = \max\{t_{(j)} \mid \hat{S}(t_{(j)}) \geq 0.55\},$$

and

$$\hat{l}(50) = \min\{t_{(j)} \mid \hat{S}(t_{(j)}) \leq 0.45\},$$

where $t_{(j)}$ is the j'th ordered death time, $j = 1, 2, \ldots, 9$. Using Table 2.3, $\hat{u}(50) =$

75 and $\hat{l}(50) = 97$, and so

$$\hat{f}\{\hat{t}(50)\} = \frac{\hat{S}(75) - \hat{S}(97)}{97 - 75} = \frac{0.5594 - 0.3729}{22} = 0.0085.$$

Then, the standard error of the median is given by

$$\text{s.e.}\,\{\hat{t}(50)\} = \frac{1}{0.0085} \times 0.1452 = 17.13.$$

A 95% confidence interval for the median discontinuation time has limits of

$$93 \pm 1.96 \times 17.13,$$

and so the required interval estimate for the median ranges from 59 to 127 days.

2.5 Comparison of two groups of survival data

The simplest way of comparing the survival times obtained from two groups of individuals is to plot the corresponding estimates of the two survivor functions on the same axes. The resulting plot can be quite informative, as the following example shows.

Example 2.10 Prognosis for women with breast cancer
Data on the survival times of women with breast cancer, grouped according to whether or not sections of a tumour were positively stained with HPA, were given in Example 1.2. The Kaplan–Meier estimate of the survivor function for each of the two groups of survival times is plotted in Figure 2.9.

This figure shows that the estimated survivor function for those women with negatively stained tumours is always greater than that for women with positively stained tumours. This means that at any time t, the estimated probability of survival beyond t is greater for women with negative staining, suggesting that the result of the HPA staining procedure might be a useful prognostic indicator. In particular, those women whose tumours are positively stained appear to have a poorer prognosis than those with negatively stained tumours.

There are two possible explanations for an observed difference between two estimated survivor functions, such as those in Example 2.10. One explanation is that there is a real difference between the survival times of the two groups of individuals, so that those in one group have a different survival experience from those in the other. An alternative explanation is that there are no real differences between the survival times in each group, and that the difference that has been observed is merely the result of chance variation. To help distinguish between these two possible explanations, we use a procedure known as the *hypothesis test*. Because the concept of the hypothesis test has a central role in the analysis of survival data, the underlying basis for this procedure is described in detail in the following section.

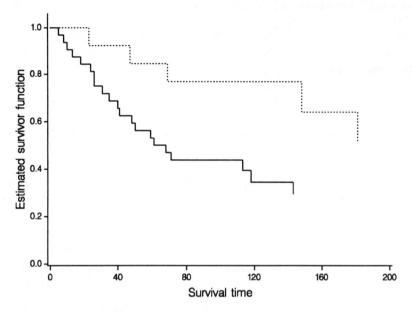

Figure 2.9 Kaplan–Meier estimate of the survivor functions for women with tumours which were positively stained (—) and negatively stained (· · ·).

2.5.1 Hypothesis testing

The hypothesis test is a procedure which enables us to assess the extent to which an observed set of data are consistent with a particular hypothesis, known as the *working* or *null hypothesis*. A null hypothesis generally represents a simplified view of the data-generating process, and is typified by hypotheses which specify that there is no difference between two groups of survival data, or that there is no relationship between survival time and explanatory variables such as age or serum cholesterol level. The null hypothesis is then the hypothesis that will be adopted, and subsequently acted upon, unless the data indicate that it is untenable.

The next step is to formulate a *test statistic* which measures the extent to which the observed data depart from the null hypothesis. In general, the test statistic is so constructed that the larger the value of the statistic, the greater the departure from the null hypothesis. Hence, if the null hypothesis is that there is no difference between two groups, relatively large values of the test statistic will be interpreted as evidence against this null hypothesis.

Once the value of the test statistic has been obtained from the observed data, we calculate the probability of obtaining a value as extreme or more extreme than the observed value, when the null hypothesis is true. This quantity summarizes the strength of the evidence in the sample data against the null hypothesis, and is known as the *probability value*, or P-*value* for short. If the P-value is large, we would conclude that it is quite likely that the observed data would have been obtained when the null hypothesis was true, and that there is no evidence to reject the null hypothesis. On the other hand, if the P-value is small, this would be interpreted as evidence against the null hypothesis; the smaller the P-value, the stronger the evidence.

In order to obtain the P-value for a hypothesis test, the test statistic must have a probability distribution that is known, or at least approximately known, when the null hypothesis is true. This probability distribution is referred to as the *null distribution* of the test statistic. More specifically, consider a test statistic W, which is such that the larger the observed value of the test statistic, w, the greater the deviation of the observed data from that expected under the null hypothesis. If W has a continuous probability distribution, the P-value is then $P(W \geq w) = 1 - F(w)$, where $F(w)$ is the distribution function of W, under the null hypothesis, evaluated at w.

In some applications, the most natural test statistic is one for which large positive values correspond to departures from the null hypothesis in one direction, while large negative values correspond to departures in the opposite direction. For example, suppose that patients suffering from a particular illness have been randomized to receive either a standard treatment or a new treatment, and their survival times are recorded. In this situation, a null hypothesis of interest will be that there is no difference in the survival experience of the patients in the two treatment groups. The extent to which the data are consistent with this null hypothesis might then be summarized by a test statistic for which positive values indicate that the new treatment is superior to the standard, while negative values indicate that the standard treatment is superior. When departures from the null hypothesis in either direction are equally important, the null hypothesis is said to have a *two-sided alternative*, and the hypothesis test itself is referred to as a two-sided test.

If W is a test statistic for which large positive or large negative observed values lead to rejection of the null hypothesis, a new test statistic, such as $|W|$ or W^2, can be defined, so that only large positive values of the new statistic indicate that there is evidence against the null hypothesis. For example, suppose that W is a test statistic which under the null hypothesis has a standard normal distribution. If w is the observed value of W, the appropriate P-value is $P(W \leq -|w|) + P(W \geq |w|)$, which in view of the symmetry of the standard normal distribution, is $2P(W \geq |w|)$. Alternatively, we can make use of the result that if W has a standard normal distribution, W^2 has a chi-squared distribution on one degree of freedom, written χ_1^2. Thus a P-value for the two-sided hypothesis test based on the statistic W is the probability that a χ_1^2 random variable exceeds w^2. The required P-value can therefore be found using tables of the standard normal or chi-squared distribution functions.

When interest centres on departures in a particular direction, the hypothesis test is said to be *one-sided*. For example, in comparing the survival times of two groups of patients where one group receives a standard treatment and the other group a new treatment, it might be argued that the new treatment cannot possibly be inferior to the standard. Then, the only relevant alternative to the null hypothesis of no treatment difference is that the new treatment is superior. If positive values of the test statistic W reflect the superiority of the new treatment, the P-value is then $P(W \geq w)$. If W has a standard normal distribution, this P-value is half of that which would have been obtained for the corresponding two-sided alternative hypothesis.

A one-sided hypothesis test can only be appropriate when there is no interest whatsoever in departures from the null hypothesis in the opposite direction to that specified in the one-sided alternative. For example, consider again the comparison of a new treatment with a standard treatment, and suppose that the observed value of the test statistic is either positive or negative, depending on whether the new treatment is superior or inferior to the standard. If the alternative to the null hypothesis of no treatment difference is that the new treatment is superior, a large negative value of the test statistic would not be regarded as evidence against the null hypothesis. Instead, it would be assumed that this large negative value is simply the result of chance variation. Generally speaking, the use of one-sided tests can rarely be justified in medical research, and so two-sided tests will be used throughout this book.

If a P-value is smaller than some value α, we say that the hypothesis is rejected at the $100\alpha\%$ *level of significance*. The observed value of the test statistic is then said to be significant at this level. But how do we decide on the basis of the P-value whether or not a null hypothesis should actually be rejected? Traditionally, P-values of 0.05 or 0.01 have been used in reaching a decision about whether or not a null hypothesis should be rejected, so that if $P < 0.05$, for example, the null hypothesis is rejected at the 5% significance level. Guidelines such as these are not hard and fast rules and should not be interpreted rigidly. For example, there is no practical difference between a P-value of 0.046 and 0.056, even though only the former indicates that the observed value of the test statistic is significant at the 5% level.

Instead of reporting that a null hypothesis is rejected or not rejected at some specified significance level, a more satisfactory policy is to report the actual P-value. This P-value can then be interpreted as a measure of the strength of evidence against the null hypothesis, using a vocabulary which depends on the range within which the P-value lies. Thus, if $P > 0.1$, there is said to be no evidence to reject the null hypothesis; if $0.05 < P \leq 0.1$, there is slight evidence against the null hypothesis; if $0.01 < P \leq 0.05$, there is moderate evidence against the null hypothesis; if $0.001 < P \leq 0.01$, there is strong evidence against the null hypothesis, and if $P \leq 0.001$, the evidence against the null hypothesis is overwhelming.

An alternative to quoting the exact P-value associated with a hypothesis test, is to compare the observed value of the test statistic with those values which would correspond to particular P-values, when the null hypothesis is true. Values of the test statistic which lead to rejection of the null hypothesis at particular levels of significance can be found from tables of the percentage points of the null distribution of that statistic. In particular, if W is a test statistic which has a standard normal distribution, for a two-sided test, the upper $\alpha/2$-point of the distribution, depicted in Figure 2.5, is the value of the test statistic for which the P-value is α. For example, values of the test statistic of 1.96, 2.58 and 3.29 correspond to P-values of 0.05, 0.01 and 0.001. Thus, if the observed value of W were between 1.96 and 2.58, we would declare that $0.01 < P < 0.05$. On the other hand, if the null distribution of W is chi-squared on one degree of freedom, the upper α-point of the distribution is the value of the test statistic which would give a P-value of α. Then, values of the test statistic of 3.84, 6.64 and 10.83 correspond to P-values of 0.05, 0.01 and 0.001 respectively. Notice that these values are simply the squares of those for the standard normal distribution, which they must be in view of the fact that the square of a standard normal random variable has a chi-squared distribution on one degree of freedom.

For commonly encountered probability distributions, such as the normal and chi-squared, percentage points are tabulated in many introductory text books on statistics, or in statistical tables such as those of Lindley and Scott (1984). Statistical software packages used in computer-based statistical analyses of survival data usually provide the exact P-values associated with hypothesis tests as a matter of course. Note that when these are rounded off to, say, three decimal places, a P-value of 0.000 should be interpreted as $P < 0.001$.

In deciding on a course of action, such as whether or not to reject the hypothesis that there is no difference between two treatments, the statistical evidence summarized in the P-value for the hypothesis test will be just one ingredient of the decision making process. In addition to the statistical evidence, there will also be scientific evidence to consider. This may, for example, concern whether the size of the treatment effect is clinically important. In particular, in a large trial, a difference between two treatments that is significant at, say, the 5% level may be found when the magnitude of the treatment effect is so small that it does not indicate a major scientific breakthrough. On the other hand, a new formulation of a treatment may prolong life by a factor of two, and yet, because of small sample sizes used in the study, may not appear to be significantly different from the standard.

Rather than report findings in terms of the results of a hypothesis testing procedure, it is more informative to provide an estimate of the size of any treatment difference, supported by a confidence interval for this difference. Unfortunately, the non-parametric approaches to the analysis of survival data being considered in this chapter do not lend themselves to this approach. We will therefore return to this theme in subsequent chapters when we consider models for survival data.

In the comparison of two groups of survival data, there are a number of methods which can be used to quantify the extent of between-group differences. Two non-parametric procedures will now be considered, namely the log-rank test and the Wilcoxon test.

Table 2.6 Number of deaths at the j'th death time in each of two groups of individuals.

Group	Number of deaths at $t_{(j)}$	Number surviving beyond $t_{(j)}$	Number at risk just before $t_{(j)}$
I	d_{1j}	$n_{1j} - d_{1j}$	n_{1j}
II	d_{2j}	$n_{2j} - d_{2j}$	n_{2j}
Total	d_j	$n_j - d_j$	n_j

2.5.2 The log-rank test

In order to construct the log-rank test, we begin by considering separately each death time in two groups of survival data. These groups will be labelled Group I and Group II. Suppose that there are r distinct death times, $t_{(1)} < t_{(2)} < \cdots < t_{(r)}$, across the two groups, and that at time $t_{(j)}$, d_{1j} individuals in Group I and d_{2j} individuals in Group II die, for $j = 1, 2, \ldots r$. Unless two or more individuals in a group have the same recorded death time, the values of d_{1j} and d_{2j} will either be zero or unity. Suppose further that there are n_{1j} individuals at risk of death in the first group just before time $t_{(j)}$, and that there are n_{2j} at risk in the second group. Consequently, at time $t_{(j)}$, there are $d_j = d_{1j} + d_{2j}$ deaths in total out of $n_j = n_{1j} + n_{2j}$ individuals at risk. The situation is summarized in Table 2.6.

Now consider the null hypothesis that there is no difference in the survival experiences of the individuals in the two groups. One way of assessing the validity of this hypothesis is to consider the extent of the difference between the observed number of individuals in the two groups who die at each of the death times, and the numbers expected under the null hypothesis. Information about the extent of these differences can then be combined over each of the death times.

If the marginal totals in Table 2.6 are regarded as fixed, and the null hypothesis that survival is independent of group is true, the four entries in this table are solely determined by the value of d_{1j}, the number of deaths at $t_{(j)}$ in Group I. We can therefore regard d_{1j} as a random variable, which can take any value in the range from 0 to the minimum of d_j and n_{1j}. In fact, d_{1j} has a distribution known as the *hypergeometric distribution*, according to which the probability that the random variable associated with the number of deaths in the first group takes the value d_{1j} is

$$\frac{\dbinom{d_j}{d_{1j}} \dbinom{n_j - d_j}{n_{1j} - d_{1j}}}{\dbinom{n_j}{n_{1j}}}. \tag{2.18}$$

In this formula, the expression

$$\dbinom{d_j}{d_{1j}}$$

represents the number of different ways in which d_{1j} times can be chosen from d_j times and is read as '$d_{1j} \ C \ d_j$'. It is given by

$$\begin{pmatrix} d_j \\ d_{1j} \end{pmatrix} = \frac{d_j!}{d_{1j}!(d_j - d_{1j})!},$$

where $d_j!$, read as 'd_j *factorial*', is such that

$$d_j! = d_j \times (d_j - 1) \times \cdots \times 2 \times 1.$$

The other two terms in expression (2.18) are interpreted in a similar manner.

The mean of the hypergeometric random variable d_{1j} is given by

$$e_{1j} = n_{1j}d_j/n_j, \tag{2.19}$$

so that e_{1j} is the expected number of individuals who die at time $t_{(j)}$ in Group I. This value has immediate intuitive appeal, since under the null hypothesis that the probability of death at time $t_{(j)}$ does not depend on the group that an individual is in, the probability of death at $t_{(j)}$ is d_j/n_j. Multiplying this by n_{1j}, gives e_{1j} as the expected number of deaths in Group I at $t_{(j)}$.

The next step is to combine the information from the individual 2×2 tables for each death time to give an overall measure of the deviation of the observed values of d_{1j} from their expected values. The most straightforward way of doing this is to sum the differences $d_{1j} - e_{1j}$ over the total number of death times, r, in the two groups. The resulting statistic is given by

$$U_L = \sum_{j=1}^{r}(d_{1j} - e_{1j}). \tag{2.20}$$

Notice that this is $\sum d_{1j} - \sum e_{1j}$, which is the difference between the total observed and expected numbers of deaths in Group I. This statistic will have zero mean, since $\mathrm{E}\,(d_{1j}) = e_{1j}$. Moreover, since the death times are independent of one another, the variance of U_L is simply the sum of the variances of the d_{1j}. Now, since d_{1j} has a hypergeometric distribution, the variance of d_{1j} is given by

$$v_{1j} = \frac{n_{1j}n_{2j}d_j(n_j - d_j)}{n_j^2(n_j - 1)}, \tag{2.21}$$

so that the variance of U_L is

$$\mathrm{var}\,(U_L) = \sum_{j=1}^{r}v_{1j} = V_L, \tag{2.22}$$

say. Furthermore, it can be shown that U_L has an approximate normal distribution, when the number of death times is not too small. It then follows that $U_L/\sqrt{V_L}$ has a normal distribution with zero mean and unit variance, denoted $N(0,1)$. We therefore write

$$\frac{U_L}{\sqrt{V_L}} \sim N(0,1),$$

where the symbol '$\sim$' is read as 'is distributed as'. The square of a standard normal random variable has a chi-squared distribution on one degree of freedom, denoted χ_1^2, and so we have that

$$\frac{U_L^2}{V_L} \sim \chi_1^2. \qquad (2.23)$$

This method of combining information over a number of 2×2 tables was proposed by Mantel and Haenszel (1959), and is known as the Mantel–Haenszel procedure. In fact, the test based on this statistic has various names, including *Mantel–Cox* and *Peto–Mantel–Haenszel*, but it is probably best known as the log-rank test.

The statistic $W_L = U_L^2/V_L$ summarizes the extent to which the observed survival times in the two groups of data deviate from those expected under the null hypothesis of no group differences. The larger the value of this statistic, the greater the evidence against the null hypothesis. Because the null distribution of W is approximately chi-squared with one degree of freedom, the P-value associated with the test statistic can be obtained from the distribution function of a chi-squared random variable. Alternatively, percentage points of the chi-squared distribution can be used to identify a range within which the P-value lies. An illustration of the log-rank test is presented below in Example 2.11.

Example 2.11 Prognosis for women with breast cancer
In this example, we return to the data on the survival times of women with breast cancer, grouped according to whether a section of the tumour was positively or negatively stained. In particular the null hypothesis that there is no difference in the survival experience of the two groups will be examined using the log-rank test. The required calculations are laid out in Table 2.7.

We begin by ordering the observed death times across the two groups of women; these times are given in column 1 of Table 2.7. The numbers of women in each group who die at each death time and the numbers who are at risk at each time are then calculated. These values are d_{1j}, n_{1j}, d_{2j} and n_{2j} given in columns 2–5 of the table. Columns 6 and 7 contain the total numbers of deaths and the total numbers of women at risk over the two groups, at each death time. The final two columns give the values of e_{1j} and v_{1j}, computed from equations (2.19) and (2.21) respectively. Summing the entries in columns 2 and 8 gives $\sum d_{1j}$ and $\sum e_{1j}$, from which the log-rank statistic can be calculated from $U_L = \sum d_{1j} - \sum e_{1j}$. The value of $V_L = \sum v_{1j}$ can be obtained by summing the entries in the final column. We find that $U_L = 5 - 9.565 = -4.565$ and $V_L = 5.929$, and so the value of the log-rank test statistic is $W_L = (-4.565)^2/5.929 = 3.515$.

The corresponding P-value is calculated from the probability that a chi-squared variate on one degree of freedom is greater than or equal to 3.515, and is 0.061, written $P = 0.061$. This P-value is sufficiently small to cast doubt on the null hypothesis that there is no difference between the survivor functions for the two groups of women. In fact the evidence against the null hypothesis is nearly significant at the 6% level. We therefore conclude that the data do provide some evidence that the prognosis of a breast cancer patient is dependent on the result of the stain-

Table 2.7 Calculation of the log-rank statistic for the data from Example 1.2.

Death time	d_{1j}	n_{1j}	d_{2j}	n_{2j}	d_j	n_j	e_{1j}	v_{1j}
5	0	13	1	32	1	45	0.2889	0.2054
8	0	13	1	31	1	44	0.2955	0.2082
10	0	13	1	30	1	43	0.3023	0.2109
13	0	13	1	29	1	42	0.3095	0.2137
18	0	13	1	28	1	41	0.3171	0.2165
23	1	13	0	27	1	40	0.3250	0.2194
24	0	12	1	27	1	39	0.3077	0.2130
26	0	12	2	26	2	38	0.6316	0.4205
31	0	12	1	24	1	36	0.3333	0.2222
35	0	12	1	23	1	35	0.3429	0.2253
40	0	12	1	22	1	34	0.3529	0.2284
41	0	12	1	21	1	33	0.3636	0.2314
47	1	12	0	20	1	32	0.3750	0.2344
48	0	11	1	20	1	31	0.3548	0.2289
50	0	11	1	19	1	30	0.3667	0.2322
59	0	11	1	18	1	29	0.3793	0.2354
61	0	11	1	17	1	28	0.3929	0.2385
68	0	11	1	16	1	27	0.4074	0.2414
69	1	11	0	15	1	26	0.4231	0.2441
71	0	9	1	15	1	24	0.3750	0.2344
113	0	6	1	10	1	16	0.3750	0.2344
118	0	6	1	8	1	14	0.4286	0.2449
143	0	6	1	7	1	13	0.4615	0.2485
148	1	6	0	6	1	12	0.5000	0.2500
181	1	5	0	4	1	9	0.5556	0.2469
Total	5						9.5652	5.9289

ing procedure.

2.5.3 The Wilcoxon test

The Wilcoxon test of the null hypothesis that there is no difference in the survivor functions for two groups of survival data is very similar to the log-rank test. The Wilcoxon test is based on the statistic

$$U_W = \sum_{j=1}^{r} n_j(d_{1j} - e_{1j}),$$

where, as in the previous section, d_{1j} is the number of deaths at time $t_{(j)}$ in the first group and e_{1j} is as defined in equation (2.19). The difference between U_W and U_L is that in the Wilcoxon test, each difference $d_{1j} - e_{1j}$ is weighted by n_j, the total number of individuals at risk at time $t_{(j)}$. The effect of this is to give less weight to differences between d_{1j} and e_{1j} at those times when the total number of individuals who are still alive is small, that is, at the longest survival times. This statistic is therefore less sensitive than the log-rank statistic to deviations of d_{1j} from e_{1j} in the tail of the distribution of survival times.

The variance of the Wilcoxon statistic U_W is given by

$$V_W = \sum_{j=1}^{r} n_j^2 v_{1j},$$

where v_{1j} is given in equation (2.21), and so the Wilcoxon test statistic is

$$W_W = U_W^2 / V_W, \tag{2.24}$$

which has a chi-squared distribution on one degree of freedom when the null hypothesis is true. The Wilcoxon test is therefore conducted in the same manner as the log-rank test.

Example 2.12 Prognosis for women with breast cancer
For the data on the survival times of women with tumours which were positively or negatively stained, the value of the Wilcoxon statistic is $U_W = -159$, and the variance of the statistic is $V_W = 6048.136$. The value of the chi-squared statistic, U_W^2 / V_W, is 4.180, and the corresponding P-value is 0.041. This is slightly smaller than the P-value for the log-rank test, and on the basis of this result, we would declare that the difference between the two groups is significant at the 5% level.

2.5.4 Comparison of the log-rank and Wilcoxon tests

Of the two tests, the log-rank test is the more suitable when the alternative to the null hypothesis of no difference between two groups of survival times is that the hazard of death at any given time for an individual in one group is proportional to the hazard at that time for a similar individual in the other group. This is the assumption of *proportional hazards*, which underlies a number of methods for analysing survival data. For other types of departure from the null hypothesis, the Wilcoxon test is more appropriate than the log-rank test for comparing the two survivor functions.

In order to help decide which test is the more suitable in any given situation, we make use of the result that if the hazard functions are proportional, the survivor functions for the two groups of survival data do not cross one another. To show this, suppose that $h_1(t)$ is the hazard of death at time t for an individual in Group I, and $h_2(t)$ is the hazard at that same time for an individual in Group II. If these two hazards are proportional, then we can write $h_1(t) = \psi h_2(t)$, where ψ is a constant that does not depend on the time t. Integrating both sides of this expression, multiplying by -1 and exponentiating gives

$$\exp\left\{-\int_0^t h_1(u)\,du\right\} = \exp\left\{-\int_0^t \psi h_2(u)\,du\right\}. \qquad (2.25)$$

Now, from equation (1.5),

$$S(t) = \exp\left\{-\int_0^t h(u)\,du\right\},$$

and so if $S_1(t)$ and $S_2(t)$ are the survivor functions for the two groups of survival data, from equation (2.25),

$$S_1(t) = [S_2(t)]^\psi.$$

Since the survivor function takes values between zero and unity, this result shows that $S_1(t)$ is greater than or less than $S_2(t)$, according to whether ψ is less than or greater than unity, at any time t. This means that if two hazard functions are proportional, the true survivor functions do not cross. This is a necessary, but not a sufficient condition for proportional hazards.

An informal assessment of the likely validity of the proportional hazards assumption can therefore be made from a plot of the estimated survivor functions for two groups of survival data, such as that shown in Figure 2.9. If the two estimated survivor functions do not cross, the assumption of proportional hazards may be justified, and the log-rank test is appropriate. Of course, sample-based estimates of survivor functions may cross even though the corresponding true hazard functions are proportional, and so some care is needed in the interpretation of such graphs.

In summary, unless a plot of the estimated survival functions shows that there is good reason to doubt the proportional hazards assumption the log-rank test should be used to test the hypothesis of equality of two survivor functions.

Example 2.13 Prognosis for women with breast cancer
From the graph of the two estimated survivor functions in Figure 2.9, we see that the survivor function for the negatively stained women always lies above that for the positively stained women. This suggests that the proportional hazards assumption is appropriate, and that the log-rank test is more appropriate than the Wilcoxon test. However, in this example, there is very little difference between the results of the two hypothesis tests.

2.6 * Comparison of three or more groups of survival data

Both the log-rank and the Wilcoxon tests can be extended to enable three or more groups of survival data to be compared. Suppose that the survival distributions of g groups of survival data are to be compared, for $g \geq 2$. We then define analogues of the U-statistics for comparing the observed numbers of deaths in groups $1, 2, \ldots g - 1$ with their expected values. In an obvious extension of the notation used in Section 2.4, we obtain

$$U_{Lk} = \sum_{j=1}^{r} \left(d_{kj} - \frac{n_{kj}d_j}{n_j} \right),$$

$$U_{Wk} = \sum_{j=1}^{r} n_j \left(d_{kj} - \frac{n_{kj}d_j}{n_j} \right),$$

for $k = 1, 2, \ldots g - 1$. These quantities are then expressed in the form of a vector with $(g - 1)$ components, which we denote by $\mathbf{U}_L$ and $\mathbf{U}_W$.

We also need expressions for the variances of the U_{Lk} and U_{Wk}, and for the covariance between pairs of values. In particular, the covariance between U_{Lk} and $U_{Lk'}$ is given by

$$V_{Lkk'} = \sum_{j=1}^{r} \frac{n_{kj}d_j(n_j - d_j)}{n_j(n_j - 1)} \left(\delta_{kk'} - \frac{n_{k'j}}{n_j} \right),$$

for $k, k' = 1, 2, \ldots, g - 1$, where $\delta_{kk'}$ is such that

$$\delta_{kk'} = \begin{cases} 1 & \text{if } k = k', \\ 0 & \text{otherwise.} \end{cases}$$

These terms are then assembled in the form of a *variance–covariance matrix*, $\mathbf{V}_L$, which is a symmetric matrix that has the variances of the U_{Lk} down the diagonal, and covariance terms in the off-diagonals. For example, in the comparison of three groups of survival data, this matrix would be given by

$$\mathbf{V}_L = \begin{pmatrix} V_{L11} & V_{L12} \\ V_{L12} & V_{L22} \end{pmatrix},$$

where V_{L11} and V_{L22} are the variances of U_{L1} and U_{L2}, respectively, and V_{L12} is their covariance.

Similarly, the variance–covariance matrix for the Wilcoxon statistic is the matrix $\mathbf{V}_W$, whose (k, k')'th element is

$$V_{Wkk'} = \sum_{j=1}^{r} n_j^2 \frac{n_{kj}d_j(n_j - d_j)}{n_j(n_j - 1)} \left(\delta_{kk'} - \frac{n_{k'j}}{n_j} \right),$$

for $k, k' = 1, 2, \ldots, g - 1$.

Finally, in order to test the null hypothesis of no group differences, we make use of the result that the test statistic $\mathbf{U}_L' \mathbf{V}_L^{-1} \mathbf{U}_L$, or $\mathbf{U}_W' \mathbf{V}_W^{-1} \mathbf{U}_W$, has a chi-squared distribution on $(g - 1)$ degrees of freedom, when the null hypothesis is true.

A number of well-known statistical packages for the analysis of survival data incorporate this methodology; details are given in Chapter 11. Furthermore, because the interpretation of the resulting chi-squared statistic is straightforward, an example will not be given here.

2.7 Stratified tests

In many circumstances, there is a need to compare two or more sets of survival data after taking account of the other variables recorded on each individual. As an illustration, consider a multi-centred clinical trial in which two forms of chemotherapy are to be compared in terms of their effect on the survival times of lung cancer patients. Information on the survival times of patients in each treatment group will be available from each centre. The resulting data are then said to be *stratified* by centre.

Individual log-rank or Wilcoxon tests based on the data from each centre will be informative, but a test that combines information about the treatment difference in each centre would provide a more precise summary of the treatment effect. A similar situation would arise in attempting to test for treatment differences when patients are stratified according to variables such as age-group, sex, performance status and other potential risk factors for the disease under study.

In situations such as those described above, a stratified version of the log-rank or Wilcoxon test may be employed. Essentially, this involves calculating the values of the U and V statistics for each *stratum*, and then combining these values over the strata. In this section, the *stratified log-rank test* will be described, but a stratified version of the Wilcoxon test can be obtained in a similar manner. An equivalent analysis based on a model for the survival times is described in Section 10.1.1 of Chapter 10.

Let U_{Lk} be the value of the log-rank statistic to compare two treatment groups, computed from the k'th of s strata using equation (2.20). Also, denote the variance of the statistic for the k'th stratum by V_{Lk}, where V_{Lk} would be computed for each stratum using equation (2.21). The stratified log-rank test is then based on the statistic

$$W_S = \frac{\left(\sum_{k=1}^{s} U_{Lk}\right)^2}{\sum_{k=1}^{s} V_{Lk}}, \tag{2.26}$$

which has a chi-squared distribution on one degree of freedom (1 d.f.) under the null hypothesis that there is no treatment difference. Comparing the observed value of this statistic with percentage points of the chi-squared distribution enables the hypothesis of no overall treatment difference to be tested.

Example 2.14 Survival times of melanoma patients
The aim of a study carried out by the University of Oklahoma Health Sciences Centre was to compare two immunotherapy treatments for their ability to prolong the life of patients suffering from melanoma, a highly malignant tumour occurring in the skin. For each patient, the tumour was surgically removed before allocation to *Bacillus Calmette-Guerin* (BCG) vaccine or to a vaccine based on the bacterium *corynebacterium parvum* (*c.parvum*).

The survival times of the patients in each treatment group were further classified according to the age-group of the patient. The data, which were given in Lee (1992), are shown in Table 2.8. An asterisk against a survival time indicates that

Table 2.8 Survival times of melanoma patients in two treatment groups, stratified by age-group.

21 – 40		41 – 60		61 –	
BCG	c. parvum	BCG	c. parvum	BCG	c. parvum
19	27*	34*	8	10	25*
24*	21*	4	11*	5	8
8	18*	17*	23*		11*
17*	16*		12*		
17*	7		15*		
34*	12*		8*		
	24		8*		
	8				
	8*				

Table 2.9 Values of the log-rank statistic for each age-group.

Age-group	U_L	V_L	W_L
21 – 40	−0.2571	1.1921	0.055
41 – 60	0.4778	0.3828	0.596
61 –	1.0167	0.6497	1.591
Total	1.2374	2.2246	

the observation is censored.

These data are analysed by first computing the log-rank statistics for comparing the survival times of patients in the two treatment groups, separately for each age-group. The resulting values of the U, V and W statistics, found using equations (2.20), (2.22) and (2.23), are summarized in Table 2.9.

The values of the W-statistic are quite similar for the three age-groups, suggesting that the treatment effect is consistent over the three age groups. Moreover, none of them are significantly large at the 10% level.

To carry out a stratified log-rank test on these data, we calculate the W_S-statistic defined in equation (2.26). Using the results in Table 2.9,

$$W_S = \frac{1.2374^2}{2.2246} = 0.688.$$

The observed value of W_S is not significant when compared with percentage points of the chi-squared distribution on 1 d.f. We therefore conclude that after allowing for the different age-groups, there is no significant difference between the survival times of patients treated with the BCG vaccine and those treated with c. parvum.

For comparison, when the division of the patients into the different age-groups is ignored, the log-rank test for comparing the two groups of patients leads to $W_L = 0.756$. The fact that this is so similar to the value that allows for age-group differences suggests that it is not necessary to stratify the patients by age.

The stratified log-rank test can be extended to compare more than two treatment groups, but the resulting formulae render it unsuitable for hand calculation. The methodology can be implemented using the major statistical packages for survival analysis. However, this method of taking account of additional variables is not as flexible as that based on a modelling approach, introduced in the next chapter.

2.8 Log-rank test for trend

Consider the situation where three or more groups of survival data are to be compared, and where these groups are ordered in some way. For example, the groups may correspond to increasing doses of a treatment, the stage of a disease, or the age-group of an individual. In comparing these groups using the log-rank test described in previous sections, it can happen that the analysis does not lead to a significant difference between the groups, even though the hazard of death increases or decreases across the groups. Indeed, a test that uses information about the ordering of the groups is more likely to lead to a trend being identified as significant than a standard log-rank test.

The log-rank test for trend across g ordered groups is based on the statistic

$$U_T = \sum_{k=1}^{g} w_k (d_{k.} - e_{k.}), \qquad (2.27)$$

where w_k is a *code* assigned to the k'th group, $k = 1, 2, \ldots, g$, and

$$d_{k.} = \sum_{j=1}^{r_k} d_{kj}, \qquad e_{k.} = \sum_{j=1}^{r_k} e_{kj},$$

are the observed and expected numbers of deaths in the k'th group, where the summation is over the r_k death times in that group. Note that the dot subscript in the notation $d_{k.}$ and $e_{k.}$ stands for summation over the subscript that the dot replaces. The codes are often taken to be equally spaced to correspond to a linear trend across the groups. For example, if there are three groups, the codes might be taken to be 1, 2, and 3, although the equivalent choice of -1, 0, and 1 does simplify the calculations somewhat. The variance of U_T is given by

$$V_T = \sum_{k=1}^{g} (w_k - \bar{w})^2 e_{k.}, \qquad (2.28)$$

where $\bar{w}$ is a weighted sum of the quantities w_k, in which the expected numbers

Table 2.10 Values of w_k and the observed and expected numbers of deaths in the three age-groups.

Age-group	w_k	$d_{k.}$	$e_{k.}$
21 – 40	−1	2	3.1871
41 – 60	0	1	1.1949
61 –	1	2	0.6179

of deaths, $e_{k.}$, are the weights, that is,

$$\bar{w} = \frac{\sum_{k=1}^{g} w_k e_{k.}}{\sum_{k=1}^{g} e_{k.}}.$$

The statistic $W_T = U_T^2/V_T$ then has a chi-squared distribution on 1 d.f. under the hypothesis of no trend across the g groups.

Example 2.15 Survival times of melanoma patients
The log-rank test for trend will be illustrated using the data from Example 2.14 on the survival times of patients suffering from melanoma. For the purpose of this illustration, only the data from those patients allocated to the BCG vaccine will be used. The log-rank statistic for comparing the survival times of the patients in the three age-groups turns out to be 3.739. When compared to percentage points of the chi-squared distribution on 2 d.f., this is not significant (P = 0.154).

We now use the log-rank test for trend to examine whether there is a linear trend over age. For this, we will take the codes w_k to be equally spaced, with values −1, 0 and 1. Some of the calculations required for the log-rank test for trend are summarized in Table 2.10. The log-rank test for trend is based on the statistic in equation (2.27), the value of which is

$$U_T = (d_{3.} - e_{3.}) - (d_{1.} - e_{1.}) = 2.5692.$$

Using the values of the expected numbers of deaths in each group given in Table 2.10, the weighted mean of the w_k's is given by

$$\bar{w} = \frac{e_{3.} - e_{1.}}{e_{1.} + e_{3.}} = 0.5138.$$

The three values of $(w_k - \bar{w})^2$ are 0.2364, 0.2640 and 2.2917, and from equation (2.28), $V_T = 2.4849$. Finally, the test statistic is

$$W_T = \frac{U_T^2}{V_T} = 2.656,$$

which is just about significant at the 10% level (P = 0.103) when judged against a chi-squared distribution on 1 d.f. We therefore conclude that there is slight evidence of a linear trend across the age-groups.

An alternative method of examining whether there is a trend across the levels of an ordered categorical variable, based on a modelling approach to the analysis of survival data, is described and illustrated in Section 3.6.2 of the next chapter.

2.9 Further reading

The life-table, which underpins the calculation of the life-table estimate of the survivor function, is widely used in the analysis of data from epidemiological studies. Fuller details of this application can be found in Armitage and Berry (1987), and books on statistical methods in demography and epidemiology, such as Pollard, Yusuf and Pollard (1990) and Kahn and Sempos (1989).

The product-limit estimate of the survivor function has been in use since the early 1900s. Kaplan and Meier (1958) derived the estimate using the method of maximum likelihood, which is why the estimate now bears their name. The properties of the Kaplan–Meier estimate of the survivor function have been further explored by Breslow and Crowley (1974) and Meier (1975).

The expression for the standard error of the Kaplan–Meier estimate was first given by Greenwood (1926). Although Section 2.1.4 shows how a confidence interval for the value of the survivor function at particular times can be found using Greenwood's formula, alternative procedures are needed for the construction of confidence bands for the complete survivor function. Hall and Wellner (1980) and Efron (1981) have shown how such bands can be computed. These procedures are also described by Harris and Albert (1991), whose book is accompanied by a diskette which contains SAS macros for computing the confidence bands using both methods.

Methods for constructing confidence intervals for the median survival time are described by Brookmeyer and Crowley (1982), Emerson (1982), Nair (1984), Simon and Lee (1982) and Slud, Byar and Green (1984). Simon (1986) emphasizes the importance of confidence intervals in reporting the results of clinical trials, and includes an illustration of a method described in Slud, Byar and Green (1984).

The formulation of the hypothesis testing procedure in the frequentist approach to inference is covered in many statistical texts. See, for example, Altman (1991) and Armitage and Berry (1987) for non-technical presentations of the ideas in a medical context.

The log-rank test results from the work of Mantel and Haenszel (1959) and Mantel (1966). The name log-rank is due to the fact that the test statistic can also be derived from the ranks of the survival times in the two groups. See Peto and Peto (1972) and Lawless (1982) for details of this. A thorough review of the hypergeometric distribution, used in the derivation of the log-rank test in Section 2.4.2, is included in Johnson and Kotz (1969).

The log-rank test for trend is derived from the test for trend in a $2 \times k$ contingency table, given in Armitage and Berry (1987). The test is also described by Altman (1991). Peto *et al.* (1976, 1977) give a non-mathematical account of the log-rank test and its extensions.

3

Modelling survival data

The non-parametric methods described in Chapter 2 can be useful in the analysis of a single sample of survival data, or in the comparison of two or more groups of survival times. However, in most medical studies which give rise to survival data, supplementary information will also be recorded on each individual. A typical example would be a clinical trial to compare the survival times of patients who receive one or other of two treatments. In such a study, demographic variables such as the age and sex of the patient, the values of physiological variables such as serum haemoglobin level and heart rate, and factors that are associated with the lifestyle of the patient, such as smoking history and dietary habits, may all have an impact on the time that the patient survives. Accordingly, the values of these variables, which are referred to as *explanatory variables*, would be recorded at the outset of the study. The resulting data set would then be more complex than those considered in Chapter 2, and the methods described in that chapter would generally be unsuitable.

In order to explore the relationship between the survival experience of a patient and explanatory variables, an approach based on statistical modelling can be used. Indeed, the particular model which is developed in this chapter both unifies and extends the non-parametric procedures of Chapter 2.

3.1 Modelling the hazard function

Through a modelling approach to the analysis of survival data, we can explore how the survival experience of a group of patients depends on the values of one or more explanatory variables, whose values have been recorded for each patient at the time origin. For example, in the study on multiple myeloma, given as Example 1.3, the aim is to determine which of seven explanatory variables have an impact on the survival time of the patients. In Example 1.4 on the survival times of patients in a clinical trial involving two treatments for prostatic cancer, the primary aim is to identify whether patients in the two treatment groups have a different survival experience. Because additional variables such as the age of the patient and the size of their tumour are likely to influence survival time, it will be important to take account of these variables when assessing the extent of any treatment difference.

In the analysis of survival data, interest centres on the risk or hazard of death at any time after the time origin of the study. As a consequence, in survival analysis the hazard function is modelled directly. The resulting models are somewhat different in form from linear models encountered in regression analysis and in the analysis of data from designed experiments, where the dependence of the mean response, or some function of it, on certain explanatory variables is modelled. However, many of the principles and procedures used in linear modelling carry over to the modelling of survival data.

There are two broad reasons for modelling survival data. One objective of the modelling process is to determine which combination of potential explanatory variables affect the form of the hazard function. In particular, the effect that the treatment has on the hazard of death can be studied, as can the extent to which other explanatory variables affect the hazard function. Another reason for modelling the hazard function is to obtain an estimate of the hazard function itself for an individual. This may be of interest in its own right, but in addition, from the relationship between the survivor function and hazard function described by equation (1.5), an estimate of the survivor function can be found. This will in turn lead to an estimate of quantities such as the median survival time, which will be a function of the explanatory variables in the model. The median survival time could then be estimated for current or future patients with particular values of these explanatory variables. The resulting estimate could be particularly useful in devising a treatment regime or counselling the patient about their prognosis.

The basic model for survival data to be considered in this chapter is the *proportional hazards model*. This model was proposed by Cox (1972) and has also come to be known as the *Cox regression model*. Although the model is based on the assumption of proportional hazards, introduced in Section 2.5.4, no particular from of probability distribution is assumed for the survival times. The model is therefore referred to as a *semi-parametric model*. We now go on to develop the model for the comparison of the hazard functions for individuals in two groups.

3.1.1 A model for the comparison of two groups

Suppose that patients are randomized to receive either a standard treatment or a new treatment, and let $h_S(t)$ and $h_N(t)$ be the hazards of death at time t for patients on the standard treatment and new treatment, respectively. According to a simple model for the survival times of the two groups of patients, the hazard at time t for a patient on the new treatment is proportional to the hazard at that same time for a patient on the standard treatment. This *proportional hazards model* can be expressed in the form

$$h_N(t) = \psi h_S(t), \tag{3.1}$$

for any non-negative value of t, where ψ is a constant. An implication of this assumption is that the corresponding true survivor functions for individuals on the new and standard treatments do not cross, as previously shown in Section 2.5.4.

The value of ψ is the ratio of the hazards of death at any time for an individual on the new treatment relative to an individual on the standard treatment, and so ψ is known as the *relative hazard* or *hazard ratio*. If $\psi < 1$, the hazard of death at t is smaller for an individual on the new drug, relative to an individual on the standard. The new treatment is then an improvement on the standard. On the other hand, if $\psi > 1$, the hazard of death at t is greater for an individual on the new drug, and the standard treatment is superior.

An alternative way of expressing the model in equation (3.1) leads to a model which can more easily be generalized. Suppose that survival data are available on n individuals and denote the hazard function for the i'th of these by $h_i(t)$, $i = 1, 2, \ldots, n$. Also, write $h_0(t)$ for the hazard function for an individual on the standard treatment. The hazard function for an individual on the new treatment is then $\psi h_0(t)$. The relative hazard ψ cannot be negative, and so it is convenient to set $\psi = \exp(\beta)$. The parameter β is then the logarithm of the hazard ratio, that is, $\beta = \log \psi$, and any value of β in the range $(-\infty, \infty)$ will lead to a positive value of ψ. Note that positive values of β are obtained when the hazard ratio, ψ, is greater than unity, that is, when the new treatment is inferior to the standard.

Now let X be an *indicator variable* which takes the value zero if an individual is on the standard drug and unity if an individual is on the new drug. If x_i is the value of X for the i'th individual in the study, $i = 1, 2, \ldots, n$, the hazard function for this individual can be written

$$h_i(t) = e^{\beta x_i} h_0(t), \tag{3.2}$$

where $x_i = 1$ if the i'th individual is on the new treatment and $x_i = 0$ otherwise. This is the *proportional hazards model* for the comparison of two treatment groups.

3.1.2 The general proportional hazards model

The model of the previous section is now generalized to the situation where the hazard of death at a particular time depends on the values $x_1, x_2, \ldots, x_p$ of p explanatory variables $X_1, X_2, \ldots, X_p$. The values of these variables will be assumed to have been recorded at the time origin of the study. An extension of the model to cover the situation where the values of one or more of the explanatory variables change over time will be considered in Section 5.8 of Chapter 5, and in fuller detail in Chapter 7.

The set of values of the explanatory variables in the proportional hazards model will be represented by the vector $\mathbf{x}$, so that $\mathbf{x} = (x_1, x_2, \ldots, x_p)'$. Let $h_0(t)$ be the hazard function for an individual for whom the values of all the explanatory variables that make up the vector $\mathbf{x}$ are zero. The function $h_0(t)$ is called the *baseline hazard function*. The hazard function for the i'th individual can then be written as

$$h_i(t) = \psi(\mathbf{x}_i)h_0(t),$$

where $\psi(\mathbf{x}_i)$ is a function of the values of the vector of explanatory variables for the i'th individual. The function $\psi(\ )$ can be interpreted as the hazard at time t for an individual whose vector of explanatory variables is $\mathbf{x}_i$, relative to the hazard for an individual for whom $\mathbf{x} = \mathbf{0}$.

Again, since the relative hazard $\psi(\mathbf{x}_i)$ cannot be negative, it is convenient to write this as $\exp(\eta_i)$, where η_i is a linear combination of the p explanatory variables in $\mathbf{x}_i$. Therefore,

$$\eta_i = \beta_1 x_{1i} + \beta_2 x_{2i} + \cdots + \beta_p x_{pi},$$

so that $\eta_i = \sum_{j=1}^{p} \beta_j x_{ji}$. In matrix notation, $\eta_i = \beta' \mathbf{x}_i$, where β is the vector of coefficients of the explanatory variables $x_1, x_2, \ldots, x_p$ in the model. The quantity η_i is called the *linear component* of the model, but it is also known as the *risk score* or *prognostic index* for the i'th individual. The general proportional hazards model then becomes

$$h_i(t) = \exp(\beta_1 x_{1i} + \beta_2 x_{2i} + \cdots + \beta_p x_{pi})h_0(t). \tag{3.3}$$

Since this model can be re-expressed in the form

$$\log\left\{\frac{h_i(t)}{h_0(t)}\right\} = \beta_1 x_{1i} + \beta_2 x_{2i} + \cdots + \beta_p x_{pi},$$

the proportional hazards model may also be regarded as a linear model for the logarithm of the hazard ratio. There are other possible forms for $\psi(\mathbf{x}_i)$, but the choice $\psi(\mathbf{x}_i) = \exp(\beta' \mathbf{x}_i)$ leads to the most commonly used model for survival data.

Notice that there is no constant term in the linear component of the proportional hazards model. If a constant term β_0, say, were included, the baseline hazard function could simply be rescaled by dividing $h_0(t)$ by $\exp(\beta_0)$, and the constant term would cancel out. Moreover, we have made no assumptions concerning the actual form of the baseline hazard function $h_0(t)$. Indeed, we will see later that the β-coefficients in this proportional hazards model can be estimated without making any such assumptions. Of course, we will often need to estimate $h_0(t)$ itself, and we will see how this can be done in Section 3.8.

3.2 The linear component of the proportional hazards model

There are two types of variable on which a hazard function may depend, namely *variates* and *factors*. A variate is a variable which takes numerical values which are often on a continuous scale of measurement, such as age or systolic blood pressure. A factor is a variable which takes a limited set of values, which are known as the *levels* of the factor. For example, sex is a factor with two levels, and type of tumour might be a factor whose levels correspond to different histologies, such as squamous, adeno or small cell.

We now consider how variates, factors, and terms that combine factors and variates, can be incorporated in the linear component of a proportional hazards model.

3.2.1 Including a variate

Variates, either alone or in combination, are readily incorporated in a proportional hazards model. Each variate appears in the model with a corresponding β-coefficient. As an illustration, consider a situation in which the hazard function depends on two variates X_1 and X_2. The value of these variates for the i'th individual will be x_{1i} and x_{2i}, respectively, and the proportional hazards model for the i'th of n individuals is written

$$h_i(t) = \exp(\beta_1 x_{1i} + \beta_2 x_{2i}) h_0(t).$$

In models such as this, the baseline hazard function, $h_0(t)$, is the hazard function for an individual for whom all the variates included in the model take the value zero.

3.2.2 Including a factor

Suppose that the dependence of the hazard function on a single factor A is to be modelled, where A has a levels. The model for an individual for whom the level of A is j will then need to incorporate the term α_j which represents the effect due to the j'th level of the factor. The terms $\alpha_1, \alpha_2, \ldots, \alpha_a$ are known as the *main effects* of the factor A. According to the proportional hazards model, the hazard function for an individual with factor A at level j is $\exp(\alpha_j) h_0(t)$. Now, the baseline hazard function $h_0(t)$ has been defined to be the hazard for an individual with values of all explanatory variables equal to zero. To be consistent with this definition, one of the α_j must be taken to be zero. One possibility is to adopt the constraint $\alpha_1 = 0$, which corresponds to taking the baseline hazard to be the hazard for an individual for whom A is at the first level. This is the constraint that will be used in the sequel.

Models which contain terms corresponding to factors can be expressed as linear combinations of explanatory variables by defining *indicator* or *dummy variables* for each factor. This procedure will be required when using computer software for survival analysis that does not allow factors to be fitted directly. If the constraint $\alpha_1 = 0$ is adopted, the term α_j can be included in the model by defining $a - 1$ indicator variables $X_2, X_3, \ldots, X_a$, which take the values shown in the table below.

Level of A	X_2	X_3	$\cdots$	X_a
1	0	0	$\cdots$	0
2	1	0	$\cdots$	0
3	0	1	$\cdots$	0
a	0	0	$\cdots$	1

The term α_j can then be incorporated in the linear part of the proportional hazards model by including the $a - 1$ explanatory variables $X_2, X_3, \ldots, X_a$ with coefficients $\alpha_2, \alpha_3, \ldots, \alpha_a$. In other words, the term α_j in the model is replaced by $\alpha_2 x_2 + \alpha_3 x_3 + \cdots + \alpha_a x_a$, where x_j is the value of X_j for an individual for whom A is at level j, $j = 2, 3, \ldots, a$. There are then $a - 1$ parameters associated with the main effect of the factor A, and A is said to have $a - 1$ *degrees of freedom*.

3.2.3 Including an interaction

When terms corresponding to more than one factor are to be included in the model, sets of indicator variables can be defined for each factor in a manner similar to that shown above. In this situation, it may also be appropriate to include a term in the model which corresponds to individual effects for each combination of levels of two or more factors. Such effects are known as *interactions*.

For example, suppose that the two factors are the sex of a patient and grade of tumour. If the effect of grade of tumour on the hazard of death was different in patients of each sex, we would say that there was an interaction between these two factors. The hazard function would then depend on the combination of levels of these two factors.

In general, if A and B are two factors, and the hazard of death depends on the combination of levels of A and B, then A and B are said to *interact*. If A and B have a and b levels respectively, the term which represents an interaction between these two factors is denoted by $(\alpha\beta)_{jk}$, for $j = 1, 2, \ldots, a$ and $k = 1, 2, \ldots, b$.

In statistical modelling, an important principle is that an interaction term should only be included in a model when the corresponding main effects are also present. Thus the term $(\alpha\beta)_{jk}$ is only included when the model contains both α_j and β_k, which correspond to the main effects of A and B. We will return to this point when we consider model building strategy in Section 3.5.

In order to include the term $(\alpha\beta)_{jk}$ in the model, products of indicator variables associated with the main effects are calculated. For example, if A and B have 2 and 3 levels respectively, indicator variables U_2 and V_2, V_3 are defined as in the following tables.

Level of A	U_2
1	0
2	1

Level of B	V_2	V_3
1	0	0
2	1	0
3	0	1

Let u_j and v_k be the values of U_j and V_k for a given individual, for $j = 2, k = 2, 3$. The term $(\alpha\beta)_{jk}$ is then fitted by including variates formed from the products of U_j and V_k in the model. The corresponding value of the product for a given individual is $u_j v_k$. The coefficient of this product is denoted $(\alpha\beta)_{jk}$, and so the term $(\alpha\beta)_{jk}$ is fitted as

$$(\alpha\beta)_{22} u_2 v_2 + (\alpha\beta)_{23} u_2 v_3.$$

There are therefore two parameters associated with the interaction between A and B. In general, if A and B have a and b levels respectively, the two-factor interaction AB has $(a-1)(b-1)$ parameters associated with it, in other words AB has $(a-1)(b-1)$ degrees of freedom. Furthermore, the term $(\alpha\beta)_{jk}$ is equal to zero whenever either A or B are at the first level, that is, when either $j = 1$ or $k = 1$.

3.2.4 Including a mixed term

Another type of term that might be needed in a model is a mixed term formed from a factor and a variate. Terms of this type would be used when the coefficient of a variate in a model was likely to be different for each level of a factor.

As an example, consider a contraceptive trial in which the time to the onset of a period of amenorrhoea, the prolonged absence of menstrual bleeding, is being modelled. The hazard of an amenorrhoea may be related to the weight of a woman, but the coefficient of this variate may differ according to the level of a factor associated with the number of previous pregnancies that the woman has experienced.

The dependence of the coefficient of a variate X on the level of a factor A would be depicted by including the term $\alpha_j x$ in the linear component of the proportional hazards model, where x is the value of X for a given individual for whom the factor A is at the j'th level, $j = 1, 2, \ldots, a$. To include such a term, indicator variables U_j, say, are defined for the factor A, and each of these is multiplied by the value of X for each individual. The resulting values of the products $U_j X$ are $u_j x$, and the coefficient of $u_j x$ in the model is α_j, where j indexes the level of the factor A.

If the same definition of indicator variables in the previous discussion were used, α_1, the coefficient of X for individuals at the first level of A, would be zero. It is then essential to include the variate X in the model as well as the products, for otherwise the dependence on X for individuals at the first level of A would not be modelled. An illustration should make this clearer.

Suppose that there are nine individuals in a study, on each of whom the value of a variate X and the level of a factor A have been recorded. We will take A to have three levels, where A is at the first level for the first three individuals, at the second level for the next three, and at the third level for the final three. In order to model the dependence of the coefficient of the variate X on the level of A, two indicator variables, U_2 and U_3 are defined as in the following table.

Individual	Level of A	X	U_2	U_3	U_2X	U_3X
1	1	x_1	0	0	0	0
2	1	x_2	0	0	0	0
3	1	x_3	0	0	0	0
4	2	x_4	1	0	x_4	0
5	2	x_5	1	0	x_5	0
6	2	x_6	1	0	x_6	0
7	3	x_7	0	1	0	x_7
8	3	x_8	0	1	0	x_8
9	3	x_9	0	1	0	x_9

Explanatory variables formed as the products U_2X and U_3X, given in the last two columns of this table, would then be included in the linear component of the model, together with the variate X. Let the coefficients of the values of the products U_2X and U_3X be α'_2 and α'_3, respectively, and let the coefficient of the value of the variate X in the model be β. Then, the model contains the terms $\beta x + \alpha'_2(u_2x) + \alpha'_3(u_3x)$. From the above table, $u_2 = 0$ and $u_3 = 0$ for individuals at level 1 of A, and so the coefficient of x for these individuals is just β. For those at level 2 of A, $u_2 = 1$ and $u_3 = 0$, and the coefficient of x is $\beta + \alpha'_2$. Similarly, at level 3 of A, $u_2 = 0$ and $u_3 = 1$, and the coefficient of x is $\beta + \alpha'_3$.

Notice that if the term βx is omitted from the model, the coefficient of x for individuals 1, 2 and 3 would be zero. There would then be no information about the relationship between the hazard function and the variate X for individuals at the first level of the factor A.

The manipulation described in the preceding paragraphs can be avoided by defining the indicator variables in a different way. If a factor A has a levels, and it is desired to include the term $\alpha_j x$ in a model, without necessarily including the term βx, a indicator variables $Z_1, Z_2, \ldots, Z_a$ can be defined for A, where $Z_j = 1$ at level j of A and zero otherwise. The corresponding values of these products for an individual, $z_1x, z_2x, \ldots, z_ax$, are included in the model with coefficients $\alpha_1, \alpha_2, \ldots, \alpha_a$. These are the coefficients of x for each level of A.

Now, if the variate X is included in the model, along with the a products of the form Z_jX, there will be $a + 1$ terms corresponding to the a coefficients. It will not then be possible to obtain unique estimates of each of these α-coefficients, and the model is said to be *overparametrized*. This overparametrization can be dealt with by forcing one of the $a + 1$ coefficients to be zero. In particular, taking $\alpha_1 = 0$ would be equivalent to a redefinition of the indicator variables, in which z_1 is taken to be zero. This then leads to the same formulation of the model that has already been discussed.

The application of these ideas in the analysis of actual data sets will be illustrated in Section 3.4, after we have seen how the proportional hazards model can be fitted. A commentary on how different statistical packages for the analysis of survival data can be used to fit terms involving factors is included in Chapter 11.

3.3 Fitting the proportional hazards model

Fitting the proportional hazards model given in equation (3.3) to an observed set of survival data entails estimating the unknown coefficients of the explanatory variables $X_1, X_2, \ldots, X_p$ in the linear component of the model, $\beta_1, \beta_2, \ldots, \beta_p$. The baseline hazard function, $h_0(t)$, may also need to be estimated. It turns out that these two components of the model can be estimated separately. The β's are estimated first and these estimates are then used to construct an estimate of the baseline hazard function. This is an important result, since it means that in order to make inferences about the effects of p explanatory variables $X_1, X_2, \ldots, X_p$ on the relative hazard, $h_i(t)/h_0(t)$, we do not need an estimate of $h_0(t)$. Methods for estimating $h_0(t)$ will therefore be deferred until Section 3.8.

The β-coefficients in the proportional hazards model, which are the unknown parameters in the model, can be estimated using the *method of maximum likelihood*. To operate this method, we first obtain the *likelihood* of the sample data. This is the joint probability of the observed data, regarded as a function of the unknown parameters in the assumed model. For the proportional hazards model, this is a function of the observed survival times and the unknown β-parameters in the linear component of the model. Estimates of the β's are then those values which are the most likely on the basis of the observed data. These *maximum likelihood estimates* are therefore the values which maximize the likelihood function. From a computational viewpoint, it is more convenient to maximize the logarithm of the likelihood function. Furthermore, approximations to the variance of maximum likelihood estimates can be obtained from the second derivatives of the log-likelihood function. Details will not be given here, but Appendix A contains a summary of relevant results from the theory of maximum likelihood estimation.

Suppose that data are available for n individuals, amongst whom there are r distinct death times and $n - r$ right-censored survival times. We will for the moment assume that only one individual dies at each death time, so that there are no *ties* in the data. The treatment of ties will be discussed in Section 3.3.2. The r ordered death times will be denoted by $t_{(1)} < t_{(2)} < \cdots < t_{(r)}$, so that $t_{(j)}$ is the j'th ordered death time. The set of individuals who are at risk at time $t_{(j)}$ will be denoted by $R(t_{(j)})$, so that $R(t_{(j)})$ is the set of individuals who are alive and uncensored at a time just prior to $t_{(j)}$. The quantity $R(t_{(j)})$ is called the *risk set*.

Cox (1972) showed that the relevant likelihood function for the proportional hazards model in equation (3.3) is given by

$$L(\beta) = \prod_{j=1}^{r} \frac{\exp(\beta' \mathbf{x}_{(j)})}{\sum_{l \in R(t_{(j)})} \exp(\beta' \mathbf{x}_l)}, \tag{3.4}$$

in which $\mathbf{x}_{(j)}$ is the vector of covariates for the individual who dies at the j'th ordered death time, $t_{(j)}$. The summation in the denominator of this likelihood function is the sum of the values of $\exp(\beta' \mathbf{x})$ over all individuals who are at risk at time $t_{(j)}$. Notice that the product is taken over the individuals for whom death times have been recorded. Individuals for whom the survival times are censored do not contribute to the numerator of the log-likelihood function, but they do enter into the summation over the risk sets at death times that occur before a censored time. Moreover, the likelihood function depends only on the ranking of the death times, since this determines the risk set at each death time. Consequently, inferences about the effect of explanatory variables on the hazard function depend only on the rank order of the survival times.

Now suppose that the data consist of n observed survival times, denoted by $t_1, t_2, \ldots, t_n$, and that δ_i is a censoring indicator which is zero if the i'th survival time t_i, $i = 1, 2, \ldots, n$, is right-censored, and unity otherwise. The likelihood function in equation (3.4) can then be expressed in the form

$$\prod_{i=1}^{n} \left[\frac{\exp(\beta' \mathbf{x}_i)}{\sum_{l \in R(t_i)} \exp(\beta' \mathbf{x}_l)} \right]^{\delta_i},$$

where $R(t_i)$ is the risk set at time t_i. The corresponding log-likelihood function is given by

$$\log L(\beta) = \sum_{i=1}^{n} \delta_i \left\{ \beta' \mathbf{x}_i - \log \sum_{l \in R(t_i)} \exp(\beta' \mathbf{x}_l) \right\}. \tag{3.5}$$

The maximum likelihood estimates of the β-parameters in the proportional hazards model can be found by maximizing this log-likelihood function using numerical methods. This maximization is generally accomplished using the *Newton–Raphson procedure* described below in Section 3.3.3.

Fortunately, most of the major statistical packages have facilities which enable the proportional hazards model to be fitted, and so it will not be necessary to write computer programs for this. Such software also gives the standard errors of the parameter estimates in the fitted model. Details of the facilities offered by the more widely available software packages are given in Chapter 11. Most of the calculations in this book have been carried out using the package SAS, with the SAS procedure proc phreg being used to fit the proportional hazards model of equation (3.3).

The justification for using equation (3.4) as a likelihood function, and further details on the structure of the likelihood function are given in Section 3.3.1. The treatment of tied survival times is then discussed in Section 3.3.2 and the Newton–

Raphson procedure is outlined in Section 3.3.3. These three sections can be omitted without loss of continuity.

3.3.1 * Likelihood function for the model

The basis of the argument used in the construction of a likelihood function for the proportional hazards model is that intervals between successive death times convey no information about the effect of explanatory variables on the hazard of death. This is because the baseline hazard function has an arbitrary form, and so it is conceivable that $h_0(t)$, and hence $h(t)$, is zero in those time intervals in which there are no deaths. This in turn means that these intervals give no information about the values of the β-parameters. We therefore consider the probability that the i'th individual dies at some time $t_{(j)}$, conditional on $t_{(j)}$ being one of the observed set of r death times $t_{(1)}, t_{(2)}, \ldots, t_{(r)}$. If the vector of explanatory variables for the individual who dies at $t_{(j)}$ is $\mathbf{x}_{(j)}$, this is

$$\text{P[individual with variables } \mathbf{x}_{(j)} \text{ dies at } t_{(j)} \mid \text{one death at } t_{(j)}]. \qquad (3.6)$$

Next, from the result that the probability of an event A, given that an event B has occurred, is given by

$$P(A \mid B) = P(A \text{ and } B)/P(B),$$

the probability in expression (3.6) becomes

$$\frac{\text{P[individual with variables } \mathbf{x}_{(j)} \text{ dies at } t_{(j)}]}{\text{P[one death at } t_{(j)}]}.$$

The numerator of the above expression is simply the hazard of death at time $t_{(j)}$ for the individual whose vector of explanatory variates is $\mathbf{x}_{(j)}$. If it is the i'th individual who dies at $t_{(j)}$, this hazard function can be written $h_i(t_{(j)})$. The denominator is the sum of the hazards of death at time $t_{(j)}$ over all individuals who are at risk of death at this time. This is the sum of the values $h_l(t_{(j)})$ over those individuals indexed by l in the risk set at time $t_{(j)}$, $R(t_{(j)})$. Consequently, the conditional probability in expression (3.6) becomes

$$\frac{h_i(t_{(j)})}{\sum_{l \in R(t_{(j)})} h_l(t_{(j)})},$$

and on using equation (3.3), the baseline hazard function in the numerator and denominator cancels out, and we are left with

$$\frac{\exp(\beta' \mathbf{x}_{(j)})}{\sum_{l \in R(t_{(j)})} \exp(\beta' \mathbf{x}_l)}.$$

Finally, taking the product of these conditional probabilities over the r death times gives the likelihood function in equation (3.4).

The likelihood function that has been obtained is not a true likelihood, since it does not make direct use of the actual censored and uncensored survival times.

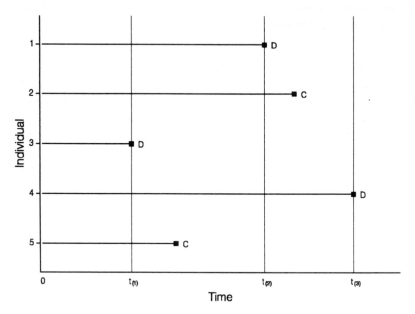

Figure 3.1   Survival times of five individuals.

For this reason it is referred to as a *partial likelihood function*.

In order to throw more light on the structure of the partial likelihood, consider a sample of survival data from five individuals, numbered from 1 to 5. The survival data are illustrated in Figure 3.1. The observed survival times of individuals 2 and 5 will be taken to be right-censored, and the three ordered death times are denoted $t_{(1)} < t_{(2)} < t_{(3)}$. Then, $t_{(1)}$ is the death time of individual 3, $t_{(2)}$ is that of individual 1, and $t_{(3)}$ that of individual 4.

The risk set at each of the three ordered death times consists of the individuals who are alive and uncensored just prior to each death time. Hence, the risk set $R(t_{(1)})$ consists of all five individuals, risk set $R(t_{(2)})$ consists of individuals 1, 2 and 4, while risk set $R(t_{(3)})$ only includes individual 4. Now write $\psi(i) = \exp(\beta' x_i)$, $i = 1, 2, \ldots, 5$, for the risk score for the i'th individual, where x_i is the vector of explanatory variables for that individual. The numerators of the partial likelihood function for times $t_{(1)}, t_{(2)}$ and $t_{(3)}$, respectively, are $\psi(3), \psi(1)$ and $\psi(4)$, since individuals 3, 1 and 4, respectively, die at the three ordered death times. The partial likelihood function over the three death times is then

$$\frac{\psi(3)}{\psi(1) + \psi(2) + \psi(3) + \psi(4) + \psi(5)} \times \frac{\psi(1)}{\psi(1) + \psi(2) + \psi(4)} \times \frac{\psi(4)}{\psi(4)}.$$

It turns out that standard results used in maximum likelihood estimation carry over without modification to maximum partial likelihood estimation. In particular, the results given in Appendix A for the variance–covariance matrix of the estimates of the β's can be used, as can distributional results associated with likelihood ratio testing, discussed in Section 3.4.

3.3.2 * Treatment of ties

The proportional hazards model for survival data assumes that the hazard function is continuous, and under this assumption, tied survival times are not possible. Of course, survival times are usually recorded to the nearest day, month or year, and so tied survival times can arise as a result of this rounding process. Indeed, Examples 1.2, 1.3 and 1.4 in Chapter 1 all contain tied observations.

In addition to the possibility of more than one death at a given time, there might also be one or more censored observations at a death time. When there are both censored survival times and deaths at a given time, the censoring is assumed to occur after all the deaths. Potential ambiguity concerning which individuals should be included in the risk set at that death time is then resolved and tied censored observations present no further difficulties in the computation of the likelihood function using equation (3.4). Accordingly, we only need consider how tied survival times can be handled in fitting the proportional hazards model.

In order to accommodate tied observations, the likelihood function in equation (3.4) has to be modified in some way. The appropriate likelihood function in the presence of tied observations has been given by Kalbfleisch and Prentice (1980). However, this likelihood has a very complicated form, and will not be reproduced here. In addition, the computation of this likelihood function can be very time consuming, particularly when there are a relatively large number of ties at one or more death times. Fortunately, there are a number of approximations to the likelihood function which have computational advantages over the exact method. But before these are given, some additional notation needs to be developed.

Let s_j be the vector of sums of each of the p covariates for those individuals who die at the j'th death time, $t_{(j)}$, $j = 1, 2, \ldots, r$. If there are d_j deaths at $t_{(j)}$, the h'th element of s_j is $s_{hj} = \sum_{k=1}^{d_j} x_{hjk}$, where x_{hjk} is the value of the h'th explanatory variable, $h = 1, 2, \ldots, p$, for the k'th of d_j individuals, $k = 1, 2, \ldots, d_j$ who die at the j'th death time, $j = 1, 2, \ldots r$.

The simplest approximation to the likelihood function is that due to Breslow (1974), who proposed the approximate likelihood

$$\prod_{j=1}^{r} \frac{\exp(\beta' s_j)}{\left[\sum_{l \in R(t_{(j)})} \exp(\beta' x_l) \right]^{d_j}}. \tag{3.7}$$

In this approximation, the d_j deaths at time $t_{(j)}$ are considered to be distinct and to occur sequentially. The probabilities of all possible sequences of deaths are then summed to give the likelihood in equation (3.7). Apart from a constant of proportionality, this is also the approximation suggested by Peto (1972). This likelihood is quite straightforward to compute, and is an adequate approximation when the number of tied observations at any one death time is not too large. For these reasons, this method is usually implemented in statistical software for survival analysis, and will be used in the examples given in this book.

Efron (1977) proposed

$$\prod_{j=1}^{r} \frac{\exp(\beta' s_j)}{\prod_{k=1}^{d_j} \left[\sum_{l \in R(t_{(j)})} \exp(\beta' s_l) - (k-1)d_k^{-1} \sum_{l \in D(t_{(j)})} \exp(\beta' x_l) \right]} \quad (3.8)$$

as an approximate likelihood for the proportional hazards model, where $D(t_{(j)})$ is the set of all individuals who die at time $t_{(j)}$. This is a closer approximation to the appropriate likelihood function than that due to Breslow, although in practice, both approximations often give similar results.

Cox (1972) suggested the approximation

$$\prod_{j=1}^{r} \frac{\exp(\beta' s_j)}{\sum_{l \in R(t_{(j)}; d_j)} \exp(\beta' s_l)}, \quad (3.9)$$

where the notation $R(t_{(j)}; d_j)$ denotes a set of d_j individuals drawn from $R(t_{(j)})$, the risk set at $t_{(j)}$. The summation in the denominator is the sum over all possible sets of d_j individuals sampled from the risk set without replacement. The approximation in expression (3.9) is based on a model for the situation where the the time-scale is discrete, so that under this model, tied observations are permissible. Now, the hazard function for an individual with vector of explanatory variables x_i, $h_i(t)$, is the probability of death in the unit time interval $(t, t+1)$, conditional on survival to time t. A discrete version of the proportional hazards model of equation (3.3) is the model

$$\frac{h_i(t)}{1 - h_i(t)} = \frac{h_0(t)}{1 - h_0(t)} \exp(\beta' x_i),$$

for which the likelihood function is that given in equation (3.9). In fact, in the limit as the width of the discrete time intervals becomes zero, this model tends to the proportional hazards model of equation (3.3).

When there are no ties, that is, when $d_j = 1$ for each death time, the approximations in equations (3.7), (3.8), and (3.9) all reduce to the likelihood function in equation (3.4).

3.3.3 * The Newton–Raphson procedure

Let $u(\beta)$ be the $p \times 1$ vector of first derivatives of the log-likelihood function in equation (3.5) with respect to the β-parameters. This quantity is known as the *vector of efficient scores*. Also, let the $p \times p$ matrix $I(\beta)$ be the matrix of negative second derivatives of the log-likelihood, so that the (j, k)'th element of $I(\beta)$ is

$$-\frac{\partial^2 \log L(\beta)}{\partial \beta_j \partial \beta_k}.$$

The matrix $I(\beta)$ is known as the *observed information matrix*.

According to the Newton–Raphson procedure, an estimate of the vector of β-parameters at the $(s + 1)$'th cycle of the iterative procedure, $\hat{\beta}_{s+1}$, is

$$\hat{\beta}_{s+1} = \hat{\beta}_s + \mathbf{I}^{-1}(\hat{\beta}_s)\mathbf{u}(\hat{\beta}_s),$$

for $s = 0, 1, 2, \ldots$, where $\mathbf{u}(\hat{\beta}_s)$ is the vector of efficient scores and $\mathbf{I}^{-1}(\hat{\beta}_s)$ is the inverse of the information matrix, both evaluated at $\hat{\beta}_s$. The process can be started by taking $\hat{\beta}_0 = \mathbf{0}$. The process is terminated when the change in the log-likelihood function is sufficiently small, or when the largest of the relative changes in the values of the parameter estimates is sufficiently small.

When the iterative procedure has converged, the variance–covariance matrix of the parameter estimates can be approximated by the inverse of the information matrix, evaluated at $\hat{\beta}$, that is, $\mathbf{I}^{-1}(\hat{\beta})$. The square roots of the diagonal elements of this matrix are then the standard errors of the estimated values of $\beta_1, \beta_2, \ldots, \beta_p$.

3.4 Confidence intervals and hypothesis tests for the β's

When a statistical package is used to fit a proportional hazards model, the parameter estimates that are provided are usually accompanied by their standard errors. These standard errors can be used to obtain approximate confidence intervals for the unknown β-parameters. In particular, a $100(1 - \alpha)\%$ confidence interval for a parameter β is the interval with limits $\hat{\beta} \pm z_{\alpha/2}$ s.e. $(\hat{\beta})$, where $\hat{\beta}$ is the estimate of β and $z_{\alpha/2}$ is the upper $\alpha/2$-point of the standard normal distribution.

If a $100(1 - \alpha)\%$ confidence interval for β does not include zero, this is evidence that the value of β is non-zero. More specifically, the null hypothesis that $\beta = 0$ can be tested by calculating the value of the statistic $\hat{\beta}/$ s.e. $(\hat{\beta})$. The observed value of this statistic is then compared to percentage points of the standard normal distribution in order to obtain the corresponding P-value. Equivalently, the square of this statistic can be compared with percentage points of a chi-squared distribution on one degree of freedom. This procedure is sometimes called a *Wald test*. Indeed, the P-values for this test are often given alongside parameter estimates and their standard errors in computer output.

When attempting to interpret the P-value for a given parameter, β_j, say, it is important to recognize that the hypothesis that is being tested is that $\beta_j = 0$ in the presence of all other terms that are in the model. For example, suppose that a model contains the three explanatory variables X_1, X_2, X_3, and that their coefficients are $\beta_1, \beta_2, \beta_3$. The test statistic $\hat{\beta}_2/$ s.e. $(\hat{\beta}_2)$ is then used to test the null hypothesis that $\beta_2 = 0$ in the presence of β_1 and β_3. If there was no evidence to reject this hypothesis, we would conclude that X_2 was not needed in the model in the presence of X_1 and X_3.

In general, the individual estimates $\hat{\beta}_1, \hat{\beta}_2, \ldots$ in a proportional hazards model are not all independent of one another. This means that the results of testing separate hypotheses about the β-parameters in a model may not be easy to interpret. For example, consider again the situation where there are three explanatory variables X_1, X_2, X_3. If $\hat{\beta}_1$ and $\hat{\beta}_2$ were not found to be significantly different from zero when compared with their standard errors, we could not conclude that only X_3 need be included in the model. This is because the coefficient of X_1, for example, could well change when X_2 is excluded from the model, and vice versa. This would certainly happen if X_1 and X_2 were correlated.

Because of the difficulty in interpreting the results of tests concerning the coefficients of the explanatory variables in a model, alternative methods for comparing different proportional hazards models are required. It turns out that the methods to be described in Section 3.5 are much more satisfactory than the Wald tests. Little attention should therefore be paid to the results of these tests given in computer-based analyses of survival data.

3.4.1 Standard errors and confidence intervals for hazard ratios

We have seen that in situations where there are two groups of survival data, the parameter β is the logarithm of the ratio of the hazard of death at time t for individuals in one group relative to those in the other. Hence the hazard ratio itself is $\psi = e^\beta$. The corresponding estimate of the hazard ratio is $\hat{\psi} = \exp(\hat{\beta})$, and the standard error of $\hat{\psi}$ can be obtained from the standard error of $\hat{\beta}$ using the result given as equation (2.8) in Chapter 2. From this result, the approximate variance of $\hat{\psi}$, a function of $\hat{\beta}$, is

$$\left\{ \exp(\hat{\beta}) \right\}^2 \operatorname{var}(\hat{\beta}),$$

that is $\hat{\psi}^2 \operatorname{var}(\hat{\beta})$, and so the standard error of $\hat{\psi}$ is given by

$$\text{s.e.}(\hat{\psi}) = \hat{\psi}\,\text{s.e.}(\hat{\beta}). \tag{3.10}$$

Generally speaking, a confidence interval for the true hazard ratio will be more informative than the standard error of the estimated hazard ratio. A $100(1 - \alpha)\%$ confidence interval for the true hazard ratio, ψ, can be found simply by exponentiating the confidence limits for β. An interval estimate obtained in this way is preferable to one found using $\hat{\psi} \pm z_{\alpha/2}\,\text{s.e.}(\hat{\psi})$. This is because the distribution of the logarithm of the estimated hazard ratio will be more closely approximated by a normal distribution than that of the hazard ratio itself.

The construction of a confidence interval for a hazard ratio is illustrated in Example 3.1 below. Fuller details on the interpretation of the parameters in the linear component of a proportional hazards model are given in Section 3.7.

3.4.2 Two examples

In this section, the results of fitting a proportional hazards model to data from two of the examples introduced in Chapter 1 are given.

Example 3.1 Prognosis for women with breast cancer
Data on the survival times of breast cancer patients, classified according to whether

or not sections of their tumours were positively stained, were first given in Example 1.2. The variable which indexes the result of the staining process can be regarded as a factor with two levels. From the arguments given in Section 3.2.1, this factor can be fitted by using an indicator variable X to denote the staining result, where $X = 0$ corresponds to negative staining and $X = 1$ to positive staining. Under the proportional hazards model, the hazard of death at time t for the i'th woman, for whom the value of the indicator variable is x_i, is

$$h_i(t) = e^{\beta x_i} h_0(t),$$

where x_i is zero or unity. The baseline hazard function $h_0(t)$ is then the hazard function for a women with a negatively stained tumour. This is essentially the model considered in Section 3.1.1 and given in equation (3.2).

In the group of women whose tumours were positively stained, there are two who die at 26 months. To cope with this tie, the Breslow approximation to the likelihood function will be used. This model is fitted by finding that value of β, $\hat{\beta}$, which maximizes the likelihood function in equation (3.7). The maximum likelihood estimate of β is $\hat{\beta} = 0.908$. The standard error of this estimate is also obtained from statistical packages for fitting the Cox regression model, and turns out to be given by s.e. $(\hat{\beta}) = 0.501$.

The quantity e^β is the ratio of the hazard function for a woman with $X = 1$ to that for a woman with $X = 0$, so that β is the logarithm of the ratio of the hazard of death at time t for positively stained relative to negatively stained women. The estimated value of this hazard ratio is $e^{0.908} = 2.48$. Since this is greater than unity, we conclude that a woman who has a positively stained tumour will have a greater risk of death at any given time than a comparable women whose tumour was negatively stained. Positive staining therefore indicates a poorer prognosis for a breast cancer patient.

The standard error of the hazard ratio can be found from the standard error of $\hat{\beta}$, using the result in equation (3.10). Since the estimated relative hazard is $\hat{\psi} = \exp(\hat{\beta}) = 2.480$, and the standard error of $\hat{\beta}$ is 0.501, the standard error of $\hat{\psi}$ is given by

$$\text{s.e.} (\hat{\psi}) = 2.480 \times 0.501 = 1.242.$$

We can go further and construct a confidence interval for this hazard ratio. The first step is to obtain a confidence interval for the logarithm of the hazard ratio, β. For example, a 95% confidence interval for β is the interval from $\hat{\beta} - 1.96$ s.e. $(\hat{\beta})$ to $\hat{\beta} + 1.96$ s.e. $(\hat{\beta})$, that is, the interval from -0.074 to 1.890. Exponentiating these confidence limits gives $(0.93, 6.62)$ as a 95% confidence interval for the hazard ratio itself. Notice that this interval barely includes unity, suggesting that there is evidence that the two groups of women have a different survival experience.

Example 3.2 Survival of multiple myeloma patients
Data on the survival times of 48 patients suffering from multiple myeloma were given in Example 1.3. The data base also contains the values of seven other variables which were recorded for each patient. For convenience, the values of the

variable which describes the sex of a patient have been redefined to be zero and
unity for males and females respectively. The variables are then as follows:

AGE: age of the patient
SEX: sex of the patient (0 = male, 1 = female)
BUN: blood urea nitrogen
CA: serum calcium
HB: serum haemoglobin
PC: percentage of plasma cells
BJ: Bence–Jones protein (0 = absent, 1 = present)

The sex of the patient and the variable associated with the occurrence of Bence–
Jones protein are factors with two levels. These terms are fitted using the indicator
variables SEX and BJ. The proportional hazards model for the i'th individual is
then

$$h_i(t) = \exp(\beta_1 AGE_i + \beta_2 SEX_i + \beta_3 BUN_i + \beta_4 CA_i + \beta_5 HB_i$$
$$+ \beta_6 PC_i + \beta_7 BJ_i)h_0(t),$$

where the subscript i on an explanatory variable denotes the value of that variable
for the i'th individual. The baseline hazard function is the hazard function for
an individual for whom the values of all seven of these variables are zero. This
function therefore corresponds to a male aged zero, who has zero values of BUN,
CA, HB and PC, and no Bence–Jones protein. In view of the obvious difficulty
in interpreting this function, it might be more sensible to redefine the variables
AGE, BUN, CA, HB and PC by subtracting values for an average patient.
For example, if we took $AGE - 60$ in place of AGE, the baseline hazard would
correspond to a male aged 60 years. This procedure also avoids the introduction of
a function that describes the hazard of individuals whose ages are rather different
from the age range of patients in the study. Although this leads to a baseline hazard
function that has a more natural interpretation, it will not affect inference about
the influence of the explanatory variables on the hazard of death. For this reason,
the untransformed variables will be used in this example.

On fitting the model, the estimates of the coefficients of the explanatory variables and their standard errors are found to be those shown in Table 3.1. We see
from Table 3.1 that some of the estimates are close to zero. Indeed, if individual
95% confidence intervals are calculated for the coefficients of the seven variables,
only those for BUN and HB exclude zero. This suggests that the hazard function
does not depend on all seven explanatory variables.

However, we cannot deduce from this that BUN and HB are the relevant variables, since the estimates of the coefficients of the seven explanatory variables in
the fitted model are not independent of one another. This means that if one of the
seven explanatory variables were excluded from the model, the coefficients of the
remaining six might be different from those in Table 3.1. For example, if BUN
is omitted, the estimated coefficients of the six remaining explanatory variables,
AGE, SEX, CA, HB, PC and BJ, turn out to be -0.009, -0.301, -0.036,

Table 3.1 Estimated values of the coefficients of the explanatory variables on fitting a proportional hazards model to the data from Example 1.3.

Variable	$\hat{\beta}$	s.e. $(\hat{\beta})$
AGE	−0.019	0.028
SEX	−0.251	0.402
BUN	0.021	0.006
CA	0.013	0.132
HB	−0.135	0.069
PC	−0.002	0.007
BJ	−0.640	0.427

−0.140, −0.001, and −0.420, respectively. Comparison with the values shown in Table 3.1 shows that there are differences in the estimated coefficients of each of these six variables, although in this case the differences are not very great.

In general, to determine on which of the seven explanatory variables the hazard function depends, a number of different models will need to be fitted, and the results compared. Methods for comparing the fit of alternative models, and strategies for model building are considered in subsequent sections of this chapter.

3.5 Comparing alternative models

In a modelling approach to the analysis of survival data, a model is developed for the dependence of the hazard function on one or more explanatory variables. In this development process, proportional hazards models with linear components that contain different sets of terms are fitted, and comparisons made between them.

As a specific example, consider the situation where there are two groups of survival times, corresponding to individuals who receive either a new treatment or a standard. The common hazard function under the model for no treatment difference can be taken to be $h_0(t)$. This model is a special case of the general proportional hazards model in equation (3.3) in which there are no explanatory variables in the linear component of the model. This model is therefore referred to as the *null model*.

Now let X be an indicator variable that takes the value zero for individuals receiving the standard treatment and unity otherwise. Under a proportional hazards model, the hazard function for an individual for whom X takes the value x is $e^{\beta x} h_0(t)$. The hazard functions for individuals on the standard and new treatments are then $h_0(t)$ and $e^{\beta} h_0(t)$, respectively. The difference between this model and the null model is that the linear component of the latter contains the additional term βx. Since $\beta = 0$ corresponds to no treatment effect, the extent of any treatment difference can be investigated by comparing these two proportional hazards models for the observed survival data.

More generally, suppose that two models are contemplated for a particular data set, Model (1) and Model (2), say, where Model (1) contains a subset of the terms in Model (2). Model (1) is then said to be *parametrically nested* within Model (2). Specifically, suppose that the p explanatory variables $X_1, X_2, \ldots, X_p$ are fitted in Model (1), so that the hazard function under this model can be written

$$\exp\{\beta_1 x_1 + \beta_2 x_2 + \cdots + \beta_p x_p\} h_0(t).$$

Also suppose that the $p + q$ explanatory variables $X_1, X_2, \ldots, X_p, X_{p+1}, \ldots, X_{p+q}$ are fitted in Model (2), so that the hazard function under this model is

$$\exp\{\beta_1 x_1 + \cdots + \beta_p x_p + \beta_{p+1} x_{p+1} + \cdots + \beta_{p+q} x_{p+q}\} h_0(t).$$

Model (2) then contains the q additional explanatory variables $X_{p+1}, X_{p+2}, \ldots, X_{p+q}$. Because Model (2) has a larger number of terms than Model (1), Model (2) must be a better fit to the observed data. The statistical problem is then to determine whether the additional q terms in Model (2) significantly improve the explanatory power of the model. If not, they might be omitted, and Model (1) would be deemed to be adequate.

In the discussion of Example 3.2, we saw that when there are a number of explanatory variables of possible relevance, the effect of each term cannot be studied independently of the others. The effect of any given term therefore depends on the other terms currently included in the model. For example, in Model (1), the effect of any of the p explanatory variables on the hazard function depends on the $p - 1$ variables that have already been fitted, and so the effect of X_p is said to be *adjusted* for the remaining $p - 1$ variables. In particular, the effect of X_p is adjusted for $X_1, X_2, \ldots, X_{p-1}$, but we also speak of the effect of X_p *eliminating* or *allowing for* $X_1, X_2, \ldots, X_{p-1}$. Similarly, when the q variables $X_{p+1}, X_{p+2}, \ldots, X_{p+q}$ are added to Model (1), the effect of these variables on the hazard function is said to be adjusted for the p variables that have already been fitted, $X_1, X_2, \ldots, X_p$.

3.5.1 The statistic $-2 \log \hat{L}$

In order to compare alternative models fitted to an observed set of survival data, a statistic which measures the extent to which the data are fitted by a particular model is required. Since the likelihood function summarizes the information that the data contain about the unknown parameters in a given model, a suitable summary statistic is the value of the likelihood function when the parameters are replaced by their maximum likelihood estimates. This is the maximized likelihood under an assumed model, and can be computed from equation (3.4) by replacing the β's by their maximum likelihood estimates under the model. For a given set of data, the larger the value of the maximized likelihood, the better is the agreement between the model and the observed data.

For reasons given in the sequel, it is more convenient to use minus twice the logarithm of the maximized likelihood in comparing alternative models. If the maximized likelihood for a given model is denoted by $\hat{L}$, the summary measure of agreement between the model and the data is $-2\log\hat{L}$. From Section 3.3.1, $\hat{L}$ is in fact the product of a series of conditional probabilities, and so this statistic will be less than unity. In consequence, $-2\log\hat{L}$ will always be positive, and for a given data set, the smaller the value of $-2\log\hat{L}$, the better the model.

The statistic $-2\log\hat{L}$ cannot be used on its own as a measure of model adequacy. The reason for this is that the value of $\hat{L}$, and hence of $-2\log\hat{L}$, is dependent upon the number of observations in the data set. Thus if, after fitting a model to a set of data, additional data became available to which the fit of the model was the same as that to the original data, the value of $-2\log\hat{L}$ for the enlarged data set would be different from that of the original data. Accordingly the value of $-2\log\hat{L}$ is only useful when making comparisons between models fitted to the same data.

3.5.2 Comparing nested models

Consider again Model (1) and Model (2) defined above, and let the value of the maximized log-likelihood function for each model be denoted by $\hat{L}(1)$ and $\hat{L}(2)$, respectively. The two models can then be compared on the basis of the difference between the values of $-2\log\hat{L}$ for each model. In particular, a large difference between $-2\log\hat{L}(1)$ and $-2\log\hat{L}(2)$ would lead to the conclusion that the q variates in Model (2) that are additional to those in Model (1) do improve the adequacy of the model. Naturally, the amount by which the value of $-2\log\hat{L}$ changes when terms are added to a model will depend on which terms have already been included. In particular, the difference in the values of $-2\log\hat{L}(1)$ and $-2\log\hat{L}(2)$, that is, $-2\log\hat{L}(1) + 2\log\hat{L}(2)$, will reflect the combined effect of adding the variables $X_{p+1}, X_{p+2}, \ldots, X_{p+q}$ to a model that already contains $X_1, X_2, \ldots, X_p$. This is said to be the change in the value of $-2\log\hat{L}$ due to fitting $X_{p+1}, X_{p+2}, \ldots, X_{p+q}$, adjusted for $X_1, X_2, \ldots, X_p$.

The statistic $-2\log\hat{L}(1) + 2\log\hat{L}(2)$, can be written as

$$-2\log\{\hat{L}(1)/\hat{L}(2)\},$$

and this is the log-likelihood ratio statistic for testing the null hypothesis that the q parameters $\beta_{p+1}, \beta_{p+2}, \ldots, \beta_{p+q}$ in Model (2) are all zero. From results associated with the theory of likelihood ratio testing (see Appendix A), this statistic has an asymptotic chi-squared distribution, under the null hypothesis that the coefficients of the additional variables are zero. The number of degrees of freedom of this chi-squared distribution is equal to the difference between the number of independent β-parameters being fitted under the two models. Hence, in order to compare the value of $-2\log\hat{L}$ for Model (1) and Model (2), we use the fact that the statistic $-2\log\hat{L}(1) + 2\log\hat{L}(2)$ has a chi-squared distribution on q degrees of freedom, under the null hypothesis that $\beta_{p+1}, \beta_{p+2}, \ldots, \beta_{p+q}$ are all zero. If the observed value of the statistic is not significantly large, the two models will be

Table 3.2 Values of $-2 \log \hat{L}$ on fitting proportional hazards models to the data from Example 1.2.

Variables in model	$-2 \log \hat{L}$
none	173.968
X	170.096

adjudged to be equally suitable. Then, other things being equal, the more simple model, that is, the one with fewer terms, would be preferred. On the other hand, if the values of $-2 \log \hat{L}$ for the two models are significantly different, we would argue that the additional terms are needed and the more complex model would be adopted.

Some texts, and some software packages, ascribe degrees of freedom to the quantity $-2 \log \hat{L}$. However, the value of $-2 \log \hat{L}$ for a particular model does not have a chi-squared distribution, and so the quantity cannot be considered to have an associated number of degrees of freedom. Additionally, the quantity $-2 \log \hat{L}$ is sometimes referred to as a *deviance*. This is also inappropriate, since unlike the deviance used in the context of generalized linear modelling, $-2 \log \hat{L}$ does not measure deviation from a model that is a perfect fit to the data.

Example 3.3 Prognosis for women with breast cancer
Consider again the data from Example 1.2 on the survival times of breast cancer patients. On fitting a proportional hazards model which contains no explanatory variables, that is, the null model, the value of $-2 \log \hat{L}$ is 173.968. As in Example 3.1, the indicator variable X, will be used to represent the result of the staining procedure, so that X is zero for women whose tumours are negatively stained and unity otherwise. When the variable X is included in the linear component of the model, the value of $-2 \log \hat{L}$ decreases to 170.096. The values of $-2 \log \hat{L}$ for alternative models are conveniently summarized in tabular form, as illustrated in Table 3.2.

The difference between the values of $-2 \log \hat{L}$ for the null model and the model that contains X can be used to assess the significance of the difference between the hazard functions for the two groups of women. Since one model contains one more β-parameter than the other, the difference in the values of $-2 \log \hat{L}$ has a chi-squared distribution on one degree of freedom. The difference in the two values of $-2 \log \hat{L}$ is $173.968 - 170.096 = 3.872$, which is just significant at the 5% level ($P = 0.049$). We may therefore conclude that there is evidence, significant at the 5% level, that the hazard functions for the two groups of women are different.

In Example 2.10, the extent of the difference between the survival times of the two groups of women was investigated using the log-rank test. The chi-squared value for this test was found to be 3.515 ($P = 0.061$). This value is not very different from the figure of 3.872 (P = 0.049) obtained above. The similarity of these two P-values means that essentially the same conclusions are drawn about the extent to which the data provide evidence against the null hypothesis of no group difference. From the practical viewpoint, the fact that one result is just significant at the 5% cent level, while the other is not quite significant at that level, is immaterial.

Although the model-based approach used in this example is operationally different from the log-rank test, the two procedures are in fact closely related. This relationship will be explored in greater detail in Section 3.9.

Example 3.4 Treatment of hypernephroma
In a study carried out at the University of Oklahoma Health Sciences Centre, data were obtained on the survival times of 36 patients with a malignant tumour in the kidney, or hypernephroma. The patients had all been treated with a combination of chemotherapy and immunotherapy, but additionally a nephrectomy, the surgical removal of the kidney, had been carried out on some of the patients. Of particular interest is whether the survival time of the patients depends on their age at the time of diagnosis and on whether or not they had received a nephrectomy. The data obtained in the study were given in Lee (1992). In the data set to be used as a basis for this example, the age of a patient has been classified according to whether the patient is less than 60, between 60 and 70 or greater than 70. Table 3.3 gives the survival times of the patients in months, where an asterisk denotes a censored observation.

In this example, there is a factor age-group with three levels (< 60, 60–70, > 70), and a factor associated with whether or not a nephrectomy was performed. There are a number of possible models for these data depending on whether the hazard function is related to neither, one or both of these factors. Suppose that the effect due to the j'th age-group is denoted by α_j, $j = 1, 2, 3$, and that due to nephrectomy status is denoted by ν_k, $k=1,2$. The terms α_j and ν_k may then be included in proportional hazards models for $h_i(t)$, the hazard function for the i'th individual in the study. Five possible models are as follows:

$$\text{Model (1):} \quad h_i(t) = h_0(t),$$
$$\text{Model (2):} \quad h_i(t) = \exp\{\alpha_j\}h_0(t),$$
$$\text{Model (3):} \quad h_i(t) = \exp\{\nu_k\}h_0(t),$$
$$\text{Model (4):} \quad h_i(t) = \exp\{\alpha_j + \nu_k\}h_0(t),$$
$$\text{Model (5):} \quad h_i(t) = \exp\{\alpha_j + \nu_k + (\alpha\nu)_{jk}\}h_0(t).$$

Under Model (1), the hazard of death does not depend on either of the two factors and is the same for all 36 individuals in the study. In Models (2) and (3), the hazard depends on either the age-group or on whether or not a nephrectomy was performed, but not on both. In Model (4), the hazard depends on both factors, where the impact of nephrectomy on the hazard is independent of the age-group of the patient. Model (5) includes an interaction between age-group and nephrec-

Table 3.3 Survival times of 36 patients classified according to age-group and whether or not they have had a nephrectomy.

No nephrectomy			Nephrectomy		
< 60	60–70	> 70	< 60	60–70	> 70
9	15	12	104*	108*	10
6	8		9	26	9
21	17		56	14	18
			35	115	6
			52	52	
			68	5*	
			77*	18	
			84	36	
			8	9	
			38		
			72		
			36		
			48		
			26		
			108		
			5		

tomy, so that under this model the effect of a nephrectomy on the hazard of death depends on the age-group of the patient.

To fit the term α_j, two indicator variables $A2$ and $A3$ are defined with values shown in the table below.

Age-group	$A2$	$A3$
< 60	0	0
60–70	1	0
> 70	0	1

The term ν_k is fitted by defining a variable N which takes the value zero when no nephrectomy has been performed and unity when it has. With this choice of indicator variables, the baseline hazard function will correspond to an individual in the youngest age-group who has not had a nephrectomy.

Models that contain the term α_j are then fitted by including the variables $A2$, $A3$ in the model, while the term ν_k is fitted by including N. The interaction is fitted by including the products $A2N = A2 \times N$ and $A3N = A3 \times N$ in the model. The explanatory variables fitted and the values of $-2 \log \hat{L}$ for each of the five models under consideration are shown in Table 3.4. Some computer software for

Table 3.4 Values of $-2 \log \hat{L}$ on fitting five models to the data in Table 3.3.

Terms in model	Variables in model	$-2 \log \hat{L}$
null model	none	177.667
α_j	$A2, A3$	172.172
ν_k	N	170.247
$\alpha_j + \nu_k$	$A2, A3, N$	165.508
$\alpha_j + \nu_k + (\alpha\nu)_{jk}$	$A2, A3, N, A2N, A3N$	162.479

modelling survival data enables factors to be included in a model without the user having to define appropriate indicator variables. The values of $-2 \log \hat{L}$ in Table 3.4 could then have been obtained directly using such software.

The first step in comparing these different models is to determine if there is an interaction between nephrectomy status and age-group. To do this, Model (4) is compared with Model (5). The reduction in the value of $-2 \log \hat{L}$ on including the interaction term in the model that contains the main effects of age-group and nephrectomy status is $165.508 - 162.479 = 3.029$ on 2 d.f. This is not significant (P = 0.220) and so we conclude that there is no interaction between age-group and whether or not a nephrectomy has been performed.

We now determine whether the hazard function is related to neither, one or both of the factors age-group and nephrectomy status. The change in the value of $-2 \log \hat{L}$ on including the term α_j in the model that contains ν_k is $170.247 - 165.508 = 4.739$ on 2 d.f. This is significant at the 10% level (P = 0.094) and so there is some evidence that α_j is needed in a model that contains ν_k. The change in $-2 \log \hat{L}$ when ν_k is added to the model that contains α_j is $172.172 - 165.508 = 6.664$ on 1 d.f., which is significant at the 1% level (P = 0.010). Putting these two results together, the term α_j may add something to the model that includes ν_k, and ν_k is certainly needed in the model that contains α_j. This means that both terms are required, and that the hazard function depends on both the patient's age-group and on whether or not a nephrectomy has been carried out.

Before leaving this example, let us consider other possible results from the comparison of the five models, and how they would affect the conclusion as to which model is the most appropriate. If the term corresponding to age-group, α_j, was needed in a model in addition to the term corresponding to nephrectomy status, ν_k, and yet ν_k was not needed in the presence of α_j, the model containing just α_j, Model (2), is probably the most suitable. To make sure that α_j was needed at all, Model (2) would be further compared with Model (1), the null model. Similarly, if the term corresponding to nephrectomy status, ν_k, was needed in addition to the term corresponding to age-group, α_j, but α_j was not required in the presence of ν_k, Model (3) would probably be satisfactory. However, the significance of ν_k would be checked by comparing Model (3) with Model (1). If neither of the terms corresponding to age-group and nephrectomy status were needed in the presence of the other, a maximum of one variable would be required. To determine which of the two is necessary, Model (2) would be compared with Model (1) and Model (3) with Model (1). If both results were significant, on statistical grounds, the model which leads to the biggest reduction in the value of $-2 \log \hat{L}$ from that for the null model would be adopted. If neither Model (2) nor Model (3) was superior to Model (1), we would conclude that neither age-group nor nephrectomy status had an effect on the hazard function.

There are two further steps in the modelling approach to the analysis of survival data. First, we will need to critically examine the fit of a model to the observed data in order to ensure that the fitted proportional hazards model is indeed appropriate. Second, we will need to interpret the model, in order to quantify the effect that the explanatory variables have on the hazard function. Interpretation of parameters in a fitted model is considered in Section 3.7, while methods for assessing the adequacy of a fitted model will be considered in Chapter 5. But first, some general comments are made on possible strategies for model selection.

3.6 Strategy for model selection

An initial step in the model selection process is to identify a set of explanatory variables that have the potential for being included in the linear component of a proportional hazards model. This set will contain those variates and factors which have been recorded for each individual, but additionally terms corresponding to interactions between factors or between variates and factors may also be required.

Once a set of potential explanatory variables has been isolated, the combination of variables which are to be used in modelling the hazard function has to be determined. In practice, a hazard function will not depend on a unique combination of variables. Instead, there are likely to be a number of equally good models, rather than a single 'best' model. For this reason, it is desirable to consider a wide range of possible models.

An important principle in statistical modelling is that when a model contains an interaction term, the corresponding lower-order terms should also be included. This rule is known as the *hierarchic principle*, and means that interactions should not be fitted unless the corresponding main effects are present. Models which are not hierarchic are difficult to interpret and no attempt should be made to fit them.

The model selection strategy depends to some extent on the purpose of the study. In some applications, information on a number of variables will have been obtained, and the aim might be to determine which of them has an effect on the hazard function, as in Example 1.3 on multiple myeloma. In other situations, there may be one or more variables of primary interest, such as terms corresponding to a treatment effect. The aim of the modelling process is then to evaluate the effect of such variables on the hazard function, as in Example 1.4 on prostatic cancer. Since the other variables that have been recorded might also be expected to influence the size and precision of the treatment effect, these variables will need to be taken account of in the modelling process.

3.6.1 Variable selection procedures

We first consider the situation where all explanatory variables are on an equal footing, and the aim is to identify subsets of variables upon which the hazard function depends. When the number of potential explanatory variables, including interactions, non-linear terms and so on, is not too large, it might be feasible to fit all possible combinations of terms, paying due regard to the hierarchic principle. Alternative nested models can be compared by examining the change in the value of $-2 \log \hat{L}$ on adding terms into a model or deleting terms from a model.

Comparisons between a number of possible models, which need not necessarily be nested, can also be made on the basis of the statistic

$$AIC = -2 \log \hat{L} + \alpha q,$$

in which q is the number of unknown β-parameters in the model and α is a pre-determined constant. The value of α is usually taken to be between 2 and 6. This statistic is known as *Akaike's information criterion*; the smaller the value of this statistic, the better the model. The motivation behind this statistic is that if the only difference between two models is that one includes unnecessary covariates, the values of AIC for the two models will not be very different. Indeed, the value of AIC will tend to increase when unnecessary terms are added to the model.

Values of α between 2 and 6 can be used in computing the value of the statistic. The choice $\alpha = 3$ is roughly equivalent to using a 5% significance level in judging the difference between the values of $-2 \log \hat{L}$ for two nested models which differ by between one and three parameters. This value of α is recommended for general use.

Of course, some terms may be identified as alternatives to those in a particular model, leading to subsets that are equally suitable. The choice of the most appropriate from these should not then rest on statistical grounds alone. When there are no subject matter grounds for model choice, the model chosen for initial consideration from a set of alternatives might be the one for which the value of $-2 \log \hat{L}$ or AIC is a minimum. It will then be important to confirm that the model does fit the data using the methods for model checking described in Chapter 5.

In some applications, information might be recorded on a number of variables, all of which relate to the same general feature. For example, the variables height, weight, head circumference, arm length, body mass index (weight/height^2), and so on, are all concerned with the size of an individual. In view of inter-relationships between these variables, a model for the survival times of these individuals may not need to include each of them. It would then be appropriate to determine which variables from this group should be included in the model, although it may not matter exactly which variables are chosen.

When the number of variables is relatively large, the number of possible models that need to be fitted can be computationally expensive. In particular, if there is a pool of p potential explanatory variables, there are 2^p possible combinations of terms, so that if $p > 10$, there are more than a thousand possible combinations of explanatory variables. In this situation, automatic routines for variable selection that are available in many software packages might seem an attractive prospect. These routines are based on *forward selection*, *backward elimination* or a combination of the two known as the *stepwise procedure*.

In forward selection, variables are added to the model one at a time. At each stage in the process, the variable added is the one which gives the largest decrease in the value of $-2 \log \hat{L}$ on its inclusion. The process ends when the next candidate for inclusion in the model does not reduce the value of $-2 \log \hat{L}$ by more than a prespecified amount. This is known as the *stopping rule*. This rule is often couched in terms of the significance level of the difference in the values of $-2 \log \hat{L}$ when a variable is added to a model, so that the selection process ends when the next term for inclusion ceases to be significant at a pre-assigned level.

In backward elimination, a model that contains the largest number of variables under consideration is first fitted. Variables are then excluded one at a time. At each stage the variable omitted is the one that increases the value of $-2 \log \hat{L}$ by the smallest amount on its exclusion. The process ends when the next candidate for deletion increases the value of $-2 \log \hat{L}$ by more than a prespecified amount.

The stepwise procedure operates in the same way as forward selection. However, a variable that has been included in the model can be considered for exclusion at a later stage. Thus after adding a variable to the model, the procedure then checks whether any previously included variable can now be deleted. These decisions are again made on the basis of prespecified stopping rules.

These automatic routines have a number of disadvantages. Typically, they lead to the identification of one particular subset, rather than a set of equally good ones. The subsets found by these routines often depend on the variable selection process that has been used, that is, whether it is forward selection, backward elimination or the stepwise procedure, and generally tend not to take any account of the hierarchic principle. They also depend on the stopping rule that is used to determine whether a term should be included in or excluded from a model. For all these reasons, these automatic routines have a limited role in model selection, and should certainly not be used uncritically.

Instead of using automatic variable selection procedures, the following general strategy for model selection is recommended.

1. The first step is to fit models that contain each of the variables one at a time. The values of $-2 \log \hat{L}$ for these models are then compared with that for the null model to determine which variables on their own significantly reduce the value of this statistic.

2. The variables which appear to be important from Step 1 are then fitted together. In the presence of certain variables, others may cease to be important. Consequently, those variables which do not significantly increase the value of

$-2 \log \hat{L}$ when they are omitted from the model can now be discarded. We therefore compute the change in the value of $-2 \log \hat{L}$ when each variable on its own is omitted from the set. Only those that lead to a significant increase in the value of $-2 \log \hat{L}$ are retained in the model. Once a variable has been dropped, the effect of omitting each of the remaining variables in turn should be examined.

3. Variables which were not important on their own, and so were not under consideration is Step 2, may become important in the presence of others. These variables are therefore added to the model from Step 2, one at a time, and any that reduce $-2 \log \hat{L}$ significantly are retained in the model. This process may result in terms in the model determined at Step 2 ceasing to be significant.

4. A final check is made to ensure that no term in the model can be omitted without significantly increasing the value of $-2 \log \hat{L}$, and that no term not included significantly reduces $-2 \log \hat{L}$.

When using this selection procedure, rigid application of a particular significance level should be avoided. In order to guide decisions on whether to include or omit a term, the significance level should not be too small; a level of around 10% is recommended.

In some applications, a small number of interactions and other higher-order terms, such as powers of certain variates, may need to be considered for inclusion in a model. Such terms would be added to the model identified in Step 3 above, after ensuring that any terms necessitated by the hierarchic principle have already been included in the model. If any higher-order term leads to a significant reduction in the value of $-2 \log \hat{L}$, that term would be included in the model.

The procedure outlined above is now illustrated in an example.

Example 3.5 Survival of multiple myeloma patients
The analysis of the data on the survival times of multiple myeloma patients in Example 3.2 suggested that not all of the seven explanatory variabies AGE, SEX, BUN, CA, HB, PC, BJ, are needed in a proportional hazards model. We now determine the most appropriate subsets of these variables. In this example, transformations of the original variables and interactions between them will not be considered. We will further assume that there are no medical grounds for including particular variables in a model. A summary of the values of $-2 \log \hat{L}$ for all models that are to be considered is given in Table 3.5.

The first step is to fit the null model and models which contain each of the seven explanatory variables on their own. Of these variables, BUN leads to the largest reduction in $-2 \log \hat{L}$, reducing the value of the statistic from 215.940 to 207.453. This reduction of 8.487 is significant at the 1% level (P = 0.004) when compared with percentage point of the chi-squared distribution on 1 d.f. The reduction in $-2 \log \hat{L}$ on adding HB to the null model is 4.872, which is also significant at the 5% level (P = 0.027). The only other variable which on its own has some explanatory power is BJ, which leads to a reduction in $-2 \log \hat{L}$ which is nearly

Table 3.5 Values of $-2\log\hat{L}$ for models fitted to the data from Example 1.3.

Variables in model	$-2\log\hat{L}$
none	215.940
AGE	215.817
SEX	215.906
BUN	207.453
CA	215.494
HB	211.068
PC	215.875
BJ	213.890
$HB + BUN$	202.938
$HB + BJ$	209.829
$BUN + BJ$	203.641
$BUN + HB + BJ$	200.503
$HB + BUN + AGE$	202.669
$HB + BUN + SEX$	202.553
$HB + BUN + CA$	202.937
$HB + BUN + PC$	202.773

significant at the 15% level (P $= 0.152$). Although this P-value is relatively high, we will for the moment keep BJ under consideration for inclusion in the model.

The next step is to fit the model that contains BUN, HB and BJ, which leads to a value of $-2\log\hat{L}$ of 200.503. The effect of omitting each of the three variables in turn from this model is shown in Table 3.5. In particular, when BUN is omitted, the increase in $-2\log\hat{L}$ is 9.326, when HB is omitted the increase is 3.138, and when BJ is omitted it is 2.435. Each of these changes in the value of $-2\log\hat{L}$ can be compared with percentage points of a chi-squared distribution on 1 d.f. Since BJ does not appear to be needed in the model in the presence of HB and BUN, this variable will not be further considered for inclusion.

If either HB or BUN is excluded from the model that contains both of these variables, the increase in $-2\log\hat{L}$ is 4.515 and 8.130, respectively. Both of these increases are significant at the 5% level, and so neither HB nor BUN can be excluded from the model without significantly increasing the value of the $-2\log\hat{L}$ statistic.

Finally, we look to see if any of variables AGE, SEX, CA and PC should be included in the model that contains BUN and HB. Table 3.5 shows that when any of these four variables is added, the reduction in $-2\log\hat{L}$ is less than 0.5, and so none of them need to be included in the model. We therefore conclude that the most satisfactory model is that containing BUN and HB.

We now turn to studies where there are variables of primary importance, such as a treatment effect. Here we proceed as follows.

1. The important prognostic variables are first selected, ignoring the treatment effect. Models with all possible combinations of the variables can be fitted when their number is not too large. Alternatively, the variable selection process might follow similar lines to those described previously in Steps 1–4.

2. The treatment effect is then included in the model. In this way, any differences between the two groups that arise as a result of differences between the distributions of the prognostic variables in each treatment group are not attributed to the treatment.

3. If the possibility of interactions between the treatment and other explanatory variables has not been discounted, these must be considered before the treatment effect can be interpreted.

It will often be interesting to fit a model that contains the treatment effect alone. This enables the effect that the prognostic variables have on the magnitude of the treatment effect to be evaluated.

In this discussion on strategies for model selection, the use of statistical criteria to guide the selection process has been emphasized. In addition, due account must be taken of the application area. In particular, on subject area grounds, it may be inappropriate to include particular combinations of variables. On the other hand, there might be some variables which it is not sensible to omit from the model, even if they appear not be needed in modelling a given data set. There is always a need for non-statistical considerations in model building.

Example 3.6 Comparison of two treatments for prostatic cancer
In the data from Example 1.4 on the survival times of 38 prostatic cancer patients, there are four prognostic variables which might have an effect on the survival times. These are the age of the patient in years (AGE), serum haemoglobin level (SHB), tumour size ($SIZE$) and Gleason index ($INDEX$). All possible combinations of these variates are fitted in a proportional hazards model and the values of $-2 \log \hat{L}$ computed. These values are shown in Table 3.6, together with the values of Akaike's information criterion, computed with $\alpha = 3$.

The two most important explanatory variables when considered separately are $SIZE$ and $INDEX$. From the change in the value of $-2 \log \hat{L}$ on omitting either of them from a model that contains both, we deduce that both variables are needed in a proportional hazards model. The value of $-2 \log \hat{L}$ is only reduced by a very small amount when AGE and SHB are added to the model that contains $SIZE$ and $INDEX$. We therefore conclude that only $SIZE$ and $INDEX$ are important prognostic variables.

From the values of Akaike's information criterion in Table 3.6 the model with $SIZE$ and $INDEX$ leads to the smallest value of the statistic, confirming that this is the most suitable model of those tried. Notice also that there are no other combinations of explanatory variables that lead to similar values of the AIC-

Table 3.6 Values of $-2 \log \hat{L}$ and AIC for models fitted to the data from Example 1.4.

Variables in model	$-2 \log \hat{L}$	AIC
none	36.349	36.349
AGE	36.269	39.269
SHB	36.196	39.196
$SIZE$	29.042	32.042
$INDEX$	29.127	32.127
$AGE + SHB$	36.151	42.151
$AGE + SIZE$	28.854	34.854
$AGE + INDEX$	28.760	34.760
$SHB + SIZE$	29.019	35.019
$SHB + INDEX$	27.981	33.981
$SIZE + INDEX$	23.533	29.533
$AGE + SHB + SIZE$	28.852	37.852
$AGE + SHB + INDEX$	27.893	36.893
$AGE + SIZE + INDEX$	23.269	32.269
$SHB + SIZE + INDEX$	23.508	32.508
$AGE + SHB + SIZE + INDEX$	23.231	35.231

statistic, which shows that there are no obvious alternatives to using $SIZE$ and $INDEX$ in the model.

We now consider the treatment effect. Let $TREAT$ be a variable which takes the value zero for individuals allocated to the placebo, and unity for those allocated to DES. When $TREAT$ is added to the model that contains $SIZE$ and $INDEX$, the value of $-2 \log \hat{L}$ is reduced to 22.572. This reduction of 0.961 on 1 d.f. is not significant (P = 0.327). This indicates that there is no treatment effect, but first we ought to examine whether the coefficients of the two explanatory variables in the model depend on treatment. To do this, we form the products $TSIZE = TREAT \times SIZE$ and $TINDEX = TREAT \times INDEX$, and add these to the model that contains $SIZE$, $INDEX$ and $TREAT$. When $TSIZE$ and $TINDEX$ are added to the model, $-2 \log \hat{L}$ is reduced to 20.829 and 20.792, respectively. On adding both of these mixed terms, $-2 \log \hat{L}$ becomes 19.705. The reductions in $-2 \log \hat{L}$ on adding these terms to the model are not significant, and so there is no evidence that the treatment effect depends on $SIZE$ and $INDEX$. This means that our original interpretation of the size of the treatment effect is valid, and that on the basis of these data, treatment with DES does not appear to affect the hazard of death. The estimated size of this treatment effect will be considered later in Example 3.10.

Before leaving this example, we note that when either $TSIZE$ or $TINDEX$ is added to the model, their estimated coefficient, and that of $TREAT$, become large. The standard errors of these estimates are also very large. In particular, in the model that contains $SIZE$, $INDEX$, $TREAT$ and $TSIZE$, the estimated coefficient of $TREAT$ is -11.28 with a standard error of 18.50. For the model that contains $SIZE$, $INDEX$, $TREAT$ and $TINDEX$, the coefficients of $TREAT$ and $TINDEX$ are respectively -161.52 and 14.66, while the standard errors of these estimates are 18476 and 1680 respectively! This is evidence of *overfitting*.

In an overfitted model, the estimated values of some of the β-coefficients will be highly dependent on the actual data. A very slight change to the values of one of these variables could have a large impact on the estimate of the corresponding coefficient. This is the reason for such estimates having large standard errors.

An overfitted model is one which is more complicated than is justified by the data, and does not provide a useful summary of the data. This is another reason for not including the mixed terms in the model for the hazard of death from prostatic cancer.

3.6.2 Testing for non-linearity

When the dependence of the hazard function on a variate which takes a wide range of values is to be modelled, we should consider whether the variate should be included as a linear term in the proportional hazards model.

For some variates, transformations of their original values might be used in place of the original variate. For example, if a variate takes a wide range of values, that variate might first be transformed by taking logarithms. This is particularly appropriate for variates that are strictly positive. The logarithm of a variate may also be used when the distribution of the values of the variate is highly positively skew.

When there are no a priori reasons for transforming a variate, the assumption of linearity in the variate should at least be critically examined. One possibility is to add quadratic or even cubic terms to the model, and examine the consequent reduction in the value of $-2 \log \hat{L}$. If the inclusion of such terms significantly reduces the value of this statistic, we would conclude that there is non-linearity, and incorporate the polynomial terms in the model.

In many situations, non-linearity in an explanatory variate cannot be adequately represented by including polynomial terms in a model, or by transforming the original variable. For this reason, the following procedure is recommended for general use.

The values of the variate are first grouped into four or five categories containing approximately equal numbers of observations. A factor is then defined whose levels correspond to this grouping. For example, a variate reflecting the size of a tumour could be fitted as a factor whose levels correspond to very small, small, medium and large.

More specifically, let A be a factor with m levels formed from a continuous variate, and let X be a variate that takes the value j when A is at level j, for $j = 1, 2, \ldots, m$. Linearity in the original variate will then correspond to there being a linear trend across the levels of A. This linear trend can be modelled by fitting X alone. Now, fitting the $m - 1$ terms $X, X^2, \ldots, X^{m-1}$ is equivalent to fitting A as a factor in the model, using indicator variables as in Section 3.2.1. Accordingly, the difference between the value of $-2 \log \hat{L}$ for the model that contains X, and that for the model that contains A is a measure of non-linearity across the levels of A. If this difference is not significant we would conclude that there is no non-linearity and the original variate would be fitted. On the other hand, if there is evidence of non-linearity, the factor which corresponds to the variate is fitted.

The actual form of the non-linearity can be further studied from the coefficients of the indicator variables corresponding to A. Indeed, a plot of these coefficients may help in establishing the nature of any trend across the levels of the factor A.

Example 3.7 Survival of multiple myeloma patients

In Example 3.5, we found that a proportional hazards model that contained the explanatory variables BUN and HB appeared to be appropriate for the data on the survival times of multiple myeloma patients. However, the values of BUN range from 6 to 172, and the distribution of their values across the 48 subjects is positively skewed. In order to guard against the extreme values of this variate having an undue impact on the coefficient of BUN, logarithms of this variable might be used in the modelling process. Let the variable $LBUN$ be the logarithm of BUN. When $LBUN$ is used in the model selection process in place of BUN, the strategy adopted in Example 3.5 leads to a similar conclusion, in that the model need only contain $LBUN$ and HB. We will use $LBUN$ rather than BUN in this particular example.

We now consider whether there is any evidence of non-linearity in the values of serum haemoglobin level, and examine whether a quadratic term is needed in the proportional hazards model that contains $LBUN$ and HB. When the term HB^2 is added to this model, the value of $-2 \log \hat{L}$ is reduced from 208.175 to 208.032. This reduction of 0.143 on 1 d.f. is clearly not significant, which suggests that a linear term in HB is sufficient.

An alternative way of examining the extent of non-linearity is to use a factor to model the effect of serum haemoglobin level on the hazard function. Suppose that a factor with four levels is defined, where level 1 corresponds to values of HB less than or equal to 7, level 2 to values between 7 and 10, level 3 to values between 10 and 13 and level 4 to values greater than 13. This choice of levels corresponds roughly to the quartiles of the distribution of the values of HB. This factor can be fitted by defining three indicator variables, $HB2$, $HB3$, and $HB4$, which take the values shown in the following table.

Level of factor (X)	Value of HB	$HB2$	$HB3$	$HB4$
1	$HB \leq 7$	0	0	0
2	$7 < HB \leq 10$	1	0	0
3	$10 < HB \leq 13$	0	1	0
4	$HB > 13$	0	0	1

When a model containing $LBUN$, $HB2$, $HB3$ and $HB4$ is fitted, the value of $-2 \log \hat{L}$ is 206.755. Although the change in this statistic on adding the indicator variables $HB2$, $HB3$ and $HB4$ to the model that contains $LBUN$ alone is significant at the 5% level, none of the individual coefficients of the indicator variables are significantly different from zero. This makes it difficult to identify any pattern across the factor levels, but there is no suggestion of anything more than a linear trend.

A linear trend across the levels of the factor corresponding to haemoglobin level can be modelled by fitting the variate X, which takes values 1, 2, 3, 4, according to the factor level. When the model containing $LBUN$ and X is fitted, $-2 \log \hat{L}$ is 209.387, and the change in the value of $-2 \log \hat{L}$ due to any non-linearity is $209.387 - 206.755 = 2.632$ on 2 d.f. This is not significant when compared with percentage points of the chi-squared distribution on 2 d.f. We therefore conclude that the effect of haemoglobin level on the hazard of death in this group of patients is adequately modelled using the linear term HB.

3.7 Interpretation of parameter estimates

When the proportional hazards model is used in the analysis of survival data, the coefficients of the explanatory variables in the model can be interpreted as logarithms of the ratio of the hazard of death to the baseline hazard. This means that estimates of this hazard ratio, and corresponding confidence intervals, can easily be found from the fitted model. The interpretation of parameters corresponding to different types of term in the proportional hazards model is described in the following sections.

3.7.1 Models with a variate

Suppose that a proportional hazards model contains a single continuous variable X, so that the hazard function for the i'th of n individuals, for whom X takes the value x_i, is

$$h_i(t) = e^{\beta x_i} h_0(t).$$

The coefficient of x_i in this model can then be interpreted as the logarithm of a hazard ratio. Now consider the ratio of the hazard of death for an individual for whom the value $x + 1$ is recorded on X, relative to one for whom the value x is obtained. This is

$$\frac{\exp\{\beta(x+1)\}}{\exp(\beta x)} = e^{\beta},$$

and so $\hat{\beta}$ in the fitted proportional hazards model is the estimated change in the logarithm of the hazard ratio when the value of X is increased by one unit.

Using a similar argument, the estimated change in the log-hazard ratio when the value of the variable X is increased by r units is $r\hat{\beta}$, and the corresponding estimate of the hazard ratio is $\exp(r\hat{\beta})$. The standard error of the estimated log-hazard ratio will be r s.e. $(\hat{\beta})$, from which confidence intervals for the true hazard ratio can be derived.

The above argument shows that when a continuous variable X is included in a proportional hazards model, the hazard ratio when the value of X is changed by r units does not depend on the actual value of X. For example, if X refers to the age of an individual, the hazard ratio for an individual aged 70, relative to one aged 65, would be the same as that for an individual aged 20, relative to one aged 15. This feature is a direct result of fitting X as a linear term in the proportional hazards model. If there is doubt about the assumption of linearity, a factor whose levels correspond to different sets of values of X can be fitted. The linearity assumption can then be checked using the procedure described in Section 3.6.2.

3.7.2 Models with a factor

When individuals fall into one of m groups, $m \geq 2$, which correspond to categories of an explanatory variable, the groups can be indexed by the levels of a factor. Under a proportional hazards model, the hazard function for an individual in the j'th group, $j = 1, 2, \ldots, m$, is given by

$$h_j(t) = \exp(\gamma_j)h_0(t),$$

where γ_j is the effect due to the j'th level of the factor, and $h_0(t)$ is the baseline hazard function. This model is overparametrized, and so, as in Section 3.2.1, we take $\gamma_1 = 0$. The baseline hazard function then corresponds to the hazard of death at time t for an individual in the first group. The ratio of the hazards at time t for an individual in the j'th group, $j \geq 2$, relative to an individual in the first group is then $\exp(\gamma_j)$. Consequently, the parameter γ_j is the logarithm of this relative hazard, that is,

$$\gamma_j = \log\{h_j(t)/h_0(t)\}.$$

A model which contains the terms γ_j, $j = 1, 2, \ldots, m$, with $\gamma_1 = 0$, can be fitted by defining $m - 1$ indicator variables, $X_2, X_3, \ldots, X_m$, as shown in Section 3.2.1. Fitting this model leads to estimates $\hat{\gamma}_2, \hat{\gamma}_3, \ldots, \hat{\gamma}_m$, and their standard errors. The estimated logarithm of the relative hazard for an individual in group j, relative to an individual in group 1, is then $\hat{\gamma}_j$.

A $100(1 - \alpha)\%$ confidence interval for the true log-hazard ratio is the interval from $\hat{\gamma}_j - z_{\alpha/2}$ s.e. $(\hat{\gamma}_j)$ to $\hat{\gamma}_j + z_{\alpha/2}$ s.e. $(\hat{\gamma}_j)$, where $z_{\alpha/2}$ is the upper $\alpha/2$-point

Table 3.7 Parameter estimates and their standard errors on fitting a proportional hazards model to data from Example 3.4.

Parameter	Estimate	s.e.
α_2	-0.065	0.498
α_3	1.824	0.682

of the standard normal distribution. A corresponding confidence interval for the hazard ratio itself is obtained by exponentiating these confidence limits.

Example 3.8 Treatment of hypernephroma
Data on the survival times of patients with hypernephroma were given in Table 3.3. In this example, we will only consider the data from those patients on whom a nephrectomy has been performed, given in columns 3–6 of Table 3.3. The survival times of this set of patients are classified according to their age-group. If the effect due to the j'th age-group is denoted by α_j, $j = 1, 2, 3$, the proportional hazards model for the hazard at time t for a patient in the j'th age-group is such that

$$h_j(t) = \exp(\alpha_j)h_0(t).$$

This model can be fitted by defining two indicator variables $A2$ and $A3$, where $A2$ is unity if the patient is aged between 60 and 70, and $A3$ is unity if the patient is more than 70 years of age, as in Example 3.4. This corresponds to taking $\alpha_1 = 0$.

The value of $-2 \log \hat{L}$ for the null model is 128.901, and when the term α_j is added, the value of this statistic reduces to 122.501. This reduction of 6.400 on 2 d.f. is significant at the 5% level ($P = 0.041$), and so we conclude that the hazard function does depend on which age-group the patient is in.

The coefficients of the indicator variables $A2$ and $A3$ are estimates of α_2 and α_3 respectively, and are given in Table 3.7. Since the constraint $\alpha_1 = 0$ has been used, $\hat{\alpha}_1 = 0$. The hazard ratio for a patient aged 60–70, relative to one aged less than 60, is $e^{-0.065} = 0.94$, while that for a patient whose age is greater than 70, relative to one aged less than 60, is $e^{1.824} = 6.20$. These results suggest that the hazard of death at any given time is greatest for patients who are older than 70, but that there is little difference in the hazard functions for patients in the other two age-groups.

The standard error of the parameter estimates in Table 3.7 can be used to obtain confidence intervals for the true hazard ratios. A 95% confidence interval for the log-hazard ratio for a patient whose age is between 60 and 70, relative to one aged less than 60, is the interval with limits $-0.065 \pm (1.96 \times 0.498)$, that is, the interval $(-1.041, 0.912)$. The corresponding 95% confidence interval for the hazard ratio itself is $(0.35, 2.49)$. This confidence interval includes unity, which suggests that the hazard function for an individual whose age is between 60 and

70 is similar to that of a patient aged less than 60. Similarly, a 95% confidence interval for the hazard for a patient aged greater than 70, relative to one aged less than 60, is found to be $(1.63, 23.59)$. This interval does not include unity, and so an individual whose age is greater than 70 has a significantly greater hazard of death, at any given time, than patients aged less than 60.

In some applications, the hazard ratio relative to the level of a factor other than the first may be required. In these circumstances, the levels of the factor, and associated indicator variables, could be redefined so that some other level of the factor corresponds to the required baseline level, and the model re-fitted. The required estimates can also be found directly from the estimates obtained when the first level of the original factor is taken as the baseline, although this is more difficult.

The hazard functions for individuals at levels j and j' of the factor are respectively $\exp(\alpha_j)h_0(t)$ and $\exp(\alpha_{j'})h_0(t)$, and so the hazard ratio for an individual at level j, relative to one at level j', is $\exp(\alpha_j - \alpha_{j'})$. The log-hazard ratio is then $\alpha_j - \alpha_{j'}$, which is estimated by $\hat{\alpha}_j - \hat{\alpha}_{j'}$.

To obtain the standard error of this estimate, we use the result that the variance of the difference $\hat{\alpha}_j - \hat{\alpha}_{j'}$ is given by

$$\text{var}\left\{\hat{\alpha}_j - \hat{\alpha}_{j'}\right\} = \text{var}\left\{\hat{\alpha}_j\right\} + \text{var}\left\{\hat{\alpha}_{j'}\right\} - 2\,\text{cov}\left\{\hat{\alpha}_j, \hat{\alpha}_{j'}\right\}.$$

In view of this, an estimate of the covariance between $\hat{\alpha}_j$ and $\hat{\alpha}_{j'}$, as well as estimates of their variance, will be needed to compute s.e. $(\hat{\alpha}_j - \hat{\alpha}_{j'})$. The calculations are illustrated in Example 3.9.

Example 3.9 Treatment of hypernephroma
Consider again the data used in Example 3.4, and suppose that an estimate of the hazard ratio for an individual aged greater than 70, relative to one aged between 60 and 70, is required. Using the estimates in Table 3.7, the estimated log-hazard ratio is $\hat{\alpha}_3 - \hat{\alpha}_2 = 1.824 + 0.065 = 1.889$, and so the estimated hazard ratio is $e^{1.889} = 6.61$. This suggests that the hazard of death at any given time for someone aged greater than 70 is more than six and a half times that for someone aged between 60 and 70.

The variance of $\hat{\alpha}_3 - \hat{\alpha}_2$ is

$$\text{var}\,(\hat{\alpha}_3) + \text{var}\,(\hat{\alpha}_2) - 2\,\text{cov}\,(\hat{\alpha}_3, \hat{\alpha}_2),$$

and the required variances and covariance are found from the variance–covariance matrix of the parameter estimates. This matrix can be obtained from statistical packages used to fit the Cox regression model. The matrix is found to be

$$\begin{array}{c} A2 \\ A3 \end{array} \left(\begin{array}{cc} 0.2484 & 0.0832 \\ 0.0832 & 0.4649 \end{array} \right),$$
$$\qquad\quad A2 \qquad A3$$

from which var $(\hat{\alpha}_2) = 0.2484$, var $(\hat{\alpha}_3) = 0.4649$, and cov $(\hat{\alpha}_2, \hat{\alpha}_3) = 0.0832$. Of course, the variances are simply the squares of the standard errors in Table 3.7.

It then follows that

$$\text{var}\,(\hat{\alpha}_3 - \hat{\alpha}_2) = 0.4649 + 0.2484 - (2 \times 0.0832) = 0.5469,$$

and so the standard error of $\hat{\alpha}_2 - \hat{\alpha}_3$ is 0.740. Consequently a 95% confidence interval for the log-hazard ratio is$(0.440, 3.338)$ and that for the hazard ratio itself is $(1.55, 8.18)$.

An easier way of obtaining the estimated value of the hazard ratio for an individual who is aged greater than 70, relative to one aged between 60 and 70, and the standard error of the estimate, is to refine the levels of the factor associated with age-group. Suppose that the data are now arranged so that the first level of the factor corresponds to the age range 60–70, level 2 corresponds to patients aged greater than 70 and level 3 to those aged less than 60. Choosing indicator variables to be such that the effect due to the first level of the redefined factor is set equal to zero leads to the variables $AA2$ and $AA3$ defined in the table below.

Age-group	$AA2$	$AA3$
< 60	0	1
60–70	0	0
> 70	1	0

The estimated log-hazard ratio is now simply the estimated coefficient of $AA2$, and its standard error can be read directly from standard computer output.

The manner in which the coefficients of indicator variables are interpreted is crucially dependent upon the coding that has been used for them. In particular, when a proportional hazards model is fitted using a statistical package that enables factors to be fitted directly, it is essential to know how indicator variables used within the package have been defined. For example, suppose that individuals fall into one of m groups and that the coding used for the $m - 1$ indicator variables, $X_2, X_3, \ldots, X_m$, corresponding to an m-level factor A, is as follows.

Level of A	X_2	X_3	$\cdots$	X_m
1	-1	-1	$\cdots$	-1
2	1	0	$\cdots$	0
3	0	1	$\cdots$	0
$\cdots$				
m	0	0	$\cdots$	1

With this choice of indicator variables, the sum of the main effects of A, $\sum_{j=1}^{m} \alpha_j$, is equal to zero. A proportional hazards model that contains this factor can be expressed in the form

$$h_j(t) = \exp(\alpha_2 x_2 + \alpha_3 x_3 + \cdots + \alpha_m x_m) h_0(t),$$

where x_j is the value of X_j for an individual for whom the factor A is at the j'th level, $j = 2, 3, \ldots, m$. The hazard of death at a given time for an individual at the first level of the factor is

$$\exp\{-(\alpha_2 + \alpha_3 + \cdots + \alpha_m)\},$$

while that for an individual at the j'th level of the factor is

$$\exp(\alpha_j),$$

for $j \geq 2$. The ratio of the hazard for an individual in group j, $j \geq 2$, relative to an individual in the first group, is then

$$\exp(\alpha_j + \alpha_2 + \alpha_3 + \cdots + \alpha_m).$$

For example, if $m = 4$ and $j = 3$, the hazard ratio is $\exp(\alpha_2 + 2\alpha_3 + \alpha_4)$, and the variance of the corresponding estimated log-hazard ratio is

$$\text{var}(\hat{\alpha}_2) + 4\,\text{var}(\hat{\alpha}_3) + \text{var}(\hat{\alpha}_4) + 4\,\text{cov}(\hat{\alpha}_2, \hat{\alpha}_3)$$
$$+ 4\,\text{cov}(\hat{\alpha}_3, \hat{\alpha}_4) + 2\,\text{cov}(\hat{\alpha}_2, \hat{\alpha}_4).$$

Each of the terms in this expression can be found from the variance–covariance matrix of the parameter estimates after fitting a proportional hazards model, and a confidence interval for the hazard ratio obtained. Although this is reasonably straightforward, this particular coding of the indicator variables does make it much more complicated to interpret the individual parameter estimates in a fitted model.

3.7.3 Models with combinations of terms

In previous sections, we have only considered the interpretation of parameter estimates in proportional hazards models that contain a single term. More generally, a fitted model will contain terms corresponding to a number of variates, factors or combinations of the two. With suitable coding of indicator variables corresponding to factors in the model, the parameter estimates can again be interpreted as logarithms of hazard ratios.

When a model contains more than one variable, the parameter estimate associated with a particular effect is said to be adjusted for the other variables in the model, and so the estimates are log-hazard ratios, adjusted for the other terms in the model. The proportional hazards model can therefore be used to estimate hazard ratios, taking account of other variables included in the model.

When interactions between factors, or mixed terms involving factors and variates, are fitted, the estimated log-hazard ratios for a particular factor will differ according to the level of any factor, or the value of any variate, with which it interacts. In this situation, the value of any such factor level or variate will need to

Table 3.8 Estimated values of the coefficients of the explanatory variables on fitting a proportional hazards model to the data from Example 1.4.

Variable	$\hat{\beta}$	s.e. $(\hat{\beta})$
$SIZE$	0.083	0.048
$INDEX$	0.710	0.338
$TREAT$	-1.113	1.203

be made clear when the estimated hazard ratios for the factor of primary interest are presented.

Instead of giving algebraic details on how hazard ratios can be estimated after fitting models with different combinations of terms, the general approach will be illustrated in two examples. The first of these involves both factors and variates, while the second includes an interaction.

Example 3.10 Comparison of two treatments for prostatic cancer
In Example 3.6, the most important prognostic variables in the study on the survival of prostatic cancer patients were found to be size of tumour ($SIZE$) and the Gleason index of tumour stage ($INDEX$). The indicator variable $TREAT$, which represents the treatment effect, is also included in a proportional hazards model, since the aim of the study is to quantify the treatment effect. The model for the i'th individual can then be expressed in the form

$$h_i(t) = \exp\{\beta_1 SIZE_i + \beta_2 INDEX_i + \beta_3 TREAT_i\}h_0(t),$$

for $i = 1, 2, \ldots, 38$. Estimates of the β-coefficients and their standard errors on fitting this model are given in Table 3.8. The estimated log-hazard ratio for an individual on the active treatment, DES, ($TREAT = 1$) relative to an individual on the placebo ($TREAT = 0$), with the same values of $SIZE$ and $INDEX$ as the individual on DES, is $\hat{\beta}_3 = -1.113$. Consequently the estimated hazard ratio is $e^{-1.113} = 0.329$. This is the same irrespective of the actual values of $SIZE$ and $INDEX$. However, since these two explanatory variables were included in the model, the estimated hazard ratio is adjusted for these variables.

For comparison, if a model that only contains $TREAT$ is fitted, the estimated coefficient of $TREAT$ is -1.978. The estimated hazard ratio for an individual on DES relative to one on the placebo, unadjusted for $SIZE$ and $INDEX$, is $e^{-1.978} = 0.14$. This shows that unless proper account is taken of the effect of size of tumour and index of tumour grade, the extent of the treatment effect is overestimated.

Now consider the hazard ratio for an individual on a particular treatment with a given value of the variable $INDEX$ and a tumour of a given size, relative to an individual on the same treatment with the same value of $INDEX$, but whose size of tumour is one unit less. This is $e^{0.083} = 1.09$. Since this is greater than unity, we conclude that, other things being equal, the greater the size of the tumour, the greater that hazard of death at any given time. Similarly, the hazard ratio for an individual on a given treatment with a given value of $SIZE$, relative to one on the same treatment with the same value of $SIZE$, whose value of $INDEX$ is one unit less, is $e^{0.710} = 2.03$. This again means that the greater the value of the Gleason index, the greater is the hazard of death at any given time. In particular, an increase of one unit in the value of $INDEX$ doubles the hazard of death.

Example 3.11 Treatment of hypernephroma

Consider again the full set of data on survival times following treatment for hyper-nephroma, given in Table 3.3. In Example 3.4, the most appropriate proportional hazards model was found to contain terms α_j, $j = 1, 2, 3$, corresponding to age-group, and terms ν_k, $k = 1, 2$, corresponding to whether or not a nephrectomy was performed. For illustrative purposes, in this example we will consider the model that also contains the interaction between these two factors, even though it was found not to be significant. Under this model, the hazard function for an individual in the j'th age-group and the k level of nephrectomy status is

$$h(t) = \exp\{\alpha_j + \nu_k + (\alpha\nu)_{jk}\}h_0(t), \qquad (3.11)$$

where $(\alpha\nu)_{jk}$ is the term corresponding to the interaction.

Consider the hazard ratio for a patient in the j'th age-group, $j = 2, 3$, relative to an individual in the first age-group. This is

$$\frac{\exp\{\alpha_j + \nu_k + (\alpha\nu)_{jk}\}}{\exp\{\alpha_1 + \nu_k + (\alpha\nu)_{1k}\}}.$$

which reduces to

$$\exp\{\alpha_j - \alpha_1 + (\alpha\nu)_{jk} - (\alpha\nu)_{1k}\}.$$

This expression confirms that when an interaction is included in a model, the hazard ratio for one of the factors involved in the interaction will depend on the level of the other.

As in Example 3.4, the model in equation (3.11) is fitted by including the indicator variables $A2$, $A3$, and N in the model, together with the products $A2N$ and $A3N$. The estimated coefficients of these variables are then $\hat{\alpha}_j, j = 2, 3$, $\hat{\nu}_2$, and $(\hat{\alpha\nu})_{j2}, j = 2, 3$, respectively. From the coding of the indicator variables that has been used, the estimates $\hat{\alpha}_1, \hat{\nu}_1, (\hat{\alpha\nu})_{11}$ and $(\hat{\alpha\nu})_{12}$ are all zero. The estimated hazard ratio for an individual in the j'th age-group, $j = 2, 3$, relative to one in the first, when a nephrectomy has not been carried out, is then $\exp(\hat{\alpha}_j)$. The corresponding hazard ratio for an individual on whom a nephrectomy has been performed is

$$\exp\{\hat{\alpha}_j + (\hat{\alpha\nu})_{j2}\}.$$

The non-zero parameter estimates are $\hat{\alpha}_2 = 0.005$, $\hat{\alpha}_3 = 0.065$, $\hat{\nu}_2 = -1.943$, $(\hat{\alpha\nu})_{22} = -0.051$, and $(\hat{\alpha\nu})_{32} = 2.003$. The estimated hazard ratios are summa-

Table 3.9 Estimated hazard ratios on fitting a model that contains an interaction to the data from Example 3.4.

Age-group	No nephrectomy	Nephrectomy
< 60	1.000	1.000
60 − 70	1.006	0.955
> 70	1.067	7.910

rized in Table 3.9.

Inclusion of the row for which the estimated hazards ratios are 1.00 in tables such as Table 3.9 emphasizes that the hazards are relative to those for individuals in the first age-group. This table shows that individuals aged greater than 70 who have had a nephrectomy have a much increased hazard of death, relative to those in other age-groups and those who have not had a nephrectomy.

Confidence intervals for the corresponding true hazard ratios can be found using the method described in Section 3.7.2. As a further illustration, a confidence interval will be obtained for the hazard ratio for individuals who have had a nephrectomy in the second age-group relative to those in the first. The log-hazard ratio is $\hat{\alpha}_2 + (\hat{\alpha\nu})_{22}$, which has a variance of

$$\text{var}\,(\hat{\alpha}_2) + \text{var}\,\{(\hat{\alpha\nu})_{22}\} + 2\,\text{cov}\,\{\hat{\alpha}_2, (\hat{\alpha\nu})_{22}\}.$$

From the variance–covariance matrix of the parameter estimates after fitting the model in equation (3.11), $\text{var}\,(\hat{\alpha}_2) = 0.697$, $\text{var}\,\{(\hat{\alpha\nu})_{22}\} = 0.942$, and the covariance term is $\text{cov}\,\{\hat{\alpha}_2, (\hat{\alpha\nu})_{22}\} = -0.695$. Consequently, the standard error of the estimated log-hazard ratio is 0.248, and so a 95% confidence interval for the true log-hazard ratio ranges from -0.532 to 0.441. The corresponding confidence interval for the true hazard ratio is $(0.59, 1.55)$. This interval includes unity, and so the hazard ratio of 0.955 is not significantly different from unity at the 5% level. Confidence intervals for the other hazard ratios in Table 3.9 can be found in a similar manner.

3.8 * Estimating the hazard and survivor functions

So far in this chapter, we have only considered the estimation of the β-parameters in the linear component of a proportional hazards model. As we have seen, this is all that is required in order to draw inferences about the effect of explanatory variables in the model on the hazard function. Once a suitable model for a set of survival data has been identified, the hazard function, and the corresponding survivor function, can be estimated. These estimates can then be used to summarize the survival experience of individuals in the study.

Suppose that the linear component of a proportional hazards model contains p explanatory variables, $X_1, X_2, \ldots, X_p$, and that the estimated coefficients of these variables are $\hat{\beta}_1, \hat{\beta}_2, \ldots, \hat{\beta}_p$. The estimated hazard function for the i'th of n individuals in the study is then given by

$$\hat{h}_i(t) = e^{\hat{\beta}'\mathbf{x}_i} \hat{h}_0(t), \tag{3.12}$$

where $\mathbf{x}_i$ is the vector of values of the explanatory variables for the i'th individual, $i = 1, 2, \ldots, n$, and $\hat{h}_0(t)$ is the estimated baseline hazard function. Using this equation, the hazard function for an individual can be estimated once an estimate of $h_0(t)$ has been found. The relationship between the hazard, cumulative hazard and survivor functions can then be used to give estimates of the cumulative hazard function and the survivor function.

An estimate of the baseline hazard function was derived by Kalbfleisch and Prentice (1973) using an approach based on the method of maximum likelihood. Suppose that there are r distinct death times which, when arranged in increasing order, are $t_{(1)} < t_{(2)} < \cdots < t_{(r)}$, and that there are d_j deaths and n_j individuals at risk at time $t_{(j)}$. The estimated baseline hazard function at time $t_{(j)}$ is then given by

$$\hat{h}_0(t_{(j)}) = 1 - \hat{\xi}_j, \tag{3.13}$$

where $\hat{\xi}_j$ is the solution of the equation

$$\sum_{l \in D(t_{(j)})} \frac{\exp(\hat{\beta}'\mathbf{x}_l)}{1 - \hat{\xi}_j^{\exp(\hat{\beta}'\mathbf{x}_l)}} = \sum_{l \in R(t_{(j)})} \exp(\hat{\beta}'\mathbf{x}_l), \tag{3.14}$$

for $j = 1, 2, \ldots, r$. In equation (3.14), $D(t_{(j)})$ is the set of all d_j individuals who die at the j'th ordered death time, $t_{(j)}$, and as in Section 3.3, $R(t_{(j)})$ is the set of all n_j individuals at risk at time $t_{(j)}$. The estimates of the β's which form the vector $\hat{\beta}$ are those which maximize the likelihood function in equation (3.4). The derivation of this estimate of $h_0(t)$ is quite complex, and so it will not be reproduced here.

In the particular case where there are no tied death times, that is, where $d_j = 1$ for $j = 1, 2, \ldots, r$, the left-hand side of equation (3.14) will be a single term. This equation can then be solved to give

$$\hat{\xi}_j = \left(1 - \frac{\exp(\hat{\beta}'\mathbf{x}_{(j)})}{\sum_{l \in R(t_{(j)})} \exp(\hat{\beta}'\mathbf{x}_l)}\right)^{\exp(-\hat{\beta}'\mathbf{x}_{(j)})},$$

where $\mathbf{x}_{(j)}$ is the vector of explanatory variables for the individual who dies at time $t_{(j)}$.

When there are tied observations, that is, when one or more of the d_j are greater than unity, the summation on the left-hand side of equation (3.14) is the sum of a series of fractions in which $\hat{\xi}_j$ occurs in the denominators, raised to different powers. Equation (3.14) cannot then be solved explicitly, and an iterative scheme is required.

We now make the assumption that the hazard of death is constant between adjacent death times. Then, $\hat{\xi}_j$ can be regarded as an estimate of the probability that an individual survives through the interval from $t_{(j)}$ to $t_{(j+1)}$. The baseline survivor function can then be estimated by

$$\hat{S}_0(t) = \prod_{j=1}^{k} \hat{\xi}_j, \qquad (3.15)$$

for $t_{(k)} \leq t < t_{(k+1)}$, $k = 1, 2, \ldots, r - 1$, and so this estimate is a step-function. The estimated value of the baseline survivor function is zero for $t \geq t_{(r)}$ unless there are censored survival times greater than $t_{(r)}$. If this is the case, $\hat{S}_0(t)$ is undefined beyond $t_{(r)}$. The cumulative baseline hazard function is, from equation (1.7), given by $H_0(t) = -\log S_0(t)$, and so an estimate of this function is

$$\hat{H}_0(t) = -\log \hat{S}_0(t) = -\sum_{j=1}^{k} \log \hat{\xi}_j, \qquad (3.16)$$

for $t_{(k)} \leq t < t_{(k+1)}$, $k = 1, 2, \ldots, r - 1$.

The estimates of the baseline hazard, survivor and cumulative hazard functions in equations (3.13), (3.15) and (3.16) can be used to obtain the corresponding estimates for an individual with vector of explanatory variables $\mathbf{x}_i$. In particular, from equation (3.12), the hazard function is estimated by $\exp(\hat{\beta}'\mathbf{x}_i)\hat{h}_0(t)$. Next, integrating both sides of equation (3.12), we get

$$\int_0^t \hat{h}_i(u)\,du = \exp(\hat{\beta}'\mathbf{x}_i) \int_0^t \hat{h}_0(u)\,du, \qquad (3.17)$$

so that the estimated cumulative hazard function for the i'th individual is given by

$$\hat{H}_i(t) = \exp(\hat{\beta}'\mathbf{x}_i)\hat{H}_0(t). \qquad (3.18)$$

On multiplying each side of equation (3.17) by -1 and exponentiating, and making use of equation (1.5), we find that the estimated survivor function for the i'th individual is

$$\hat{S}_i(t) = [\hat{S}_0(t)]^{\exp(\hat{\beta}'\mathbf{x}_i)}, \qquad (3.19)$$

for $t_{(k)} \leq t < t_{(k+1)}$, $k = 1, 2, \ldots, r - 1$. Note that once the estimated survivor function, $\hat{S}_i(t)$, has been obtained, an estimate of the cumulative hazard function is simply $-\log \hat{S}_i(t)$.

3.8.1 The special case of no covariates

When there are no covariates, so that we have just a single sample of survival times, equation (3.14) becomes

$$\frac{d_j}{1 - \hat{\xi}_j} = n_j,$$

from which

$$\hat{\xi}_j = \frac{n_j - d_j}{n_j}.$$

Then, the estimated hazard function at time $t_{(j)}$ is $1 - \hat{\xi}_j$, which is d_j/n_j. The corresponding estimate of the survivor function from equation (3.15) is $\prod_{j=1}^{k} \hat{\xi}_j$, that is,

$$\prod_{j=1}^{k} \frac{n_j - d_j}{n_j},$$

which is the Kaplan–Meier estimate of the survivor function given earlier in equation (2.4). This shows that the estimate of the survivor function given in equation (3.15) generalizes the Kaplan–Meier estimate to the case where the hazard function depends on explanatory variables.

3.8.2 An approximation

When there are tied survival times, the estimated baseline hazard can only be found by using an iterative method to solve equation (3.14). This iterative process can be avoided by using an approximation to the summation on the left-hand side of equation (3.14).

The term

$$\hat{\xi}_j^{\exp(\hat{\beta}' \mathbf{x}_l)}$$

in the denominator of the left-hand side of equation (3.14) can be written

$$\exp[e^{\hat{\beta}' \mathbf{x}_l} \log \hat{\xi}_j],$$

and taking the first two terms in the expansion of the exponent gives

$$\exp[e^{\hat{\beta}' \mathbf{x}_l} \log \hat{\xi}_j] \approx 1 + e^{\hat{\beta}' \mathbf{x}_l} \log \hat{\xi}_j.$$

Writing $1 - \xi_j^*$ for the estimated baseline hazard at time $t_{(j)}$, obtained using this approximation, and substituting $1 + e^{\hat{\beta}' \mathbf{x}_l} \log \xi_j^*$ for $\hat{\xi}_j^{\exp(\hat{\beta}' \mathbf{x}_l)}$ in equation (3.14), we find that ξ_j^* is such that

$$-\sum_{l \in D(t_{(j)})} \frac{1}{\log \xi_j^*} = \sum_{l \in R(t_{(j)})} \exp(\hat{\beta}' \mathbf{x}_l).$$

Therefore,

$$\log \xi_j^* = \frac{-d_j}{\sum_{l \in R(t_{(j)})} \exp(\hat{\beta}' \mathbf{x}_l)},$$

and so

$$\xi_j^* = \exp\left(\frac{-d_j}{\sum_{l \in R(t_{(j)})} \exp(\hat{\beta}' \mathbf{x}_l)}\right). \tag{3.20}$$

The corresponding estimate of the survivor function is given by

$$S_0^*(t) = \prod_{j=1}^{k} \exp\left(\frac{-d_j}{\sum_{l \in R(t_{(j)})} \exp(\hat{\beta}' \mathbf{x}_l)}\right),$$

for $t_{(k)} \leq t < t_{(k+1)}$, $k = 1, 2, \ldots, r - 1$. When there are no covariates, this estimate becomes

$$\prod_{j=1}^{k} \exp(-d_j/n_j), \tag{3.21}$$

which is known as *Altshuler's estimate* of the survivor function.

A corresponding estimate of the baseline cumulative hazard function is then

$$H_0^*(t) = -\log S_0^*(t) = \sum_{j=1}^{k} \frac{d_j}{\sum_{l \in R(t_{(j)})} \exp(\hat{\beta}' \mathbf{x}_l)},$$

for $t_{(k)} \leq t < t_{(k+1)}$, $k = 1, 2, \ldots, r - 1$.

A further approximation is found from noting that the term

$$-d_j / \sum_{l \in R(t_{(j)})} \exp(\hat{\beta}' \mathbf{x}_l)$$

in the exponent of equation (3.20) will tend to be small, unless there are large numbers of ties at particular death times. Taking the first two terms of the expansion of this exponent, and denoting this new approximation to ξ_j by $\tilde{\xi}_j$ gives

$$\tilde{\xi}_j = 1 - \frac{d_j}{\sum_{l \in R(t_{(j)})} \exp(\hat{\beta}' \mathbf{x}_l)}.$$

The estimated baseline hazard function at time $t_{(j)}$ is then given by

$$\tilde{h}_0(t_{(j)}) = \frac{d_j}{\sum_{l \in R(t_{(j)})} \exp(\hat{\beta}' \mathbf{x}_l)}.$$

The cumulative hazard function is found by summing the individual hazards to give

$$\tilde{H}_0(t) = \sum_{j=1}^{k} \frac{d_j}{\sum_{l \in R(t_{(j)})} \exp(\hat{\beta}' \mathbf{x}_l)},$$

for $t_{(k)} \leq t < t_{(k+1)}$, $k = 1, 2, \ldots, r - 1$, which is the same as $H_0^*(t)$. The corresponding estimated baseline survivor function is

$$\tilde{S}_0(t) = \prod_{j=1}^{k} \left(1 - \frac{d_j}{\sum_{l \in R(t_{(j)})} \exp(\hat{\beta}' \mathbf{x}_l)} \right).$$

In the particular case where there are no covariates, the estimates $\tilde{h}_0(t)$, $\tilde{H}_0(t)$, and $\tilde{S}_0(t)$ are the same as those given in Section 3.8.1. Equations similar to equations (3.18) and (3.19) can then be used to estimate the cumulative hazard and survivor functions for an individual whose vector of explanatory variables is $\mathbf{x}_i$.

In practice, it will often be computationally advantageous to use either $S_0^*(t)$ or $\tilde{S}_0(t)$ in place of $\hat{S}_0(t)$. When the number of tied survival times is small, all three estimates will tend to be very similar. Moreover, since the estimates will tend to be used as descriptive summaries of the survival data, small differences between the estimates are unlikely to be of practical importance. Comments on which methods are used to estimate these baseline functions in statistical packages for survival analysis are included in Chapter 11.

Once an estimate of the survivor function has been obtained, the median and other percentiles of the survival time distribution can be found from tabular or graphical displays of the function for individuals with particular values of explanatory variables. The method used is very similar to that described in Section 2.3, and is illustrated in the following example.

Example 3.12 Treatment of hypernephroma
In Example 3.4, a proportional hazards model was fitted to the data on the survival times of patients with hypernephroma. The hazard function was found to depend on the age-group of a patient, and whether or not a nephrectomy had been performed. The estimated hazard function for the i'th patient was found to be

$$\hat{h}_i(t) = \exp\{0.013\,A2_i + 1.342\,A3_i - 1.411\,N_i\}\hat{h}_0(t),$$

where $A2_i$ is unity if the patient is aged between 60 and 70 and zero otherwise, $A3_i$ is unity if the patient is aged over 70 and zero otherwise, and N_i is unity if the patient has had a nephrectomy and zero otherwise. The estimated baseline hazard function is therefore the estimated hazard of death at time t, for an individual whose age is less than 60 and who has not had a nephrectomy.

In Table 3.10, the estimated baseline hazard function, $\hat{h}_0(t)$, cumulative hazard function, $\hat{H}_0(t)$, and survivor function, $\hat{S}_0(t)$, obtained using equations (3.14), (3.16) and (3.15), respectively, are tabulated. From this table, we see that the general trend is for the estimated baseline hazard function to increase with time. From the manner in which the estimated baseline hazard function has been computed, the estimates only apply at the death times of the patients in the study. However, if the assumption of a constant hazard in each time interval is made, by dividing the estimated hazard by the corresponding time interval, the risk of death per unit time can be found and plotted. Such a graph is shown in Figure 3.2. This graph shows that the risk of death per unit time is roughly constant over the duration of the study.

We now consider estimation of the median survival time, which is the smallest observed survival time for which the estimated survivor function is less than 0.5. From Table 3.10, the estimated median survival time for patients aged less than 60 who have not had a nephrectomy is 12 months.

By raising the estimate of the baseline survivor function to a suitable power, the estimated survivor functions for patients in other age-groups, and for patients who have had a nephrectomy, can be obtained through equation (3.19). Thus, the estimated survivor function for the i'th individual is given by

$$\hat{S}_i(t) = [\hat{S}_0(t)]^{\exp\{0.013A2_i + 1.342A3_i - 1.411N_i\}}.$$

For an individual aged less than 60 who has had a nephrectomy, $A2 = 0$, $A3 = 0$,

Table 3.10 Estimates of the baseline hazard and survivor functions for the data from Example 3.4.

Time interval	$\hat{h}_0(t)$	$\hat{S}_0(t)$	$\hat{H}_0(t)$
0–	0.000	1.000	0.000
5–	0.050	0.950	0.051
6–	0.104	0.852	0.161
8–	0.113	0.755	0.281
9–	0.237	0.576	0.551
10–	0.073	0.534	0.628
12–	0.090	0.486	0.721
14–	0.108	0.433	0.836
15–	0.116	0.383	0.960
17–	0.132	0.333	1.101
18–	0.285	0.238	1.436
21–	0.185	0.194	1.641
26–	0.382	0.120	2.122
35–	0.232	0.092	2.386
36–	0.443	0.051	2.971
38–	0.279	0.037	3.299
48–	0.299	0.026	3.654
52–	0.560	0.011	4.475
56–	0.382	0.007	4.957
68–	0.421	0.004	5.502
72–	0.467	0.002	6.132
84–	0.598	0.001	7.042
108–	0.807	0.000	8.689
115–	–	0.000	–

and $N = 1$, so that the estimated survivor function for this individual is

$$[\hat{S}_0(t)]^{\exp\{-1.411\}}.$$

This function is plotted in Figure 3.3, together with the estimated baseline survivor function which is for an individual in the same age-group but who has not had a nephrectomy. This figure shows that the probability of surviving beyond any given time is greater for those who have had a nephrectomy, confirming that a nephrectomy improves the prognosis for patients with hypernephroma.

Note that because of the assumption of proportional hazards, the two estimated survivor functions in Figure 3.3 cannot cross. Moreover the curve for those who have had a nephrectomy lies above that of those on whom a nephrectomy has not been performed. This is a direct consequence of the estimated hazard ratio for those who have had the operation, relative to those who have not, being less than unity.

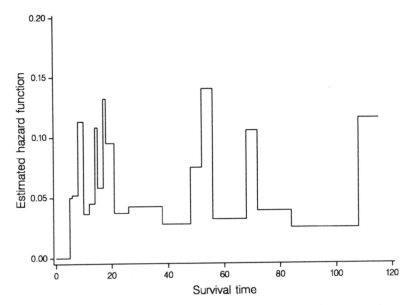

Figure 3.2 Estimated baseline hazard function, per unit time, assuming constant hazard between adjacent death times.

An estimate of the median survival time for this type of patient can be obtained from the tabulated values of the estimated survivor function or from the graph in Figure 3.3. We find that the estimated median survival time for a patient aged less than 60 who has had a nephrectomy is 36 months. Other percentiles of the distribution of survival times can be estimated using a similar approach.

In a similar manner, the survivor functions for patients in the different age-groups can be compared, either for those who have had or not had a nephrectomy. For example, for patients who have had a nephrectomy, the estimated survivor functions for patients in the three age-groups are respectively $[\hat{S}_0(t)]^{\exp\{-1.411\}}$, $[\hat{S}_0(t)]^{\exp\{-1.411+0.013\}}$ and $[\hat{S}_0(t)]^{\exp\{-1.411+1.342\}}$. These estimated survivor functions are shown in Figure 3.4. This figure clearly shows that patients aged over 70 have a poorer prognosis than those in the other two age-groups.

3.9 * Proportional hazards modelling and the log-rank test

The proportional hazards model can be used to test the null hypothesis that there is no difference between the hazard functions for two groups of survival times, as illustrated in Example 3.3. This modelling approach therefore provides an alternative to the log-rank test in this situation. However, there is a close connection between the two procedures, which is explored in greater detail in this section.

Following the notation used in Section 2.5.2, and summarized in Table 2.6, the

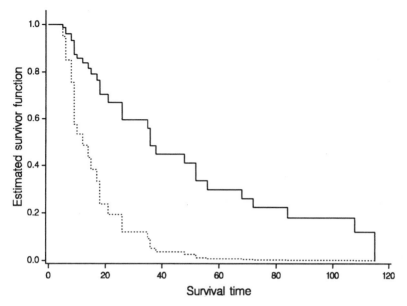

Figure 3.3 Estimated survivor functions for patients aged less than 60 with (—) and without ($\cdots$) a nephrectomy.

two groups will be labelled Group I and Group II respectively. The numbers of individuals in the two groups who die at the j'th ordered death time, $t_{(j)}$, $j = 1, 2, \ldots, r$, will be denoted by d_{1j} and d_{2j}, respectively. Similarly, the number of individuals at risk in the two groups at time $t_{(j)}$, that is, the numbers who are alive and uncensored just prior to this time, will be denoted n_{1j} and n_{2j}, respectively.

Now let X be an indicator variable which is unity when an individual is in Group I and zero when an individual is in Group II. The proportional hazards model for the i'th individual can be written

$$h_i(t) = e^{\beta x_i} h_0(t),$$

where x_i is the value of X for the i'th individual, $i = 1, 2, \ldots, n$. When there are no tied observations, that is, when $d_j = d_{1j} + d_{2j} = 1$, this model can be fitted by finding that value $\hat{\beta}$ which maximizes the likelihood function in equation (3.4). Denoting the value of X for the individual who dies at $t_{(j)}$ by $x_{(j)}$, the likelihood function is given by

$$L(\beta) = \prod_{j=1}^{r} \frac{\exp(\beta x_{(j)})}{\sum_{l=1}^{n_j} \exp(\beta x_l)}, \tag{3.22}$$

since there are $n_j = n_{1j} + n_{2j}$ individuals in the risk set, $R(t_{(j)})$, at time $t_{(j)}$, and

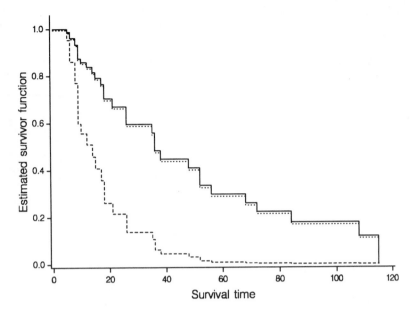

Figure 3.4 Estimated survivor functions for patients aged less than 60 (—), between 60 and 70 ($\cdots$) and greater than 70 (---), who have had a nephrectomy.

the corresponding log-likelihood function is

$$\log L(\beta) = \sum_{j=1}^{r} \beta x_{(j)} - \sum_{j=1}^{r} \log \left\{ \sum_{l=1}^{n_j} \exp(\beta x_l) \right\}.$$

Since $x_{(j)}$ is zero for individuals in Group II, the first summation in this expression is over the death times in Group I, and so is simply $d_1 \beta$, where d_1 is the total number of deaths in Group I. Also,

$$\sum_{l=1}^{n_j} \exp(\beta x_l) = n_{1j} e^{\beta} + n_{2j},$$

and so

$$\log L(\beta) = d_1 \beta - \sum_{j=1}^{r} \log \left\{ n_{1j} e^{\beta} + n_{2j} \right\}. \tag{3.23}$$

The maximum likelihood estimate of β can be found by maximizing this expression with respect to β, for which a non-linear optimization routine is required. Then the null hypotheses that $\beta = 0$ can be tested by comparing the value of $-2 \log \hat{L}(\hat{\beta})$ with $-2 \log \hat{L}(0)$. This latter quantity is simply $2 \sum_j \log n_j$.

Computation of $\hat{\beta}$ can be avoided by using a *score test* of the null hypothesis that $\beta = 0$. This test is based on the test statistic

$$\frac{u^2(0)}{i(0)},$$

where

$$u(\beta) = \frac{\partial \log L(\beta)}{\partial \beta}$$

is the efficient score, and

$$i(\beta) = -\frac{\partial^2 \log L(\beta)}{\partial \beta^2}$$

is Fisher's (observed) information function. Under the null hypothesis that $\beta = 0$, $u^2(0)/i(0)$ has a chi-squared distribution on one degree of freedom.

Now, from equation (3.23),

$$\frac{\partial \log L(\beta)}{\partial \beta} = \sum_{j=1}^{r} \left(d_{1j} - \frac{n_{1j} e^{\beta}}{n_{1j} e^{\beta} + n_{2j}} \right),$$

and

$$\frac{\partial^2 \log L(\beta)}{\partial \beta_2} = -\sum_{j=1}^{r} \frac{(n_{1j} e^{\beta} + n_{2j}) n_{1j} e^{\beta} - (n_{1j} e^{\beta})^2}{(n_{1j} e^{\beta} + n_{2j})^2}$$

$$= -\sum_{j=1}^{r} \frac{n_{1j} n_{2j} e^{\beta}}{(n_{1j} e^{\beta} + n_{2j})^2}.$$

The efficient score and information function evaluated at $\beta = 0$ are therefore given by

$$u(0) = \sum_{j=1}^{r} \left(d_{1j} - \frac{n_{1j}}{n_{1j} + n_{2j}} \right),$$

and

$$i(0) = \sum_{j=1}^{r} \frac{n_{1j} n_{2j}}{(n_{1j} + n_{2j})^2}.$$

These are simply the expressions for U_L and V_L given in equations (2.20) and (2.22) of Chapter 2, for the special case where there are no ties, that is, where $d_j = 1$ for $j = 1, 2, \ldots, r$.

When there are tied observations, the likelihood function in equation (3.22) has to be replaced by one that allows for ties. In particular, if the likelihood function in equation (3.9) is used, the efficient score and information function are exactly those given in equations (2.20) and (2.22). Hence, when there are tied survival times, the log-rank test corresponds to using the score test for the discrete proportional hazards model due to Cox (1972). In practice, the P-values that result from this score test will not usually differ much from those obtained from comparing the values of the statistic $-2 \log \hat{L}$ for the models with and without a term corresponding to the treatment effect. This was noted in the discussion of Example

3.3. Of course, one advantage of using the Cox regression model in the analysis of such data is that it leads directly to an estimate of the hazard ratio.

3.10 Further reading

Comprehensive introductions to statistical modelling in the context of linear regression analysis are given by Draper and Smith (1981) and Montgomery and Peck (1982). McCullagh and Nelder (1989) include a chapter on models for survival data in their encyclopaedic survey of generalized linear modelling. Aitkin *et al.* (1989) illustrate the theory and practice of linear modelling through the statistical package GLIM, and also include a chapter on the analysis of survival data.

Model formulation, and strategies for model selection, are discussed in books on linear regression analysis, and also in Chapter 5 of Chatfield (1988), Chapter 4 of Cox and Snell (1981), and Appendix 2 of Cox and Snell (1989). Miller (1990) describes a wide range of procedures for identifying suitable subsets of variables to use in linear regression modelling. What has come to be known as Akaike's information criterion was introduced by Akaike (1974). It is widely used in times series analysis and described in books on this subject, such as Chatfield (1989) and Cryer (1986). The hierarchic principle is fully discussed by Nelder (1977) and in Chapter 3 of McCullagh and Nelder (1989).

The proportional hazards model for survival data, in which the baseline hazard function remains unspecified, was proposed by Cox (1972). This paper introduced the notion of partial likelihood, which was subsequently considered in greater detail by Cox (1975). See also the contributions to the discussion of Cox (1972) by Kalbfleisch and Prentice (1972) and Breslow (1972).

Introductions to the proportional hazards model which are intended for medical researchers have been given by Christensen (1987), Elashoff (1983) and Tibshirani (1982). Harris and Albert (1991) include several examples of the use of this model in their textbook on survival analysis.

A detailed treatment of ties in survival data is given in Kalbfleisch and Prentice (1980) and Lawless (1982). See also Breslow (1972) and Peto (1972). The estimate of the baseline survivor function denoted by $\hat{S}_0(t)$ in Section 3.8 was introduced by Kalbfleisch and Prentice (1973) and is also described in Kalbfleisch and Prentice (1980). The estimate $\tilde{S}_0(t)$ was presented by Breslow (1972, 1974), although it was derived using a different argument to that used in Section 3.8.2.

4

The Weibull model for survival data

When the Cox regression model is used in the analysis of survival data, there is no need to assume a particular form of probability distribution for the survival times. As a result, the hazard function is not restricted to a specific functional form, and the model has flexibility and widespread applicability. On the other hand, if the assumption of a particular probability distribution for the data is valid, inferences based on such an assumption will be more precise. In particular, estimates of quantities such as relative hazards and median survival times will tend to have smaller standard errors than they would in the absence of a distributional assumption.

A probability distribution which plays a central role in the analysis of survival data is the Weibull distribution, introduced by W. Weibull in 1951 in the context of industrial reliability testing. Indeed, this distribution is as central to the parametric analysis of survival data as the normal distribution is in linear modelling. Models based on the Weibull distribution are the subject of this chapter

4.1 Models for the hazard function

Once a distributional model for survival times has been specified in terms of a probability density function, the corresponding survivor and hazard functions can be obtained from the relations

$$S(t) = 1 - \int_0^t f(u) \, \mathrm{d}u, \tag{4.1}$$

and

$$h(t) = \frac{f(t)}{S(t)} = -\frac{\mathrm{d}}{\mathrm{d}t} \{\log S(t)\}, \tag{4.2}$$

where $f(t)$ is the probability density function of the survival times. These relationships were derived in Section 1.3. An alternative approach is to specify a functional form for the hazard function, from which the survivor function and probability density functions can be determined from the equations

$$S(t) = \exp \{-H(t)\}, \tag{4.3}$$

and

$$f(t) = h(t)S(t) = -\frac{dS(t)}{dt},$$ (4.4)

where

$$H(t) = \int_0^t h(u)\, du$$

is the integrated hazard function.

4.1.1 The exponential distribution

The simplest model for the hazard function is to assume that it is constant over time. The hazard of death at any time after the time origin of the study is then the same, irrespective of the time elapsed. Under this model, the hazard function may be written

$$h(t) = \lambda,$$

for $0 \leq t < \infty$. The parameter λ is a positive constant which would be estimated by fitting the model to the observed data. From equation (4.3), the corresponding survivor function is

$$S(t) = \exp\left\{-\int_0^t \lambda\, du\right\} = e^{-\lambda t},$$

and so the implied probability density function of the survival times is

$$f(t) = \lambda e^{-\lambda t},$$ (4.5)

for $0 \leq t < \infty$. This is the probability density function of a random variable T that has an *exponential distribution* with a mean of λ^{-1}. It is sometimes convenient to write $\mu = \lambda^{-1}$, so that the hazard function is μ^{-1}, and the survival time distribution has a mean of μ. However, the former specification of the hazard function will generally be used in this book.

The median of the exponential distribution, $t(50)$, is such that $S\{t(50)\} = 0.5$, that is,

$$\exp\{-\lambda t(50)\} = 0.5,$$

so that

$$t(50) = \frac{1}{\lambda}\log 2.$$

More generally, the p'th percentile of the survival time distribution is the value $t(p)$ such that $S\{t(p)\} = 1 - (p/100)$, that is,

$$t(p) = \frac{1}{\lambda}\log\left(\frac{100}{100 - p}\right).$$

A plot of the hazard function for three values of λ, namely 1.0, 0.1 and 0.01, is given in Figure 4.1, and the corresponding probability density functions are shown in Figure 4.2. For these values of λ, the means of the corresponding exponential distributions are 1, 10 and 100, and the median survival times are 0.69, 6.93 and 69.31, respectively.

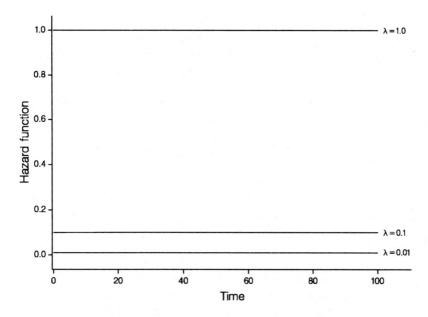

Figure 4.1 Hazard functions for exponential distributions with $\lambda=1.0$, 0.1 and 0.01.

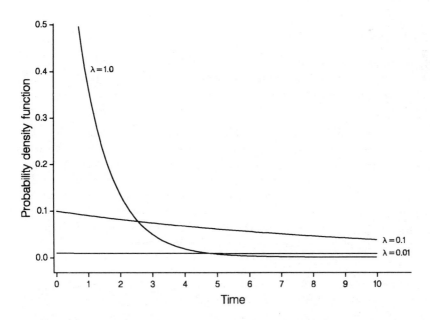

Figure 4.2 Probability density functions for exponential distributions with $\lambda=1.0$, 0.1 and 0.01.

4.1.2 * Lack of memory property

An important property of the exponential distribution is known as the *lack of memory* property. To demonstrate this property, suppose that the random variable T is associated with survival time, and that T has an exponential distribution with mean λ^{-1}. Now consider the probability that T exceeds some value t_1, where $t_1 > t_0$, conditional on the survival time being at least t_0. From the standard result for conditional probability used in Section 3.3.1,

$$P(T > t_1 \mid T > t_0) = \frac{P(T > t_1 \text{ and } T > t_0)}{P(T > t_0)}.$$

Since $t_1 > t_0$, the numerator of this expression is simply $P(T > t_1)$. The required probability is then the ratio of the probability of survival beyond t_1 to the probability of survival beyond t_0, that is, $S(t_1)/S(t_0)$. The survivor function for the exponential distribution is $S(t) = e^{-\lambda t}$, and so

$$P(T > t_1 \mid T > t_0) = \exp\{-\lambda(t_1 - t_0)\}.$$

This result means that conditional on survival to time t_0, the excess survival time beyond t_0 also has an exponential distribution with mean λ^{-1}. In other words, the distribution of the additional survival time is not affected by the knowledge that an individual has survived for a certain length of time. We will make use of this result later in Section 5.1.2 of Chapter 5.

4.1.3 The Weibull distribution

In practice, the assumption of a constant hazard function, or equivalently of exponentially distributed survival times, is rarely tenable. A more general form of hazard function is such that

$$h(t) = \lambda \gamma t^{\gamma-1}, \tag{4.6}$$

for $0 \leq t < \infty$, a function which depends on two parameters λ and γ, which are both greater than zero. In the particular case where $\gamma = 1$, the hazard function takes a constant value λ, and the survival times have an exponential distribution. For other values of γ, the hazard function increases or decreases monotonically, that is, it does not change direction. The shape of the hazard function depends critically on the value of γ, and so γ is known as the *shape parameter*, while the parameter λ is a *scale parameter*. The general form of this hazard function for different values of γ is shown in Figure 4.3. For this particular choice of hazard function, the survivor function is given by

$$S(t) = \exp\left\{-\int_0^t \lambda \gamma u^{\gamma-1}\, du\right\} = \exp(-\lambda t^\gamma). \tag{4.7}$$

The corresponding probability density function is then

$$f(t) = \lambda \gamma t^{\gamma-1} \exp(-\lambda t^\gamma),$$

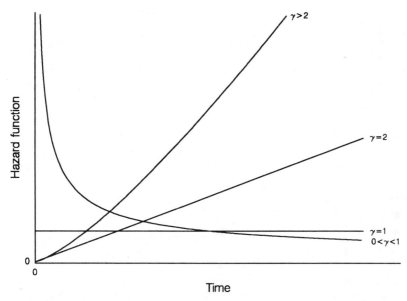

Figure 4.3 The form of the hazard function $h(t) = \lambda\gamma t^{\gamma-1}$ for different values of γ.

for $0 \leq t < \infty$, which is the density of a random variable that has a *Weibull distribution* with scale parameter λ and shape parameter γ. This distribution will be denoted $W(\lambda, \gamma)$. The right-hand tail of this distribution is longer than the left-hand one, and so the distribution is positively skewed.

The mean, or expected value, of a random variable T that has a $W(\lambda, \gamma)$ distribution can be shown to be given by

$$E(T) = \lambda^{-1/\gamma}\Gamma(\gamma^{-1} + 1),$$

where $\Gamma(x)$ is the gamma function defined by the integral

$$\Gamma(x) = \int_0^\infty u^{x-1}e^{-u}\,du.$$

The value of this integral is $(x - 1)!$, and so for integer values of x it can easily be calculated. For non-integer values of x, tables of the gamma function, such as those in Abramowitz and Stegun (1972), or suitable computer software, will be needed to compute the mean. However, since the Weibull distribution is skewed, a more appropriate and more tractable summary of the location of the distribution is the median survival time. This is the value $t(50)$ such that $S\{t(50)\} = 0.5$, so that

$$\exp\{-\lambda[t(50)]^\gamma\} = 0.5,$$

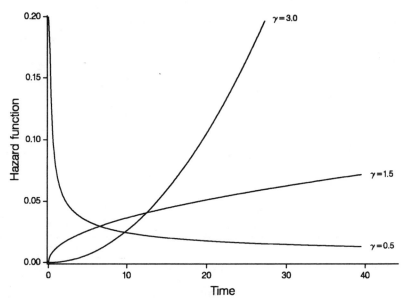

Figure 4.4 Hazard functions for a Weibull distribution with a median of 20 and $\gamma = 0.5$, 1.5 and 3.0.

and

$$t(50) = \left[\frac{1}{\lambda} \log 2 \right]^{1/\gamma}.$$

More generally, the p'th percentile of the Weibull distribution, $t(p)$, is such that

$$t(p) = \left[\frac{1}{\lambda} \log \left(\frac{100}{100 - p} \right) \right]^{1/\gamma}. \tag{4.8}$$

The median and other percentiles of the Weibull distribution are therefore much simpler to compute than the mean of the distribution.

The hazard function and corresponding probability density function for Weibull distributions with a median of 20 and shape parameters $\gamma = 0.5$, 1.5 and 3.0 are shown in Figures 4.4 and 4.5, respectively. The corresponding value of the scale parameter, λ, for these three Weibull distributions is 0.15, 0.0078 and 0.000087, respectively.

Since the Weibull hazard function can take a variety of forms depending on the value of the shape parameter γ, and appropriate summary statistics can be easily obtained, this distribution is widely used in the parametric analysis of survival data.

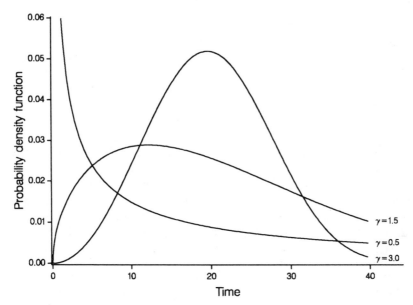

Figure 4.5 Probability density functions for a Weibull distribution with a median of 20 and $\gamma = 0.5$, 1.5 and 3.0.

4.2 Assessing the suitability of a parametric model

Prior to fitting a model based on an assumed parametric form for the hazard function, a preliminary study of the validity of this assumption should be carried out. One approach would be to estimate the hazard function using the methods outlined in Section 2.2. If the hazard function were reasonably constant over time, this would indicate that the exponential distribution might be a suitable model for the data. On the other hand, if the hazard function increased or decreased monotonically with increasing survival time, a model based on the Weibull distribution would be indicated.

A more informative way of assessing whether a particular distribution for the survival times is plausible is to compare the survivor function for the data with that of a chosen model. This is greatly helped by transforming the survivor function to produce a plot that should give a straight line if the assumed model is appropriate.

Suppose that a single sample of survival data is available, and that a Weibull distribution for the survival times is contemplated. If the survival times have a Weibull distribution with scale parameter λ and shape parameter γ, the survivor function is given by

$$S(t) = \exp\left\{-\lambda t^{\gamma}\right\}.$$

Taking the logarithm of $S(t)$, multiplying by -1 and taking logarithms a second time, gives

$$\log\{-\log S(t)\} = \log \lambda + \gamma \log t. \qquad (4.9)$$

We now substitute the Kaplan–Meier estimate of the survivor function, $\hat{S}(t)$, for $S(t)$ in equation (4.9). If the Weibull assumption is tenable, $\hat{S}(t)$ will be 'close' to $S(t)$, and a plot of $\log\{-\log \hat{S}(t)\}$ against $\log t$ would then give an approximately straight line. From equation (1.7), the cumulative hazard function, $H(t)$, is $-\log S(t)$ and so $\log\{-\log S(t)\}$ is the *log-cumulative hazard*. A plot of the values of $\log\{-\log \hat{S}(t)\}$ against $\log t$ is therefore known as a *log-cumulative hazard plot*.

If the log-cumulative hazard plot gives a straight line, the plot can be used to provide a rough estimate of the two parameters of the Weibull distribution. Specifically, from equation (4.9), the intercept and slope of the straight line will be $\log \lambda$ and γ, respectively. Thus, the slope of the line in a log-cumulative hazard plot gives an estimate of the shape parameter, and the exponential of the intercept an estimate of the scale parameter. Note that if the slope of the log-cumulative hazard plot is close to unity, the survival times could have an exponential distribution.

Example 4.1 Time to discontinuation of the use of an IUD
In Example 2.3, the Kaplan–Meier estimate of the survivor function, $\hat{S}(t)$, for the data on the time to discontinuation of an IUD was obtained. A log-cumulative hazard plot for these data, that is, a plot of $\log\{-\log \hat{S}(t)\}$ against $\log t$, is shown in Figure 4.6.

The plot indicates that there is a straight line relationship between the log-cumulative hazard and $\log t$, confirming that the Weibull distribution is an appropriate model for the discontinuation times. From the graph, the intercept of the line is approximately -6.0 and the slope is approximately 1.25. Estimates of the parameters of the Weibull distribution are therefore $\lambda^* = \exp(-6.0) = 0.002$ and $\gamma^* = 1.25$. The estimated value of γ, the shape parameter of the Weibull distribution, is quite close to unity, suggesting that the discontinuation times might be adequately modelled by an exponential distribution.

These informal estimates of λ and γ can be used to estimate the parameters of the distribution, and hence functions of these estimates, such as the median of the survival time distribution. However, this graphical approach does not lead to a measure of the precision with which the quantities have been estimated. In view of this limitation, a more formal way of fitting parametric models to survival data is developed in the next section.

4.3 Fitting a parametric model to a single sample

Parametric models can be fitted to an observed set of survival data using the method of maximum likelihood, outlined in Section 3.3. Consider first the situation where actual survival times have been observed for n individuals, so that there are no censored observations. If the probability density function of the ran-

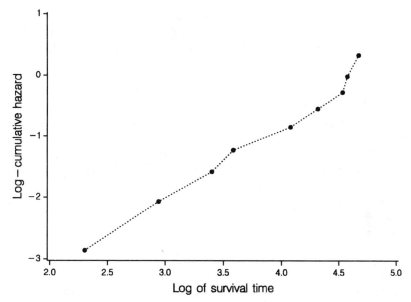

Figure 4.6   Log-cumulative hazard plot for the data from Example 1.1.

dom variable associated with survival time is $f(t)$, the likelihood of the n observations $t_1, t_2, \ldots, t_n$ is simply the product

$$\prod_{i=1}^{n} f(t_i).$$

This likelihood will be a function of the unknown parameters in the probability density function, and the maximum likelihood estimates of these parameters are those values for which the likelihood function is a maximum. In practice, it is generally more convenient to work with the logarithm of the likelihood function. Those values of the unknown parameters in the density function which maximize the log-likelihood are of course the same values which maximize the likelihood function itself.

We now consider the more usual situation where the survival data include one or more censored survival times. Specifically, suppose that r of the n individuals die at times $t_{(1)} < t_{(2)} < \cdots < t_{(r)}$ and that the survival times of the remaining $n-r$ individuals, $t_1^*, t_2^*, \ldots, t_{n-r}^*$, are right-censored. The r death times contribute a term of the form

$$\prod_{j=1}^{r} f(t_{(j)})$$

to the overall likelihood function. Naturally, we cannot ignore information about the survival experience of the $n - r$ individuals for whom a censored survival time has been recorded. If a survival time is censored at time t^*, say, we know that the lifetime of the individual is at least t^*, and the probability of this event is $P(T \geq t^*)$, which is $S(t^*)$. Thus each censored observation contributes a term of this form to the likelihood function of the n observations. The total likelihood function is therefore

$$\prod_{j=1}^{r} f(t_{(j)}) \prod_{l=1}^{n-r} S(t_l^*), \qquad (4.10)$$

in which the first product is taken over the r death times and the second over the $n - r$ censored survival times.

More compactly, suppose that the data are regarded as n pairs of observations, where the pair for the i'th individual is (t_i, δ_i), $i = 1, 2, \ldots, n$. In this notation, δ_i is an indicator variable which takes the value zero when the survival time t_i is censored and unity when t_i is an uncensored survival time. The likelihood function can then be expressed in the form

$$\prod_{i=1}^{n} \{f(t_i)\}^{\delta_i} \{S(t_i)\}^{1-\delta_i}. \qquad (4.11)$$

This function, which is equivalent to that in equation (4.10), can then be maximized with respect to the unknown parameters in the density and survivor functions.

We now consider fitting exponential and Weibull distributions to a single sample of survival data.

4.3.1* Fitting the exponential distribution

Suppose that the survival times of n individuals, $t_1, t_2, \ldots, t_n$, are assumed to have an exponential distribution with mean λ^{-1}. Further suppose that the data give the actual death times of r individuals, and that the remaining $n - r$ survival times are right-censored.

For the exponential distribution,

$$f(t) = \lambda e^{-\lambda t}, \qquad S(t) = e^{-\lambda t},$$

and on substituting into expression (4.11), the likelihood function for the n observations is given by

$$L(\lambda) = \prod_{i=1}^{n} \left(\lambda e^{-\lambda t_i}\right)^{\delta_i} \left(e^{-\lambda t_i}\right)^{1-\delta_i},$$

where δ_i is zero if the survival time of the i'th individual is censored and unity otherwise. After some simplification,

$$L(\lambda) = \prod_{i=1}^{n} \lambda^{\delta_i} e^{-\lambda t_i},$$

and the corresponding log-likelihood function is

$$\log L(\lambda) = \sum_{i=1}^{n} \delta_i \log \lambda - \lambda \sum_{i=1}^{n} t_i.$$

Since the data contain r deaths, $\sum_{i=1}^{n} \delta_i = r$ and the log-likelihood function becomes

$$\log L(\lambda) = r \log \lambda - \lambda \sum_{i=1}^{n} t_i.$$

We now need to identify the value $\hat{\lambda}$ for which the log-likelihood function is a maximum. Differentiation with respect to λ gives

$$\frac{\mathrm{d} \log L(\lambda)}{\mathrm{d}\lambda} = \frac{r}{\lambda} - \sum_{i=1}^{n} t_i,$$

and equating the derivative to zero and evaluating it at $\hat{\lambda}$ gives

$$\hat{\lambda} = r / \sum_{i=1}^{n} t_i \tag{4.12}$$

for the maximum likelihood estimator of λ.

The mean of an exponential distribution is $\mu = \lambda^{-1}$, and so the maximum likelihood estimator of μ is

$$\hat{\mu} = \hat{\lambda}^{-1} = \frac{1}{r} \sum_{i=1}^{n} t_i.$$

This estimator of μ is the total time survived by the n individuals in the data set divided by the number of deaths observed. The estimator therefore has intuitive appeal as an estimate of the mean lifetime from censored survival data.

The standard error of either $\hat{\lambda}$ or $\hat{\mu}$ can be obtained from the second derivative of the log-likelihood function, using a result from the theory of maximum likelihood estimation given in Appendix A. Differentiating $\log L(\lambda)$ a second time gives

$$\frac{\mathrm{d}^2 \log L(\lambda)}{\mathrm{d}\lambda^2} = -\frac{r}{\lambda^2},$$

and so the asymptotic variance of $\hat{\lambda}$ is

$$\mathrm{var}\,(\hat{\lambda}) = \left[-\mathrm{E} \left\{ \frac{\mathrm{d}^2 \log L(\lambda)}{\mathrm{d}\lambda^2} \right\} \right]^{-1} = \frac{\lambda^2}{r}.$$

Consequently, the standard error of $\hat{\lambda}$ is given by

$$\mathrm{s.e.}\,(\hat{\lambda}) = \hat{\lambda}/\sqrt{r}. \tag{4.13}$$

This result could be used to obtain a confidence interval for the mean survival time. In particular, the limits of a $100(1 - \alpha)\%$ confidence interval for λ are

$\hat{\lambda} \pm z_{\alpha/2}$ s.e. $(\hat{\lambda})$, where $z_{\alpha/2}$ is the upper $\alpha/2$-point of the standard normal distribution.

In presenting the results of a survival analysis, the estimated survivor and hazard functions, and the median and other percentiles of the distribution of survival times, are useful. Once an estimate of λ has been found, all these functions can be estimated using the results given in Section 4.1.1. In particular, under the assumed exponential distribution, the estimated hazard function is $\hat{h}(t) = \hat{\lambda}$ and the estimated survivor function is $\hat{S}(t) = \exp(-\hat{\lambda}t)$. In addition, the estimated median survival time is given by

$$\hat{t}(50) = \hat{\lambda}^{-1}\log 2, \tag{4.14}$$

and the estimated p'th percentile is given by

$$\hat{t}(p) = \frac{1}{\hat{\lambda}}\log\left(\frac{100}{100-p}\right). \tag{4.15}$$

The standard error of an estimate of the median survival time can be found using the result for the approximate variance of a function of a random variable given in equation (2.8) of Chapter 2. According to this result, an approximation to the variance of a function $g(\hat{\lambda})$ of $\hat{\lambda}$ is such that

$$\text{var}\,[g(\hat{\lambda})] \approx \left\{\frac{\mathrm{d}g(\hat{\lambda})}{\mathrm{d}\hat{\lambda}}\right\}^2 \text{var}\,(\hat{\lambda}). \tag{4.16}$$

Using this result, the approximate variance of the estimated p'th percentile is given by

$$\text{var}\,\{\hat{t}(p)\} \approx \left[-\frac{1}{\hat{\lambda}^2}\log\left(\frac{100}{100-p}\right)\right]^2 \text{var}\,(\hat{\lambda}).$$

On simplifying this and taking the square root, we get

$$\text{s.e.}\,\{\hat{t}(p)\} = \frac{1}{\hat{\lambda}^2}\log\left(\frac{100}{100-p}\right)\text{s.e.}\,(\hat{\lambda}),$$

and on further substituting for s.e. $(\hat{\lambda})$ from equation (4.13) and $\hat{t}(p)$ from equation (4.15), we find

$$\text{s.e.}\,\{\hat{t}(p)\} = \hat{t}(p)/\sqrt{r}. \tag{4.17}$$

In particular, the standard error of the estimated median survival time is

$$\text{s.e.}\,\{\hat{t}(50)\} = \hat{t}(50)/\sqrt{r}. \tag{4.18}$$

Confidence intervals for a true percentile are best obtained from exponentiating the confidence limits for the logarithm of the percentile. This procedure ensures that confidence limits for the percentile will be non-negative. Again making use of the result in equation (4.16), the standard error of $\log\hat{t}(p)$ is given by

$$\text{s.e.}\,\{\log\hat{t}(p)\} = \hat{t}(p)^{-1}\,\text{s.e.}\,\{\hat{t}(p)\},$$

Figure 4.7 Estimated hazard function on fitting the exponential distribution.

and after substituting for s.e. $\{\hat{t}(p)\}$ from equation (4.17), this standard error becomes

$$\text{s.e.}\,\{\log\hat{t}(p)\} = 1/\sqrt{r}.$$

Using this result, $100(1 - \alpha)\%$ confidence limits for the $100p$'th percentile are $\exp\{\log\hat{t}(p) \pm z_{\alpha/2}/\sqrt{r}\}$, that is, $\hat{t}(p)\exp\{\pm z_{\alpha/2}/\sqrt{r}\}$, where $z_{\alpha/2}$ is the upper $\alpha/2$-point of the standard normal distribution.

Example 4.2 Time to discontinuation of the use of an IUD
In this example, the data of Example 1.1 on the times to discontinuation of an IUD for 18 women are analysed under the assumption of a constant hazard of discontinuation. An exponential distribution is therefore fitted to the discontinuation times. For these data, the total of the observed and right-censored discontinuation times is 1046 days, and the number of uncensored times is 9. Therefore, using equation (4.12), $\hat{\lambda} = 9/1046 = 0.0086$, and the standard error of $\hat{\lambda}$ from equation (4.13) is s.e. $(\hat{\lambda}) = 0.0086/\sqrt{9} = 0.0029$. The estimated hazard function is therefore $\hat{h}(t) = 0.0086$, $t > 0$, and the estimated survivor function is $\hat{S}(t) = \exp(-0.0086t)$. The estimated hazard and survivor functions are shown in Figures 4.7 and 4.8 respectively.

Estimates of the median and other percentiles of the distribution of discontinuation times can be found from Figure 4.8, but more accurate estimates are obtained from equation (4.15). In particular, using equation (4.14), the median discontinu-

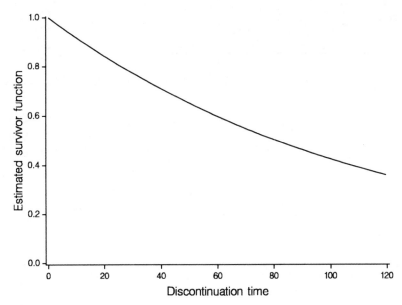

Figure 4.8 Estimated survivor function on fitting the exponential distribution.

ation time is 81 days.

An estimate of the 90'th percentile of the distribution of discontinuation times is, from equation (4.15), $\hat{t}(90) = \log 10/0.0086 = 267.61$. This means that on the assumption that the risk of discontinuing the use of an IUD is independent of time, 90% of women will have a discontinuation time of less than 268 days.

From equation (4.18), the standard error of the estimated median time to discontinuation is $80.56/\sqrt{9}$, that is, 26.85 days. The limits of a 95% confidence interval for the true median discontinuation time are

$$80.56 \exp\{\pm 1.96/\sqrt{9}\},$$

and so the interval is from 42 days to 155 days. Confidence intervals for other percentiles can be calculated in a similar manner.

4.3.2 * Fitting the Weibull distribution

The survival times of n individuals are now taken to be a censored sample from a Weibull distribution with scale parameter λ and shape parameter γ. Suppose that there are r deaths amongst the n individuals and $n - r$ right-censored survival times. We can again use expression (4.11) to obtain the likelihood of the sample data. The probability density function and survivor function of a $W(\lambda, \gamma)$ distribution are given by

$$f(t) = \lambda\gamma t^{\gamma-1}\exp(-\lambda t^{\gamma}), \qquad S(t) = \exp(-\lambda t^{\gamma}),$$

and so the likelihood of the n survival times is

$$\prod_{i=1}^{n}\left\{\lambda\gamma t_i^{\gamma-1}\exp(-\lambda t_i^{\gamma})\right\}^{\delta_i}\{\exp(-\lambda t_i^{\gamma})\}^{1-\delta_i},$$

where δ_i is zero if the i'th survival time is censored and unity otherwise. This is regarded as a function of λ and γ, the unknown parameters in the Weibull distribution, and so can be written $L(\lambda, \gamma)$. The corresponding log-likelihood function is given by

$$\log L(\lambda, \gamma) = \sum_{i=1}^{n}\delta_i\log(\lambda\gamma) + (\gamma-1)\sum_{i=1}^{n}\delta_i\log t_i - \lambda\sum_{i=1}^{n}t_i^{\gamma},$$

and noting that $\sum_{i=1}^{n}\delta_i = r$, the log-likelihood becomes

$$\log L(\lambda, \gamma) = r\log(\lambda\gamma) + (\gamma-1)\sum_{i=1}^{n}\delta_i\log t_i - \lambda\sum_{i=1}^{n}t_i^{\gamma}.$$

The maximum likelihood estimates of λ and γ are found by differentiating this function with respect to λ and γ, equating the derivatives to zero, and evaluating them at $\hat{\lambda}$ and $\hat{\gamma}$. The resulting equations are

$$\frac{r}{\hat{\lambda}} - \sum_{i=1}^{n}t_i^{\hat{\gamma}} = 0, \tag{4.19}$$

and

$$\frac{r}{\hat{\gamma}} + \sum_{i=1}^{n}\delta_i\log t_i - \hat{\lambda}\sum_{i=1}^{n}t_i^{\hat{\gamma}}\log t_i = 0. \tag{4.20}$$

From equation (4.19),

$$\hat{\lambda} = r/\sum_{i=1}^{n}t_i^{\hat{\gamma}}, \tag{4.21}$$

and on substituting for $\hat{\lambda}$ in equation (4.20), we get the equation

$$\frac{r}{\hat{\gamma}} + \sum_{i=1}^{n}\delta_i\log t_i - \frac{r}{\sum_i t_i^{\hat{\gamma}}}\sum_{i=1}^{n}t_i^{\hat{\gamma}}\log t_i = 0. \tag{4.22}$$

This is a non-linear equation in $\hat{\gamma}$ which can only be solved using an iterative numerical procedure. Once the estimate $\hat{\gamma}$ which satisfies equation (4.22) has been found, equation (4.21) can be used to obtain $\hat{\lambda}$.

In practice, a numerical procedure, such as the Newton–Raphson algorithm, is used to find the values $\hat{\lambda}$ and $\hat{\gamma}$ which maximize the likelihood function simultaneously. This procedure was described in Section 3.3.3 of Chapter 3, in connection with fitting the Cox regression model. In that section it was noted that an important by-product of the Newton–Raphson procedure is an approximation to the variance–covariance matrix of the parameter estimates, from which their standard errors can be obtained.

Once estimates of the parameters λ and γ have been found from fitting the Weibull distribution to the observed data, percentiles of the survival time distribution can be estimated using equation (4.8). The estimated p'th percentile of the distribution is

$$\hat{t}(p) = \left[\frac{1}{\hat{\lambda}} \log\left(\frac{100}{100-p}\right)\right]^{1/\hat{\gamma}}, \tag{4.23}$$

and so the estimated median survival time is given by

$$\hat{t}(50) = \left[\frac{1}{\hat{\lambda}} \log 2\right]^{1/\hat{\gamma}}. \tag{4.24}$$

The standard error of the estimated p'th percentile can be obtained using a generalization of the result in equation (4.16) to the case where the approximate variance of a function of two estimates is required. Details of the derivation are given in Appendix B, where it is shown that

$$\text{s.e.} \left\{\hat{t}(p)\right\} = \frac{\hat{t}(p)}{\hat{\lambda}\hat{\gamma}^2} \left\{\hat{\gamma}^2 \text{ var}\left(\hat{\lambda}\right) + \hat{\lambda}^2 \left(c_p - \log\hat{\lambda}\right)^2 \text{ var}\left(\hat{\gamma}\right) \right.$$
$$\left. + 2\hat{\lambda}\hat{\gamma} \left(c_p - \log\hat{\lambda}\right) \text{ cov}\left(\hat{\lambda}, \hat{\gamma}\right)\right\}^{1/2}, \tag{4.25}$$

where

$$c_p = \log\log\left(\frac{100}{100-p}\right).$$

The variances of $\hat{\lambda}$ and $\hat{\gamma}$, and their covariance are found from the variance–covariance matrix of the estimates.

As before, a confidence interval for the true value of the p'th percentile, $t(p)$, is best obtained from the corresponding interval for $\log t(p)$. The standard error of $\log\hat{t}(p)$ is

$$\text{s.e.} \{\log\hat{t}(p)\} = \frac{1}{\hat{t}(p)} \text{ s.e.} \{\hat{t}(p)\}, \tag{4.26}$$

and $100(1-\alpha)\%$ confidence limits for $\log t(p)$ are

$$\log\hat{t}(p) \pm z_{\alpha/2}\, \text{s.e.} \{\log\hat{t}(p)\}.$$

Corresponding interval estimates for $t(p)$ are found by exponentiating these limits. For example, the limits of a $100(1-\alpha)\%$ confidence interval for the median survival time, $t(50)$, are $\hat{t}(50) \exp\left[\pm z_{\alpha/2}\, \text{s.e.} \{\log\hat{t}(50)\}\right]$.

There is a substantial amount of arithmetic involved in these calculations, and care must be taken to ensure that significant figures are not lost during the course of the calculation. For this reason, it is better to perform the calculations using a suitable computer program.

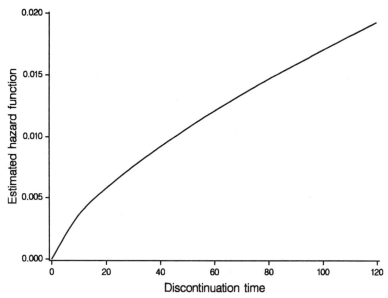

Figure 4.9   Estimated hazard function on fitting the Weibull distribution.

Example 4.3 Time to discontinuation of the use of an IUD
In Example 4.1, it was found that an exponential distribution provides a satisfactory model for the data on the discontinuation times of 18 IUD users. For comparison, a Weibull distribution will be fitted to the same data set. The distribution can be fitted using a computer package such as SAS or BMDP. From the resulting output, the estimated scale parameter of the distribution is found to be $\hat{\lambda} = 0.000454$, while the estimated shape parameter is $\hat{\gamma} = 1.676$. The standard errors of these estimates are given by s.e. $(\hat{\lambda}) = 0.000965$ and s.e. $(\hat{\gamma}) = 0.460$, respectively. Note that approximate confidence limits for γ, found using $\hat{\gamma} \pm 1.96$ s.e. $(\hat{\gamma})$, include unity, suggesting that the exponential distribution would provide a satisfactory model for the discontinuation times.

The estimated hazard and survivor functions are obtained by substituting these estimates into equations (4.6) and (4.7), whence

$$\hat{h}(t) = \hat{\lambda}\hat{\gamma}t^{\hat{\gamma}-1},$$

and

$$\hat{S}(t) = \exp\left(-\hat{\lambda}t^{\hat{\gamma}}\right).$$

These two functions are shown in Figures 4.9 and 4.10.

Although percentiles of the discontinuation time can be read from the estimated survivor function in Figure 4.10, they are better estimated using equation (4.23). Hence, under the Weibull distribution, the median discontinuation time can be

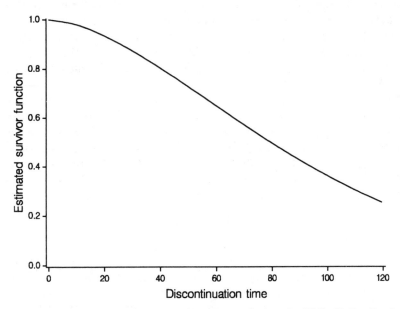

Figure 4.10 Estimated survivor function on fitting the Weibull distribution.

estimated using equation (4.24), and is given by

$$\hat{t}(50) = \left[\frac{1}{0.000454} \log 2\right]^{1/1.676} = 79.27.$$

As a check, notice that this is quite consistent with the value of the survival time corresponding to $\hat{S}(t) = 0.5$ in Figure 4.10. The standard error of this estimate, from equation (4.25) is, after much arithmetic, found to be

$$\text{s.e.}\ \{\hat{t}(50)\} = 15.795.$$

In order to obtain a 95% confidence interval for the median discontinuation time, the standard error of $\log \hat{t}(50)$ is required. From equation (4.26),

$$\text{s.e.}\ \{\log \hat{t}(50)\} = \frac{15.795}{79.272} = 0.199.$$

The required confidence limits for the log median discontinuation time are therefore $\log 79.272 \pm 1.96 \times 0.199$, that is, $(3.982, 4.763)$. The corresponding interval estimate for the true median discontinuation time is $(53.64, 117.15)$, so that there is a 95% chance that the interval from 54 days to 117 days includes the true value of the median discontinuation time. This interval is rather wide because of the small number of actual discontinuation times in the data set.

   It is interesting to compare these results with those found in Example 4.2, where the discontinuation times were modelled using an exponential distribution. The estimated median survival times are very similar, at 80.6 days for the exponential and 79.3 days for the Weibull model. However, the standard error of the estimated median survival time is 26.8 days when the times are assumed to have an exponential distribution, and only 15.8 days under the Weibull model. The median is therefore estimated more precisely when the discontinuation times are assumed to have a Weibull distribution.

Other percentiles of the survival time distribution, and accompanying standard errors and confidence intervals, can be found in a similar fashion. For example, the 90'th percentile, that is, the time beyond which 10% of those in the study continue with the use of the IUD, is 162.23 days, and 95% confidence limits for the true percentile are from 95.41 to 275.84 days. Notice that the width of the confidence interval is larger than that for the median survival time, reflecting the fact that the median is more precisely estimated than other percentiles.

4.4 A model for the comparison of two groups

We saw in Section 3.1 that a convenient general model for comparing two groups of survival times is the proportional hazards model. Here, the two groups will be labelled Group I and Group II, and X will be an indicator variable which takes the value zero if an individual is in Group I and unity if an individual is in Group II. Under the proportional hazards model, the hazard of death at time t for the i'th individual is given by

$$h_i(t) = e^{\beta x_i} h_0(t), \tag{4.27}$$

where x_i is the value of X for the i'th individual. Consequently, the hazard at time t for an individual in Group I is $h_0(t)$ and that for an individual in Group II is $\psi h_0(t)$, where $\psi = \exp(\beta)$. The quantity β is then the logarithm of the ratio of the hazard for an individual in Group II, to that of an individual in Group I.

We will now make the additional assumption that the survival times for the individuals in Group I have a Weibull distribution with scale parameter λ and shape parameter γ. Using equation (4.27), the hazard function for the individuals in this group is $h_0(t)$, where

$$h_0(t) = \lambda \gamma t^{\gamma-1}.$$

Now, also from equation (4.27), the hazard function for those in Group II is $\psi h_0(t)$, that is,

$$\psi \lambda \gamma t^{\gamma-1}.$$

This is the hazard function for a Weibull distribution with scale parameter $\psi \lambda$ and shape parameter γ. We therefore have the result that if the survival times of individuals in one group have a Weibull distribution with shape parameter γ, and the hazard of death at time t for an individual in the second group is proportional to that of an individual in the first, the survival times of those in the second group will also have a Weibull distribution with shape parameter γ. The Weibull distribution is then said to have the *proportional hazards property*. This property is another reason for the importance of the Weibull distribution in the analysis of survival data.

4.4.1 The log-cumulative hazard plot

When a single sample of survival times has a Weibull distribution $W(\lambda, \gamma)$, the log-cumulative hazard plot described in Section 4.2 will give a straight line with intercept $\log \lambda$ and slope γ. It then follows that if the survival times in a second group have a $W(\psi\lambda, \gamma)$ distribution, as they would under the proportional hazards model in equation (4.27), the log-cumulative hazard plot will give a straight line, also of slope γ, but with intercept $\log \psi + \log \lambda$. If the estimated log-cumulative hazard function is plotted against the logarithm of the survival time for individuals in two groups, parallel straight lines would mean that the assumptions of a proportional hazards model and Weibull survival times were tenable. The vertical separation of the two lines provides an estimate of $\beta = \log \psi$, the logarithm of the relative hazard.

If the two lines in a log-cumulative hazard plot are essentially straight but not parallel, this means that the shape parameter γ is different in the two groups, and the hazards are no longer proportional. If the lines are not particularly straight, the Weibull model may not be appropriate. However, if the curves can be taken to be parallel, this would mean that the proportional hazards model is valid, and the Cox regression model discussed in Chapter 3 might be more satisfactory.

Example 4.4 Prognosis for women with breast cancer
In this example, we investigate whether the Weibull proportional hazards model is likely to be appropriate for the data of Example 1.2 on the survival times of breast cancer patients. These data relate to women classified according to whether their tumours were positively or negatively stained. The Kaplan–Meier estimate of the survivor functions for the women in each group were shown in Figure 2.9. From these estimates, the log-cumulative hazards can be estimated and plotted against $\log t$. The resulting log-cumulative hazard plot is shown in Figure 4.11.

In this figure, the lines corresponding to the two staining groups are reasonably straight. This means that the assumption of Weibull distributions for the survival times of the women in each group is quite plausible. Moreover, the gradients of the two lines are very similar, which means that the proportional hazards model is valid. The vertical separation of the two lines provides an estimate of the log relative hazard. From Figure 4.11, the vertical distance between the two straight lines is approximately 1.0, and so a rough estimate of the hazard ratio is $e^{1.0} = 2.72$. Women in the positively stained group would appear to have nearly three times the risk of death at any time compared to those in the negatively stained group. More accurate estimates of the relative hazard will be obtained from fitting the Weibull model to the data of this example in Example 4.5.

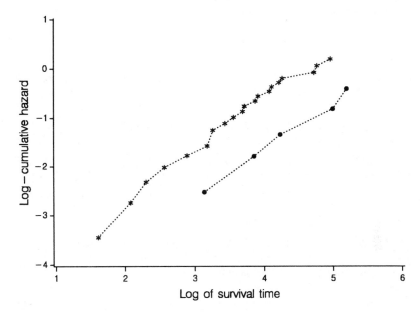

Figure 4.11 Log-cumulative hazard plot for women with tumours which were positively stained (∗) and negatively stained (•).

4.4.2 * Fitting the model

The proportional hazards model in equation (4.27) can be fitted using the method of maximum likelihood. To illustrate the process involved, we consider the situation where the survival times in each group have an exponential distribution.

Suppose that the observations from n_1 individuals in Group I can be expressed as (t_{i1}, δ_{i1}), $i = 1, 2, \ldots n_1$, where δ_{i1} takes the value zero if the survival time of the i'th individual in that group is censored, and unity if that time is a death time. Similarly, let $(t_{i'2}, \delta_{i'2})$, $i' = 1, 2, \ldots n_2$, be the observations from the n_2 individuals in Group II. For individuals in Group I, the hazard function will be taken to be λ, and the probability density function and survivor function are given by

$$f(t_{i1}) = \lambda e^{-\lambda t_{i1}}, \qquad S(t_{i1}) = e^{-\lambda t_{i1}}.$$

For those in Group II, the hazard function is $\psi\lambda$, and the probability density function and survivor function are

$$f(t_{i'2}) = \psi\lambda e^{-\psi\lambda t_{i'2}}, \qquad S(t_{i'2}) = e^{-\psi\lambda t_{i'2}}.$$

Using equation (4.11), the likelihood of the $n_1 + n_2$ observations, $L(\psi, \lambda)$, is

$$\prod_{i=1}^{n_1} \left[\lambda e^{-\lambda t_{i1}}\right]^{\delta_{i1}} \left[e^{-\lambda t_{i1}}\right]^{1-\delta_{i1}} \prod_{i'=1}^{n_2} \left[\psi\lambda e^{-\psi\lambda t_{i'2}}\right]^{\delta_{i'2}} \left[e^{-\psi\lambda t_{i'2}}\right]^{1-\delta_{i'2}},$$

which simplifies to

$$\prod_{i=1}^{n_1} \lambda^{\delta_{i1}} e^{-\lambda t_{i1}} \prod_{i'=1}^{n_2} (\psi\lambda)^{\delta_{i'2}} e^{-\psi\lambda t_{i'2}}.$$

If the numbers of actual death times in the two groups are r_1 and r_2 respectively, then $r_1 = \sum_i \delta_{i1}$ and $r_2 = \sum_{i'} \delta_{i'2}$, and the log-likelihood is given by

$$\log L(\psi, \lambda) = r_1 \log \lambda - \lambda \sum_{i=1}^{n_1} t_{i1} + r_2 \log(\psi\lambda) - \psi\lambda \sum_{i'=1}^{n_2} t_{i'2}.$$

Now write T_1 and T_2 for the total known time survived by the individuals in Groups I and II respectively. Then, T_1 and T_2 are the totals of uncensored and censored survival times in each group, and the log-likelihood function becomes

$$\log L(\psi, \lambda) = (r_1 + r_2) \log \lambda + r_2 \log \psi - \lambda(T_1 + \psi T_2).$$

In order to obtain the values $\hat{\psi}, \hat{\lambda}$ for which this function is a maximum, we differentiate with respect to ψ and λ, and set the derivatives equal to zero. The resulting equations which are satisfied by $\hat{\psi}, \hat{\lambda}$ are

$$\frac{r_2}{\hat{\psi}} - \hat{\lambda} T_2 = 0, \tag{4.28}$$

$$\frac{r_1 + r_2}{\hat{\lambda}} - (T_1 + \hat{\psi} T_2) = 0. \tag{4.29}$$

From equation (4.28),

$$\hat{\lambda} = \frac{r_2}{\hat{\psi} T_2},$$

and on substituting for $\hat{\lambda}$ in equation (4.29) we get

$$\hat{\psi} = \frac{r_2 T_1}{r_1 T_2}. \tag{4.30}$$

Then, from equation (4.28),

$$\hat{\lambda} = r_1/T_1.$$

Both of these estimates have an intuitive justification. The estimated value of λ is the reciprocal of the average time survived by individuals in Group I, while the estimated relative hazard $\hat{\psi}$ is the ratio of the average times survived by the individuals in the two groups.

The asymptotic variance–covariance matrix of the parameter estimates is the inverse of the information matrix, whose elements are found from the second derivatives of the log-likelihood function. See Appendix A. We have that

$$\frac{\mathrm{d}^2 \log L(\psi, \lambda)}{\mathrm{d}\psi^2} = -\frac{r_2}{\psi^2}, \quad \frac{\mathrm{d}^2 \log L(\psi, \lambda)}{\mathrm{d}\lambda^2} = -\frac{r_1 + r_2}{\lambda^2}, \quad \frac{\mathrm{d}^2 \log L(\psi, \lambda)}{\mathrm{d}\lambda\mathrm{d}\psi} = -T_2,$$

and the information matrix is the matrix of negative expected values of these partial derivatives. The only second derivative for which expectations need to be obtained is the derivative with respect to λ and ψ, for which $\mathrm{E}(T_2)$ is required. This is straightforward when the survival times have an exponential distribution, but as shown in Section 4.1.3, the expected value of a survival time that has a Weibull distribution is much more difficult to calculate. For this reason, the information matrix is approximated by using the observed values of the negative second partial derivatives. The observed information matrix is thus

$$\mathbf{I}(\psi, \lambda) = \begin{pmatrix} r_2/\psi^2 & T_2 \\ T_2 & (r_1 + r_2)/\lambda^2 \end{pmatrix},$$

and the inverse of this matrix is

$$\frac{1}{(r_1 + r_2)r_2 - T_2^2 \psi^2 \lambda^2} \begin{pmatrix} (r_1 + r_2)\psi^2 & -T_2\psi^2\lambda^2 \\ -T_2\psi^2\lambda^2 & r_2\lambda^2 \end{pmatrix}.$$

The standard errors of $\hat{\psi}$ and $\hat{\lambda}$ are found by substituting $\hat{\psi}$ and $\hat{\lambda}$ for ψ and λ in this matrix, and taking square roots. Thus, the standard error of $\hat{\psi}$ is given by

$$\text{s.e.}\,(\hat{\psi}) = \sqrt{\frac{(r_1 + r_2)\hat{\psi}^2}{(r_1 + r_2)r_2 - T_2^2 \hat{\psi}^2 \hat{\lambda}^2}}.$$

On substituting for $\hat{\psi}$ and $\hat{\lambda}$ in the denominator of this expression, this standard error simplifies to

$$\hat{\psi}\sqrt{\frac{r_1 + r_2}{r_1 r_2}}. \tag{4.31}$$

Similarly, the standard error of $\hat{\lambda}$ turns out to be given by

$$\text{s.e.}\,(\hat{\lambda}) = \hat{\lambda}/\sqrt{r_1}.$$

The standard error of these estimates cannot be used directly in the construction of confidence intervals for ψ and λ. The reason for this is that the values of both parameters must be positive and their estimated values will tend to have skewed distributions. This means that the assumption of normality used in constructing a confidence interval would not be justified. The distribution of the logarithm of an estimate of either ψ or λ is much more likely to be symmetric, and so confidence limits for the logarithm of the parameter are found using the standard error of the logarithm of the parameter estimate. The resulting confidence limits are then exponentiated to give an interval estimate for the parameter itself.

The standard error of the logarithm of a parameter estimate can be found using the general result given in equation (4.16). Thus the approximate variance of $\log \hat{\psi}$ is

$$\text{var}\,(\log \hat{\psi}) \approx \hat{\psi}^{-2} \,\text{var}\,(\hat{\psi}),$$

and so the standard error of $\log \hat{\psi}$ is given by

$$\text{s.e.}\,(\log \hat{\psi}) \approx \hat{\psi}^{-1}\,\text{s.e.}\,(\hat{\psi}) = \sqrt{\frac{r_1 + r_2}{r_1 r_2}}. \tag{4.32}$$

A $100(1 - \alpha)\%$ confidence interval for the logarithm of the relative hazard has limits $\log \hat{\psi} \pm z_{\alpha/2}\,\text{s.e.}\,(\log \hat{\psi})$, and confidence limits for the hazard ratio ψ are found by exponentiating these limits for $\log \psi$. If required, a confidence interval for λ can be found in a similar manner.

Example 4.5 Prognosis for women with breast cancer
The theoretical results developed in this section will now be illustrated using the data on the survival times of women with breast cancer. In Example 4.4, a Weibull proportional hazards model was found to be appropriate for these data. The relative hazard of death will therefore be estimated under this model.

From the data given in Table 1.2 of Chapter 1, the number of death times in the negatively and positively stained groups is respectively $r_1 = 5$ and $r_2 = 21$. Also, the total time survived in each group is $T_1 = 1652$ and $T_2 = 2679$ months.

From equation (4.30), the estimated hazard of death for a woman in the positively stained group, relative to one in the negatively stained group, is

$$\hat{\psi} = \frac{21 \times 1652}{5 \times 2679} = 2.59,$$

so that a woman in the positively stained group has about two and a half times the risk of death at any given time compared to a woman whose tumour was negatively stained. This is consistent with the estimated value of ψ of 2.72 from the graphical procedure used in Example 4.4.

Next, using equation (4.31), the standard error of the estimated hazard ratio is given by

$$\text{s.e.} (\hat{\psi}) = 2.59\sqrt{\frac{5 + 21}{5 \times 21}} = 1.289.$$

In order to obtain a 95% confidence interval for the true relative hazard, the standard error of $\log \hat{\psi}$ is required. Using equation (4.32), this is found to be given by $\text{s.e.} (\log \hat{\psi}) = 0.498$, and so 95% confidence limits for $\log \psi$ are $\log(2.59) \pm 1.96\,\text{s.e.} (\log \hat{\psi})$, that is, $0.952 \pm (1.96 \times 0.498)$. The confidence interval for the log relative hazard is $(-0.024, 1.927)$, and the corresponding interval estimate for the relative hazard itself is $(e^{-0.024}, e^{1.927})$, that is, $(0.98, 6.87)$. This interval only just includes unity, and suggests that women with positively stained tumours have a poorer prognosis than those whose tumours were negatively stained. This result is consistent with the result of the log-rank test in Example 2.11, where a P-value of 0.061 was obtained on testing the hypothesis of no group difference.

In practice, the analyses shown in the above example would have been carried out using computer software to fit the Weibull model to two groups of survival data. This is discussed in detail in Chapter 11. When the model in equation (4.27) is fitted using a computer package such as SAS, estimates of β, λ and γ, and their standard errors can be obtained from the resulting output. Further calculation may then be needed to obtain an estimate of the relative hazard and the standard error of the estimate. In particular, the estimated hazard ratio would be obtained as $\hat{\psi} = \exp(\hat{\beta})$ and $\text{s.e.} (\hat{\psi})$ found from the equation

$$\text{s.e.} (\hat{\psi}) = \exp(\hat{\beta})\,\text{s.e.} (\hat{\beta}),$$

a result which follows from equation (4.16).

The median and other percentiles of the survival time distributions in the two groups can be estimated from the values of $\hat{\lambda}$ and $\hat{\psi}$. For example, from equation (4.23), the estimated p'th percentile for those in Group I is found from

$$\hat{t}(p) = \left[\frac{1}{\hat{\lambda}} \log \left(\frac{100}{100 - p} \right) \right]^{1/\hat{\gamma}},$$

and that for individuals in Group II is

$$\hat{t}(p) = \left[\frac{1}{\hat{\psi}\hat{\lambda}} \log \left(\frac{100}{100 - p} \right) \right]^{1/\hat{\gamma}}.$$

An expression similar to that in (4.25) can be used to obtain the standard error of an estimated percentile for individuals in each group, once the variances and covariances of the parameter estimates in the model have been found. Specific results for the standard error of percentiles of the survival time distributions in each of the two groups will not be given. Instead, a general expression for the standard error of the p'th percentile after fitting a Weibull model is given in Appendix B.

Example 4.6 Prognosis for women with breast cancer

For the data on the survival times of two groups of breast cancer patients, the estimated value of the shape parameter of the fitted Weibull distribution is $\hat{\gamma} = 0.937$. The estimated scale parameter for women in Group I is $\hat{\lambda} = 0.00414$ and that for women in Group II is $\hat{\lambda}\hat{\psi} = 0.0105$. Putting $\hat{\gamma} = 0.937$ and $\hat{\lambda} = 0.00414$ in equation (4.24) gives 235.89 for the median survival time of those in Group I. The estimated median survival time for women in Group II is found by putting $\hat{\gamma} = 0.937$ and $\hat{\lambda} = 0.0105$ in that equation and gives 87.07 for the estimated median survival time of those women. The median survival time of women whose tumour was positively stained is about one third that of those whose tumour was negatively stained.

Using the general procedure for obtaining the standard error of the median survival time given in Appendix B, the standard errors of the two medians are found using equation (B.2) with $p = 50$, and taking $\hat{\gamma} = 0.937$ and $\hat{\lambda} = 0.00414$ and 0.0105 in turn. They turn out to be 114.126 and 20.550 respectively.

As in Section 4.3.2, 95% confidence limits for the true median survival times for each group of women are best obtained by working with the logarithm of the median. The standard error of $\log \hat{t}(50)$ is found using equation (4.26), from which

$$\text{s.e.} \{\hat{t}(50)\} = \frac{1}{\hat{t}(50)} \text{ s.e.} \{\hat{t}(50)\}.$$

Confidence limits for $\log t(50)$ are then exponentiated to give the corresponding confidence limits for $t(50)$ itself.

In this example, 95% confidence intervals for the true median survival times of the two groups of women are $(91.4, 608.9)$ and $(54.8, 138.3)$ respectively. Notice that the confidence interval for the median survival time of patients with positive staining is much narrower than that for women with negative staining. This is due to there being a relatively small number of uncensored survival times in the women whose tumours were negatively stained.

4.5 The Weibull proportional hazards model

The model in equation (4.27) for the comparison of two groups of survival data can easily be generalized to give a model that is similar in form to the Cox regression model described in Section 3.1.2. Suppose that the values $x_1, x_2, \ldots, x_p$ of p explanatory variables, $X_1, X_2, \ldots, X_p$, are recorded for each of n individuals. Under the proportional hazards model, the hazard of death at time t for the i'th individual is

$$h_i(t) = \exp(\beta_1 x_{i1} + \beta_2 x_{i2} + \cdots + \beta_p x_{ip}) h_0(t), \tag{4.33}$$

for $i = 1, 2, \ldots, n$. Although this model has a similar appearance to that given in equation (3.3), there is one fundamental difference which concerns the specification of the baseline hazard function $h_0(t)$. In the Cox regression model, the form of $h_0(t)$ is unspecified, and the shape of the function is essentially determined by the actual data. In the model being considered in this section, the survival times are assumed to have a Weibull distribution, and this imposes a particular parametric form on $h_0(t)$.

Consider an individual for whom the values of the p explanatory variables in the model of equation (4.33) are all equal to zero. The hazard function for such an individual is $h_0(t)$. If the survival time of this individual has a Weibull distribution with scale parameter λ and shape parameter γ, then their hazard function is such that

$$h_0(t) = \lambda \gamma t^{\gamma - 1}.$$

Using equation (4.33), the hazard function for the i'th individual in the study is then given by

$$h_i(t) = \exp(\beta_1 x_{1i} + \beta_2 x_{2i} + \cdots + \beta_p x_{pi}) \lambda \gamma t^{\gamma - 1}. \tag{4.34}$$

In vector notation, this is $\exp(\beta' x_i) \lambda \gamma t^{\gamma - 1}$. From the form of this hazard function we can see that the survival time of the i'th individual in the study has a Weibull distribution with scale parameter $\lambda \exp(\beta' x_i)$ and shape parameter γ. This again is a manifestation of the proportional hazards property of the Weibull distribution. This result shows that the effect of the explanatory variates in the model is to alter the scale parameter of the distribution, while the shape parameter remains constant.

The survivor function corresponding to the hazard function in equation (4.34) is found using equation (1.5), and turns out to be

$$S_i(t) = \exp\left\{-\lambda \exp(\beta' x_i) t^\gamma\right\}.$$

The probability density function can be found by differentiating the survivor function with respect to t and multiplying by -1, or from the result that $f(t) =$

$S(t)h(t)$, whence

$$f_i(t) = \lambda\gamma\exp(\beta'\mathbf{x}_i)t^{\gamma-1}\exp\left\{-\lambda\exp(\beta'\mathbf{x}_i)t^\gamma\right\}.$$

The Weibull proportional hazards model is fitted by constructing the likelihood function of the n observations using equation (4.11), and maximizing this function with respect to the unknown parameters, $\beta_1, \beta_2, \dots \beta_p, \lambda$ and γ. In practice, this is done using computer software for survival analysis. Such software generally gives the standard errors of the parameter estimates, from which confidence intervals for relative hazards and the median and other percentiles of the survival time distribution can be found.

Specifically, suppose that the estimates of the parameters in the model of equation (4.34) are $\hat{\beta}_1, \hat{\beta}_2, \dots, \hat{\beta}_p, \hat{\lambda}$ and $\hat{\gamma}$. The estimated survivor function for the i'th individual in the study, for whom the values of the explanatory variables in the model are $x_{1i}, x_{2i}, \dots, x_{pi}$, is then

$$\hat{S}_i(t) = \exp\left\{-\exp(\hat{\beta}_1 x_{1i} + \hat{\beta}_2 x_{2i} + \dots + \hat{\beta}_p x_{pi})\hat{\lambda}t^{\hat{\gamma}}\right\}. \qquad (4.35)$$

The corresponding hazard function is

$$\hat{h}_i(t) = \exp(\hat{\beta}_1 x_{1i} + \hat{\beta}_2 x_{2i} + \dots + \hat{\beta}_p x_{pi})\hat{\lambda}\hat{\gamma}t^{\hat{\gamma}-1}.$$

Both of these functions can be estimated and plotted against t, for individuals with particular values of the explanatory variables in the model.

Generalizing the result in equation (4.23) to the situation where the Weibull scale parameter is $\lambda\exp(\beta'\mathbf{x})$, we get

$$\hat{t}(p) = \left[\frac{1}{\hat{\lambda}\exp(\hat{\beta}'\mathbf{x})}\log\left(\frac{100}{100-p}\right)\right]^{1/\hat{\gamma}}$$

for the estimated p'th percentile of the survival time distribution for an individual whose vector of explanatory variables is $\mathbf{x}$. The estimated median survival time for such an individual is therefore

$$\hat{t}(50) = \left[\frac{\log 2}{\hat{\lambda}\exp(\hat{\beta}'\mathbf{x})}\right]^{1/\hat{\gamma}}. \qquad (4.36)$$

The standard error of $\hat{t}(p)$ and confidence intervals for $t(p)$ can be found after first obtaining the standard error of $\log\hat{t}(p)$. The standard error of $\log\hat{t}(p)$ is shown in equation (B.5) of Appendix B to be given by

$$\text{s.e.}\{\log\hat{t}(p)\} = \hat{\gamma}^{-1}\sqrt{(\mathbf{d}_0'\mathbf{V}\mathbf{d}_0)},$$

where $\mathbf{V}$ is the variance–covariance matrix of the estimates of $\gamma, \lambda, \beta_1, \beta_2, \dots, \beta_p$, and $\mathbf{d}_0$ is a vector whose $p + 2$ components are $\hat{\lambda}^{-1}, \hat{\gamma}^{-1}\{c_p - \log\hat{\lambda} - \hat{\beta}'\mathbf{x}\}$, $x_1, x_2, \dots, x_p$, with

$$c_p = \log\log\left(\frac{100}{100-p}\right).$$

If required, the standard error of $\hat{t}(p)$ is found using

$$\text{s.e.} \{\hat{t}(p)\} = \hat{t}(p) \, \text{s.e.} \{\log \hat{t}(p)\},$$

and as usual, confidence intervals for $t(p)$ can be found by exponentiating confidence limits for $\log t(p)$.

4.5.1 Exploratory analyses

In Sections 4.2 and 4.4.1, we saw how a log-cumulative hazard plot can be used to assess whether survival data can be modelled by a Weibull distribution, and whether the proportional hazards assumption is valid. These procedures work perfectly well when we are faced with a single sample of survival data, or data where the number of groups is small and there is a reasonably large number of individuals in each group. But in situations where there are a small number of death times distributed over a relatively large number of groups, it may not be possible to estimate the survivor function and hence the log-cumulative hazard function for each group.

As an example, consider the data on the survival times of patients with hypernephroma given in Table 3.3. Here, individuals are classified according to age-group and whether or not a nephrectomy has been performed, giving six combinations of age-group and nephrectomy status. To examine the assumption of a Weibull distribution for the survival times in each group, and the assumption of proportional hazards across the groups, a log-cumulative hazard plot would be required for each group. The number of patients in each age-group who have not had a nephrectomy is so small that the survivor function cannot be properly estimated in these groups. If there were more individuals in the study who had died and not had a nephrectomy, it would be possible to construct a log-cumulative hazard plot. If this plot featured six parallel straight lines, the Weibull proportional hazards model is likely to be satisfactory.

When a model contains continuous variables, their values will first need to be grouped before a log-cumulative hazard plot can be obtained. This may also result in there being insufficient numbers of individuals in some groups to enable the log-cumulative hazard function to be estimated.

The only alternative to using each combination of factor levels in constructing a log-cumulative hazard plot is to ignore some of the factors. However, the resulting plot can be very misleading. For example, suppose that patients are classified according to the levels of two factors, A and B. The log-cumulative hazard plot obtained by grouping the individuals according to the levels of A ignoring B, or according to the levels of B ignoring A, may not give cause to doubt the Weibull or proportional hazards assumptions. However, if the log-cumulative hazard plot is obtained for individuals at each combination of levels of A and B, the plot may not feature a series of four parallel lines. By the same token, the log-cumulative hazard plot obtained when either A or B is ignored may not show sets of parallel straight lines, but when a plot is obtained for all combinations of A and B, paral-

Table 4.1 Artificial data on the survival times of 36 patients classified according to the levels of two factors, A and B.

$A = 1$		$A = 2$	
$B = 1$	$B = 2$	$B = 1$	$B = 2$
59	10	88	25*
20	4	70*	111
71	16	54	152
33	18	139	86
25	19	31	212
25	35	59	187*
15	11	111	54
53		149	357
47		30	301
		44	195
		25	

lel lines may result. This feature is illustrated in the following example, which is based on artificial data.

Example 4.7 An artificial data set

Suppose a number of individuals are classified according to the levels of two factors, A and B, each with two levels, and that the recorded survival times are as shown in Table 4.1. As usual, an asterisk denotes a censored observation.

The log-cumulative hazard plot shown in Figure 4.12 is derived from the individuals classified according to the two levels of A, ignoring the level of factor B. The plot in Figure 4.13 is from individuals classified according to the two levels of B, ignoring the level of factor A.

From Figure 4.12 there is no reason to doubt the assumption of a Weibull distribution for the survival times for the two levels of A, and the assumption of proportional hazards is clearly tenable. However, the crossed lines on the plot shown as Figure 4.13 strongly suggest that the hazards are not proportional when individuals are classified according to the levels of B. A different picture emerges when the 37 survival times are classified according to the levels of both A and B. The log-cumulative hazard plot based on the four groups is shown in Figure 4.14. The four parallel lines show that there is no doubt about the validity of the proportional hazards assumption across the groups.

In this example, the reason why the log-cumulative hazard plot for B ignoring A is misleading is that there is an interaction between A and B. An examination of the data reveals that, on average, the difference in the survival times of patients for whom $B = 1$ and $B = 2$ is greater when $A = 2$ than when $A = 1$.

Even when a log-cumulative hazard plot gives no reason to doubt the assump-

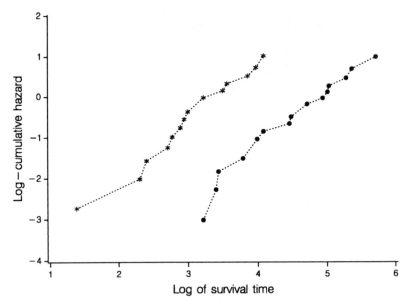

Figure 4.12 Log-cumulative hazard plot for individuals for whom $A = 1$ ($*$) and $A = 2$ ($\bullet$).

tion of a Weibull proportional hazards model, the validity of the adequacy of the fitted model will need to be examined using the methods to be described in Chapter 5. When it is not possible to use a log-cumulative hazard plot to explore whether a Weibull proportional hazards model is reasonable in advance of fitting the model, this assessment has to be delayed until after the model has been fitted.

4.6 Comparing alternative Weibull models

In order to ascertain which explanatory variables should be included in a Weibull proportional hazards model, alternative models need to be compared. Weibull models can be compared using methods analogous to those for the Cox regression model described in Section 3.4.

Suppose that one model contains a subset of the explanatory variables in another, so that the two models are nested. The two models can then be compared on the basis of the statistic $-2 \log \hat{L}$, where $\hat{L}$ is the maximized value of the likelihood function under the fitted model. For a model which contains p explanatory variables, the sample likelihood is a function of $p + 2$ unknown parameters $\beta_1, \beta_2, \ldots, \beta_p, \lambda$ and γ. The maximized likelihood is then the value of this function when these parameters take their estimates $\hat{\beta}_1, \hat{\beta}_2, \ldots, \hat{\beta}_p, \hat{\lambda}$ and $\hat{\gamma}$.

More specifically, if one model, Model (1), say, contains p explanatory variables, and another model, Model (2), contains an additional q explanatory vari-

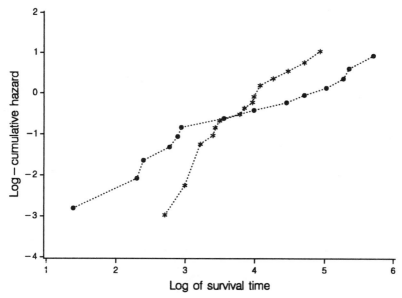

Figure 4.13 Log-cumulative hazard plot for individuals for whom $B = 1$ ($*$) and $B = 2$ ($\bullet$).

ables, the estimated hazard functions for these two models are as follows:

Model (1): $\quad h_i(t) = \exp\{\hat{\beta}_1 x_1 + \hat{\beta}_2 x_2 + \cdots + \hat{\beta}_p x_p\}\hat{\lambda}\hat{\gamma}t^{\hat{\gamma}-1}$,

Model (2): $\quad h_i(t) = \exp\{\hat{\beta}_1 x_1 + \hat{\beta}_2 x_2 + \cdots + \hat{\beta}_{p+q} x_{p+q}\}\hat{\lambda}\hat{\gamma}t^{\hat{\gamma}-1}$.

The maximized likelihoods under Model (1) and Model (2) will be denoted by $\hat{L}_1$ and $\hat{L}_2$, respectively. The difference between the values of $-2\log\hat{L}_1$ and $-2\log\hat{L}_2$, that is, $-2[\log\hat{L}_1 - \log\hat{L}_2]$, then has an approximate chi-squared distribution with q degrees of freedom, under the null hypothesis that the coefficients of the additional q variates in Model (2) are all equal to zero. If the difference between the values of $-2\log\hat{L}$ for these two models is significantly large when compared with percentage points of the chi-squared distribution, we would deduce that the extra q terms are needed in the model, in addition to the p that are already included.

The description of the modelling process in Section 3.4 applies equally well to models based on the Weibull proportional hazards model, and so will not be repeated here. However, the modelling process will be illustrated using two examples.

Example 4.8 Treatment of hypernephroma
Data on the survival times of 36 patients classified according to their age-group

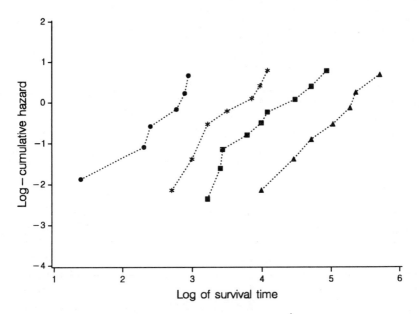

Figure 4.14 Log-cumulative hazard plot for individuals in the groups defined by the four combinations of levels of A and B.

and whether or not they have had a nephrectomy were introduced in Example 3.4 of Chapter 3. In that example, the data were analysed using the Cox proportional hazards model. Here, the analysis is repeated using the Weibull proportional hazards model. As in Example 3.4, the effect of the j'th age-group will be denoted by α_j, and that associated with whether or not a nephrectomy was performed by ν_k. There are then five possible models for the hazard function of the i'th individual, $h_i(t)$, which are as follows:

$$
\begin{aligned}
\text{Model (1):} \quad & h_i(t) = h_0(t), \\
\text{Model (2):} \quad & h_i(t) = \exp\{\alpha_j\}h_0(t), \\
\text{Model (3):} \quad & h_i(t) = \exp\{\nu_k\}h_0(t), \\
\text{Model (4):} \quad & h_i(t) = \exp\{\alpha_j + \nu_k\}h_0(t), \\
\text{Model (5):} \quad & h_i(t) = \exp\{\alpha_j + \nu_k + (\alpha\nu)_{jk}\}h_0(t).
\end{aligned}
$$

In these models, $h_0(t) = \lambda\gamma t^{\gamma-1}$ is the baseline hazard function, and the parameters λ and γ have to be estimated along with those in the linear component of the model.

These five models have the interpretations given in Example 3.4. They can be fitted by constructing indicator variables corresponding to the factors age-group and nephrectomy status, as shown in Example 3.4, but some packages will allow factors to be fitted directly. Details are given in Chapter 11.

Table 4.2 Values of $-2 \log \hat{L}$ on fitting five Weibull models to the data on hypernephroma.

Terms in model	$-2 \log \hat{L}$
null model	104.886
α_j	96.400
ν_k	94.384
$\alpha_j + \nu_k$	87.758
$\alpha_j + \nu_k + (\alpha\nu)_{jk}$	83.064

Once a Weibull proportional hazards model has been fitted to the data, values of $-2 \log \hat{L}$ can be found. These are given in Table 4.2 for the five models of interest. The reduction in the value of $-2 \log \hat{L}$ on adding the interaction term to Model (4) is 4.69 on two degrees of freedom. This reduction is just about significant at the 10% level (P = 0.096) and so there is some suggestion of an interaction between age-group and nephrectomy status. For comparison, note that when the Cox regression model was fitted in Example 3.4, the interaction was not significant (P = 0.220).

The interaction can be investigated in greater detail by examining the hazard ratios under the model. Under Model (5), the estimated hazard function for the i'th individual is

$$\hat{h}_i(t) = \exp\{\hat{\alpha}_j + \hat{\nu}_k + (\widehat{\alpha\nu})_{jk}\}\hat{h}_0(t),$$

where

$$\hat{h}_0(t) = \hat{\lambda}\hat{\gamma}t^{\hat{\gamma}-1}$$

is the estimated baseline hazard function. The logarithm of the hazard ratio for an individual in the j'th age-group, $j = 1, 2, 3$, and k'th level of nephrectomy status, $k = 1, 2$, relative to an individual in the youngest age-group who has not had a nephrectomy, is therefore

$$\hat{\alpha}_j + \hat{\nu}_k + (\widehat{\alpha\nu})_{jk} - \hat{\alpha}_1 - \hat{\nu}_1 - (\widehat{\alpha\nu})_{11}, \qquad (4.37)$$

since the baseline hazard functions cancel out.

As in Example 3.4, models can be fitted to the data by defining indicator variables $A2$ and $A3$ for age-group and N for nephrectomy status. As in that example, $A2$ is unity for an individual in the second age-group and zero otherwise, $A3$ is unity for an individual in the third age-group and zero otherwise, and N is unity if a nephrectomy has been performed and zero otherwise. Thus, fitting the term α_j corresponds to fitting the variables $A2$ and $A3$, fitting ν_k corresponds to fitting N, and fitting the interaction term $(\alpha\nu)_{jk}$ corresponds to fitting the products $A2N = A2 \times N$ and $A3N = A3 \times N$. In particular, to fit Model (5), the five variables $A2, A3, N, A2N, A3N$ are included in the model. With this choice of indi-

Table **4.3** Parameter estimates on fitting a Weibull model to the data on hypernephroma.

Parameter	Estimate
α_2	−0.085
α_3	0.115
ν_2	−2.436
$(\alpha\nu)_{22}$	0.121
$(\alpha\nu)_{32}$	2.538

Table **4.4** Hazard ratios for individuals classified by age-group and whether or not a nephrectomy has been performed.

Age-group	No nephrectomy	Nephrectomy
< 60	1.00	0.09
$60 - 70$	0.92	0.09
> 70	1.12	1.24

cator variables, $\hat{\alpha}_1 = 0$, $\hat{\nu}_1 = 0$ and $(\hat{\alpha\nu})_{jk} = 0$ when either i or j is unity. The remaining values of $\hat{\alpha}_j$, $\hat{\nu}_k$ and $(\hat{\alpha\nu})_{jk}$ are the coefficients of $A2$, $A3$, N, $A2N$, $A3N$ and are given in Table 4.3.

Computer packages such as SAS set up indicator variables internally, and so estimates such as those in the above table can be obtained directly from the output. However, to repeat an earlier warning, when packages are used to fit factors, the coding used to define the indicator variables must be known if the output is to be properly interpreted.

When the indicator variables specified above are used, the logarithm of the hazard ratio given in equation (4.37) reduces to

$$\hat{\alpha}_j + \hat{\nu}_k + (\hat{\alpha\nu})_{jk},$$

for $j = 1, 2, 3$, $k = 1, 2$. Table 4.4 gives the hazards for the individuals, relative to the baseline hazard. The baseline hazard corresponds to an individual in the youngest age-group who has not had a nephrectomy, and so a hazard ratio of unity for these individuals is recorded in Table 4.4.

This table helps to explain the interaction between age-group and nephrectomy status, in that the effect of a nephrectomy is not the same for individuals in each of the three age-groups. For patients in the two youngest age-groups, a nephrectomy substantially reduces the hazard of death at any given time. Performing a nephrectomy on patients aged over 70 does not have much effect on the risk of

Table 4.5 Median survival times for individuals classified by age-group and whether or not a nephrectomy has been performed.

Age-group	No nephrectomy	Nephrectomy
< 60	10.21	48.94
60 − 70	10.78	47.81
> 70	9.48	8.87

death. We also see that for those patients who have not had a nephrectomy, age does not much affect the hazard of death.

Estimated median survival times can be found in a similar way. Using equation (4.36), the median survival time for a patient in the j'th age-group, $j = 1, 2, 3$, and the k'th level of nephrectomy status, $k = 1, 2$, becomes

$$\hat{t}(50) = \left[\frac{\log 2}{\hat{\lambda} \exp\{\hat{\alpha}_j + \hat{\nu}_k + (\hat{\alpha\nu})_{jk}\}} \right]^{1/\hat{\gamma}}.$$

When the model containing the interaction term is fitted to the data, the estimated values of the parameters in the baseline hazard function are $\hat{\lambda} = 0.0188$ and $\hat{\gamma} = 1.5538$. Table 4.5 gives the estimated median survival times, in months, for individuals with each combination of age-group and nephrectomy status. This table shows that a nephrectomy leads to more than a four-fold increase in the median survival time in patients aged up to 70 years. The median survival time of patients aged over 70 is not much affected by the performance of a nephrectomy.

We end this example with a note of caution. For some combinations of age-group and nephrectomy status, particularly the groups of individuals who have not had a nephrectomy, the estimated hazard ratios and median survival times are based on small numbers of survival times. As a result, the standard errors of estimates of such quantities, which have not been given here, will be large.

Example 4.9 Chemotherapy in ovarian cancer patients
Surgical treatment of ovarian cancer may be followed by a course of chemotherapy. In a study of different chemotherapy treatments, Edmunson *et al.* (1979) compared the anti-tumour effects of cyclophosphamide alone and cyclophosphamide combined with adriamycin. The trial involved 26 women with minimal residual disease and who had experienced surgical excision of all tumour masses greater than 2 cm in diameter. Following surgery, the patients were further classified according to whether the residual disease was completely or partially excised. The age of the patient and their performance status were also recorded at the start of the trial. The response variable was the survival time in days following randomization to one or other of the two chemotherapy treatments. The variables in the data set are therefore as follows:

Table 4.6 Survival times of ovarian cancer patients.

Patient	TIME	CENS	TREAT	AGE	RDISEASE	PERF
1	156	1	1	66	2	2
2	1040	0	1	38	2	2
3	59	1	1	72	2	1
4	421	0	2	53	2	1
5	329	1	1	43	2	1
6	769	0	2	59	2	2
7	365	1	2	64	2	1
8	770	0	2	57	2	1
9	1227	0	2	59	1	2
10	268	1	1	74	2	2
11	475	1	2	59	2	2
12	1129	0	2	53	1	1
13	464	1	2	56	2	2
14	1206	0	2	44	2	1
15	638	1	1	56	1	2
16	563	1	2	55	1	2
17	1106	0	1	44	1	1
18	431	1	1	50	2	1
19	855	0	1	43	1	2
20	803	0	1	39	1	1
21	115	1	1	74	2	1
22	744	0	2	50	1	1
23	477	0	1	64	2	1
24	448	0	1	56	1	2
25	353	1	2	63	1	2
26	377	0	2	58	1	1

TIME: Survival time in days
CENS: Censoring indicator (0 = censored, 1 = uncensored)
TREAT: Treatment (1 = single, 2 = combined)
AGE: Age of patient in years
RDISEASE: Extent of residual disease (1 = incomplete, 2 = complete)
PERF: Performance status (1 = good, 2 = poor)

The data, which were obtained from Therneau (1986), are given in Table 4.6.

In modelling these data, the factors $TREAT$, $RDISEASE$ and $PERF$ each have two levels, and will be fitted as variates that take the values given in Table 4.6. This does of course mean that the baseline hazard function is not directly interpretable, since there can be no individual for whom the values of all these variates are zero. From both a computational and interpretive viewpoint, it is more convenient to relocate the values of the variables AGE and $TREAT$. If the variable

Table 4.7 Values of $-2\log \hat{L}$ on fitting models to the data in Table 4.6.

Variables in model	$-2\log \hat{L}$
none	59.534
AGE	43.566
$RDISEASE$	55.382
$PERF$	58.849
$AGE, RDISEASE$	41.663
$AGE, PERF$	43.518
$AGE, TREAT$	41.126
$AGE, TREAT, TREAT \times AGE$	39.708

$AGE - 50$ is used in place of AGE, and $TREAT - 1$ in place of $TREAT$, the baseline hazard then corresponds to the hazard for an individual of age 50 who has been allocated to the cyclophosphamide group. However, the original variables will be used in this example.

We begin by identifying which prognostic factors are associated with the survival times of the patients. The values of the statistic $-2\log \hat{L}$ on fitting a range of models to these data are given in Table 4.7.

When Weibull models that contain just one of AGE, $RDISEASE$ and $PERF$ are fitted, the only variable which significantly reduces the value of $-2\log \hat{L}$ from that for the null model is AGE. After fitting AGE, the variables $RDISEASE$ and $PERF$ further reduce $-2\log \hat{L}$ by 1.903 and 0.048 respectively, neither of which is significant at the 10% level. This leads us to the conclusion that AGE is the only prognostic variable that needs to be incorporated in the model.

The term associated with the treatment effect is now added to the model. The value of $-2\log \hat{L}$ is then reduced by 2.440 on 1 d.f. This reduction of 2.440 is not quite large enough for it to be significant at the 10% level (P = 0.118). There is therefore only very slight evidence of a difference in the effect of the two chemotherapy treatments on the hazard of death.

For comparison, when $TREAT$ alone is added to the null model, the value of $-2\log \hat{L}$ is reduced from 59.534 to 58.355. This reduction of 1.179 is certainly not significant when compared to percentage points of the chi-squared distribution on 1 d.f. Ignoring AGE therefore leads to an underestimate of the magnitude of the treatment effect.

To explore whether the treatment difference is consistent over age, the interaction term formed as the product of AGE and $TREAT$ is added to the model. On doing so, $-2\log \hat{L}$ is only reduced by 1.419. This reduction is nowhere near being significant and so there is no need to incorporate an interaction term in the model.

The variable $TREAT$ will be retained in the model, since interest centres on the magnitude of the treatment effect. The fitted model for the hazard of death at time t for the i'th individual is then found to be

$$\hat{h}_i(t) = \exp\{0.144\, AGE_i - 1.023\, TREAT_i\}\hat{\lambda}\hat{\gamma}t^{\hat{\gamma}-1},$$

where $\hat{\lambda} = 5.645 \times 10^{-9}$ and $\hat{\gamma} = 1.822$. In this model, $TREAT = 1$ for cyclophosphamide alone and $TREAT = 2$ for the combination of cyclophosphamide with adriamycin. The hazard for a patient on the single treatment, relative to one on the combined treatment, is therefore estimated by

$$\hat{\psi} = \exp\{(-1.023 \times 1) - (-1.023 \times 2)\} = 2.78.$$

This means that a patient receiving the single chemotherapy treatment is nearly three times more likely to die at any given time than a patient on the combined treatment. Expressed in this way, the benefits of the combined chemotherapy treatment sound to be great. However, when account is taken of the inherent variability of the data on which these results are based, this relative hazard is only significantly greater than unity at the 12% level (P = 0.118).

The median survival time can be estimated for patients of a given age on a given treatment from the equation

$$\hat{t}(50) = \left[\frac{\log 2}{\hat{\lambda}\exp(0.144\, AGE - 1.023\, TREAT)}\right]^{1/\hat{\gamma}}.$$

For example, a woman aged 60 ($AGE = 60$) who is given cyclophosphamide alone ($TREAT = 1$) has an estimated median survival time of 423 days, whereas someone of the same age on the combination of the two chemotherapy treatments has an estimated median survival time of 741 days. Confidence intervals for these estimates can be found using the method illustrated in Example 4.6.

4.7 * An alternative form of the proportional hazards model

Before ending this chapter, an alternative formulation of the Weibull proportional hazards model will be presented. One reason for the importance of this alternative form of the model is that it is adopted in statistical packages used in fitting the parametric version of the proportional hazards model.

Consider a *log-linear model* for the random variable T_i associated with the lifetime of the i'th individual in a survival study, according to which

$$\log T_i = \mu + \alpha_1 x_{1i} + \alpha_2 x_{2i} + \cdots + \alpha_p x_{pi} + \sigma\epsilon_i. \tag{4.38}$$

In this model, $\alpha_1, \alpha_2, \ldots, \alpha_p$ are the unknown coefficients of the values of p explanatory variables, $X_1, X_2, \ldots, X_p$, and μ, σ are two further parameters, known as the *intercept* and *scale parameter*, respectively. The quantity ϵ_i is a random variable used to model the departure of the values of $\log T_i$ from the linear part of the model, and ϵ is assumed to have a particular probability distribution.

Suppose that ϵ does in fact have a type of extreme value distribution known as the *Gumbel distribution*. This is an asymmetric distribution with probability density function given by

$$f(\epsilon) = \exp(\epsilon - e^\epsilon),$$

for $-\infty < \epsilon < \infty$. If we define $\xi = e^\epsilon$, then the probability density function of ξ is $e^{-\xi}$.

To show this, we use the general result that if $f_X(x)$ is the probability density function of X, the density of the random variable $Y = g(X)$ is given by

$$f_Y(y) = f_X \left\{ g^{-1}(y) \right\} / \left| \frac{dy}{dx} \right|. \tag{4.39}$$

In this result, $f_X \left\{ g^{-1}(y) \right\}$ is the density function of X with the argument of the function expressed in terms of y. Using equation (4.39), the probability density function of $\xi = e^\epsilon$ is given by

$$f(\xi) = \exp \left[\log \xi - e^{\log \xi} \right] / \xi,$$

since $d\xi/d\epsilon = e^\epsilon = \xi$, which reduces to $f(\xi) = e^{-\xi}$. The random variable $\xi = e^\epsilon$ therefore has an exponential distribution with unit mean.

Now consider the survivor function of T_i, the random variable associated with the survival time of the i'th individual. This is

$$
\begin{aligned}
S_i(t) &= P(T_i \geq t) \\
&= P(\log T_i \geq \log t) \\
&= P(\mu + \alpha' \mathbf{x}_i + \sigma \epsilon_i \geq \log t),
\end{aligned}
$$

where $\alpha' \mathbf{x}_i = \sum_{j=1}^{p} \alpha_j x_{ji}$, and $\mathbf{x}_i$ is the vector of values of the p explanatory variables for the i'th individual. Therefore,

$$S_i(t) = P \left[\epsilon_i \geq \frac{\log t - \mu - \alpha' \mathbf{x}_i}{\sigma} \right], \tag{4.40}$$

and so

$$S_i(t) = P \left[e^{\epsilon_i} \geq \exp \left\{ \frac{\log t - \mu - \alpha' \mathbf{x}_i}{\sigma} \right\} \right].$$

But e^ϵ has a unit exponential distribution, and so

$$P(e^\epsilon \geq \xi) = e^{-\xi},$$

for $\xi \geq 0$. It then follows that

$$S_i(t) = \exp \left[-\exp \left\{ \frac{\log t - \mu - \alpha' \mathbf{x}_i}{\sigma} \right\} \right]. \tag{4.41}$$

We now return to the Weibull proportional hazards model. From equation (4.35), the survivor function for the i'th individual under this model is

$$S_i(t) = \exp \left\{ -\exp(\beta_1 x_{1i} + \beta_2 x_{2i} + \cdots + \beta_p x_{pi}) \lambda t^\gamma \right\}, \tag{4.42}$$

in which λ and γ are the parameters of the Weibull baseline hazard function. There is a direct correspondence between equation (4.41) and equation (4.42), in the sense that

$$\lambda = \exp(-\mu/\sigma), \qquad \gamma = \sigma^{-1}, \qquad \beta_j = -\alpha_j/\sigma,$$

for $j = 1, 2, \ldots p$. We therefore deduce that the log-linear model where

$$\log T_i = \frac{1}{\gamma}\left[-\log \lambda - \beta'\mathbf{x}_i + \epsilon_i\right],$$

and in which ϵ has a Gumbel distribution, provides an alternative representation of the Weibull proportional hazards model. We will return to this formulation of the model in Chapter 6 in connection with the accelerated failure time model. For the moment, note that computer output from fitting the Weibull proportional hazards model in packages such as SAS and BMDP is expressed in terms of the parameters μ, σ and $\alpha_1, \alpha_2, \ldots, \alpha_p$ in equation (4.41).

In this alternative form of the model, the p'th percentile of the survival time distribution for the i'th individual is the value $t(p)$, which is such that $S_i\{t(p)\} = 1 - (p/100)$, where $S_i\{t(p)\}$ is as given in equation (4.41). Straightforward algebra leads to the result that

$$t(p) = \left\{-\log\left(\frac{100 - p}{p}\right)\right\}^\sigma \exp(\mu + \alpha'\mathbf{x}_i)$$

for that individual. When packages that provide estimates of μ, σ and $\alpha_1, \alpha_2, \ldots,$ α_p are used to fit the Weibull proportional hazards model, this expression enables percentiles to be estimated directly. In addition, the result for the standard error of the p'th percentile in Appendix B can be modified in a similar way.

4.7.1 Maximum likelihood estimation

With the alternative specification of the Weibull proportional hazards function, the likelihood function has a quite simple form. We have seen that the probability density function of $\epsilon_i = (\log T_i - \mu - \alpha'\mathbf{x}_i)/\sigma$ is given by

$$f_\epsilon(\epsilon_i) = \exp\{\epsilon_i - \exp(\epsilon_i)\}.$$

Using the result in equation (4.39) for the density of a function of a random variable, the density of $Y_i = \log T_i = \mu + \alpha'\mathbf{x}_i + \sigma\epsilon_i$ is

$$\frac{1}{\sigma}f_\epsilon\left(\frac{\log t - \mu - \alpha'\mathbf{x}_i}{\sigma}\right),$$

in which t is an observation on T_i. Hence, the probability density function of Y_i for the i'th individual is

$$f_i(y) = \frac{1}{\sigma}\exp(z_i - e^{z_i}), \tag{4.43}$$

where

$$z_i = \frac{y - \mu - \alpha'\mathbf{x}_i}{\sigma},$$

and $y = \log t$. The likelihood function based on the logarithms $y_1, y_2, \ldots, y_n$ of the n observed survival times $t_1, t_2, \ldots, t_n$ is then

$$L(\alpha, \mu, \sigma) = \prod_{i=1}^{n} \{f_i(y_i)\}^{\delta_i} \{S_i(y_i)\}^{1-\delta_i},$$

where $f_i(y_i)$ and $S_i(y_i)$ are the density and survivor functions for the i'th individual at $\log t_i$ and δ_i is the censoring indicator for the i'th observation, so that δ_i is unity if the i'th observation is uncensored and zero if it is censored. From equation (4.43), $S_i(y_i) = \exp\{-e^{z_i}\}$, and so the likelihood function becomes

$$L(\alpha, \mu, \sigma) = \prod_{i=1}^{n} \sigma^{-\delta_i} \{\exp(z_i - e^{z_i})\}^{\delta_i} \{\exp(-e^{z_i})\}^{1-\delta_i},$$

whence

$$L(\alpha, \mu, \sigma) = \prod_{i=1}^{n} \sigma^{-\delta_i} e^{\delta_i z_i} \exp(-e^{z_i}). \qquad (4.44)$$

The log-likelihood function is then

$$\log L(\alpha, \mu, \sigma) = \sum_{i=1}^{n} \{-\delta_i \log \sigma + \delta_i z_i - e^{z_i}\},$$

and the maximum likelihood estimates of the $p+2$ unknown parameters, μ, σ and $\alpha_1, \alpha_2, \ldots, \alpha_p$, are found by maximizing this function using the Newton–Raphson procedure, described in Section 3.3.3.

4.8 Further reading

The properties of the exponential and Weibull distributions are presented in Johnson and Kotz (1970). A thorough discussion of the theory of maximum likelihood estimation is included in Barnett (1982) and Cox and Hinkley (1974), and a useful summary of the main results is contained in Hinkley, Reid and Snell (1991). Numerical methods for obtaining maximum likelihood estimates, and the Newton–Raphson procedure in particular, are described by Everitt (1987) and Thisted (1988), for example.

A description of the log-linear model for survival data is contained in many of the major textbooks on survival analysis; see in particular Cox and Oakes (1984), Kalbfleisch and Prentice (1980) or Lawless (1980). Descriptions of this form of the model are also presented in computer manuals that accompany the packages SAS and BMDP.

5

Model checking in the proportional hazards model

After a model has been fitted to an observed set of survival data, the adequacy of the fitted model needs to be assessed. Indeed, the use of diagnostic procedures for model checking is an essential part of the modelling process.

In some situations, careful inspection of an observed set of data may lead to the identification of certain features, such as individuals with unusually large or small survival times. However, unless there are only one or two explanatory variables, a visual examination of the data may not be very revealing. The situation is further complicated by censoring, in that the occurrence of censored survival times make it difficult to judge aspects of model adequacy, even in the simplest of situations. Visual inspection of the data has therefore to be supplemented by more formal procedures for detecting inadequacies in a fitted model. Because methods used in assessing the adequacy of survival models have to cope with the occurrence of censored survival times, they are a little more complicated than the corresponding methods used in linear regression modelling. However, many of the procedures are easily carried out using modern computer software for survival analysis.

Once a model has been fitted, there are a number of aspects of the fit of a model that need to be studied. For example, the model must include an appropriate set of explanatory variables from those measured during the study. We therefore require procedures that allow us to check whether any variable included in the model needs to be transformed, and whether any omitted variables should actually be included in the model. It might be important to identify observed survival times that are greater than would have been anticipated, or individuals whose explanatory variables have an undue impact on particular hazard ratios. Also, if a proportional hazards model has been fitted, some means of checking the assumption of proportional hazards might be required.

Many model checking procedures are based on quantities known as *residuals*. These are values which can be calculated for each individual in the study, and have the feature that their behaviour is known, at least approximately, when the fitted model is satisfactory. A number of residuals have been proposed for use in the analysis of survival data, and we begin this chapter by reviewing some

of these. Residuals for the Cox regression model and the Weibull proportional
hazards model are considered separately.

5.1 Residuals for the Cox regression model

Throughout this section, we will suppose that the survival times of n individuals
are available, where r of these are death times and the remaining $n - r$ are right-
censored. We further suppose that a Cox regression model has been fitted to the
survival times, and that the linear component of the model contains p explanatory
variables, $X_1, X_2, \ldots, X_p$. The fitted hazard function for the i'th individual, $i =
1, 2, \ldots, n$, is therefore

$$\hat{h}_i(t) = \exp(\hat{\beta}'\mathbf{x}_i)\hat{h}_0(t),$$

where $\hat{\beta}'\mathbf{x}_i = \hat{\beta}_1 x_{1i} + \hat{\beta}_2 x_{2i} + \cdots + \hat{\beta}_p x_{pi}$ is the value of the fitted component, or
linear predictor, of the model for that individual and $\hat{h}_0(t)$ is the estimated baseline
hazard function.

5.1.1 Cox–Snell residuals

The residual which is most widely used in the analysis of survival data is the Cox–
Snell residual, so called because it is a particular example of the general definition
of residuals given by Cox and Snell (1968).

The Cox–Snell residual for the i'th individual, $i = 1, 2, \ldots, n$, is given by

$$r_{Ci} = \exp(\hat{\beta}'\mathbf{x}_i)\hat{H}_0(t_i), \tag{5.1}$$

where $\hat{H}_0(t_i)$ is the estimated cumulative baseline hazard function at time t_i, the
observed survival time of that individual. This function can be found from equation (3.16). Note that from equation (3.18), the Cox–Snell residual, r_{Ci}, is the
value of $\hat{H}_i(t_i) = -\log \hat{S}_i(t_i)$, where $\hat{H}_i(t_i)$ and $\hat{S}_i(t_i)$ are the estimated values
of the cumulative hazard and survivor functions of the i'th individual at t_i.

This residual can be derived from a general result in mathematical statistics
on the distribution of a function of a random variable. According to this result,
if T is the random variable associated with the survival time of an individual,
and $S(t)$ is the corresponding survivor function, then the random variable $Y =
-\log S(T)$ has an exponential distribution with unit mean, irrespective of the form
of $S(t)$. The proof of this result is outlined in the following paragraph, which can
be omitted without loss of continuity.

From equation (4.39) of Chapter 4, the probability density function of the random variable $Y = -\log S(T)$ is given by

$$f_Y(y) = f_T\left\{S^{-1}(e^{-y})\right\} / \left|\frac{dy}{dt}\right|, \tag{5.2}$$

where $f_T(t)$ is the probability function of T. Now,

$$\frac{dy}{dt} = \frac{d\{-\log S(t)\}}{dt} = \frac{f_T(t)}{S(t)},$$

and when the absolute value of this function is expressed in terms of y, the derivative becomes

$$\frac{f_T\{S^{-1}(e^{-y})\}}{S\{S^{-1}(e^{-y})\}} = \frac{f_T\{S^{-1}(e^{-y})\}}{e^{-y}}.$$

Finally, on substituting for the derivative in equation (5.2), we find that

$$f_Y(y) = e^{-y},$$

which, from equation (4.5) is the probability density function of an exponential random variable with unit mean.

The next and crucial step in the argument is as follows. If the model fitted to the observed data is satisfactory, then a model-based estimate of the survivor function for the i'th individual at t_i, the survival time of that individual, will be close to the corresponding true value $S_i(t_i)$. This suggests that if the correct model has been fitted, the values $\hat{S}_i(t_i)$ will have properties similar to those of $S_i(t_i)$. Then, the negative logarithms of the estimated survivor functions, $-\log \hat{S}_i(t_i)$, $i = 1, 2, \ldots, n$, will be n observations from a unit exponential distribution. These estimates are the Cox–Snell residuals.

If the observed survival time for an individual is right-censored then the corresponding value of the residual is also right-censored. The residuals will therefore be a censored sample from the unit exponential distribution, and a test of this assumption provides a test of model adequacy, to which we return in Section 5.2.

The Cox–Snell residuals r_{Ci} have properties that are quite dissimilar to those of residuals used in linear regression analysis, for example. In particular, they will not be symmetrically distributed about zero, and in fact they cannot be negative. Furthermore, since the Cox–Snell residuals are assumed to have an exponential distribution when an appropriate model has been fitted, they have a highly skew distribution and the mean and variance of the i'th residual will both be unity. One other point to note is that if the largest survival time is uncensored, the estimated value of the survivor function beyond that time is zero, and r_{Ci} is undefined for that observation.

5.1.2 Modified Cox–Snell residuals

Censored observations lead to residuals that cannot be regarded on the same footing as residuals derived from uncensored observations. We might therefore seek to modify the Cox–Snell residuals so that explicit account can be taken of censoring.

Suppose that the i'th survival time is a censored observation, t_i^*, and let t_i be the actual, but unknown, survival time, so that $t_i > t_i^*$. The Cox–Snell residual for this individual, evaluated at the censored survival time, is then given by

$$r_{Ci} = \hat{H}_i(t_i^*) = -\log \hat{S}_i(t_i^*), \quad .$$

where $\hat{H}_i(t_i^*)$ and $\hat{S}_i(t_i^*)$ are the estimated cumulative hazard and survivor functions respectively for the i'th individual at the censored survival time.

If the fitted model is correct, then the values r_{Ci} can be taken to have a unit exponential distribution. The cumulative hazard function of this distribution increases linearly with time, and so the greater the value of the survival time t_i for the i'th individual, the greater the value of the Cox–Snell residual for that individual. It then follows that the residual for the i'th individual at the actual (unknown) failure time, $\hat{H}_i(t_i)$, will be greater than the residual evaluated at the observed censored survival time.

To take account of this, Cox–Snell residuals can be modified by the addition of a positive constant Δ, which can be called the *excess residual*. Modified Cox–Snell residuals are therefore of the form

$$r'_{Ci} = \begin{cases} r_{Ci} & \text{for uncensored observations,} \\ r_{Ci} + \Delta & \text{for censored observations,} \end{cases}$$

where r_{Ci} is the Cox–Snell residual for the i'th observation, defined in equation (5.1). It now remains to identify a suitable value for Δ. For this, we use the *lack of memory property* of the exponential distribution, described in Section 4.1.2 of Chapter 4. From this result, since r_{Ci} has a unit exponential distribution, the excess residual Δ will also have a unit exponential distribution. The expected value of Δ is therefore unity, suggesting that Δ is taken to be unity, and this leads to modified Cox–Snell residuals

$$r'_{Ci} = \begin{cases} r_{Ci} & \text{for uncensored observations,} \\ r_{Ci} + 1 & \text{for censored observations.} \end{cases} \tag{5.3}$$

The i'th modified Cox–Snell residual can be expressed in an alternative form by introducing a censoring indicator, δ_i, which takes the value zero if the observed survival time of the i'th individual is censored and unity if it is uncensored. Then the modified Cox–Snell residual is given by

$$r'_{Ci} = 1 - \delta_i + r_{Ci}. \tag{5.4}$$

Note that from the definition of this type of residual, r'_{Ci} must be greater than unity for a censored observation. Also, as for the unmodified residuals, the r'_{Ci} can take any value between zero and infinity, and they will have a skew distribution.

On the basis of empirical evidence, Crowley and Hu (1977) found that the addition of unity to a Cox–Snell residual for a censored observation inflated the residual to too great an extent. They therefore suggested that the median value of the excess residual be used rather than the mean. For the unit exponential distribution, the survivor function is $S(t) = e^{-t}$, and so the median, $t(50)$, is such that $e^{-t(50)} = 0.5$, whence $t(50) = \log 2 = 0.693$. Thus a second version of the modified Cox–Snell residual has

$$r''_{Ci} = \begin{cases} r_{Ci} & \text{for uncensored observations,} \\ r_{Ci} + 0.693 & \text{for censored observations.} \end{cases} \tag{5.5}$$

However, if the proportion of censored observations is not too great, the set of residuals obtained from each of these two forms of modification will not appear too different.

5.1.3 Martingale residuals

The modified residuals r'_{Ci} defined in equation (5.4) have a mean of unity for uncensored observations. Accordingly, these residuals might be further refined by relocating the r'_{Ci} so that they have a mean of zero when an observation is uncensored. If in addition the resulting values are multiplied by -1, we obtain the residuals

$$r_{Mi} = \delta_i - r_{Ci}. \tag{5.6}$$

These residuals are known as *martingale residuals*, since they can also be derived using what are known as martingale methods. Because these methods rely heavily on probability theory and stochastic processes, this approach will not be discussed in this book. However, a comprehensive account of the martingale approach to the analysis of survival data is given by Fleming and Harrington (1991).

Martingale residuals take values between $-\infty$ and unity, with the residuals for censored observations, where $\delta_i = 0$, being negative. It can also be shown that in large samples, the martingale residuals are uncorrelated with one another and have an expected value of zero. In this respect, they have properties similar to those possessed by residuals encountered in linear regression analysis. However, the martingale residuals are not symmetrically distributed about zero.

Another way of looking at the martingale residuals is to note that the quantity r_{Mi} in equation (5.6) is the difference between the observed number of deaths for the i'th individual in the interval $(0, t_i)$ and the corresponding estimated expected number on the basis of the fitted model. To see this, note that the observed number of deaths is unity if the survival time t_i is uncensored, and zero if censored, that is δ_i. The second term in equation (5.6) is an estimate of $H_i(t_i)$, the cumulative hazard or cumulative probability of death for the i'th individual over the interval $(0, t_i)$. Since we are dealing with just one individual, this can be viewed as the expected number of deaths in that interval. This shows another similarity between the martingale residuals and residuals from other areas of data analysis.

5.1.4 Deviance residuals

Although martingale residuals share many of the properties possessed by residuals encountered in other situations, such as in linear regression analysis, they are not symmetrically distributed about zero, even when the fitted model is correct. This skewness makes plots based on the residuals difficult to interpret (see below). The deviance residuals, which were introduced by Therneau, Grambsch and Fleming (1990), are much more symmetrically distributed about zero. They are defined by

$$r_{Di} = \text{sgn}(r_{Mi}) \left[-2 \left\{ r_{Mi} + \delta_i \log(\delta_i - r_{Mi}) \right\} \right]^{1/2}, \tag{5.7}$$

where r_{Mi} is the martingale residual for the i'th individual, and the function sgn( ) is the sign function. This is the function that takes the value $+1$ if its argument is positive and -1 if negative. Thus $\text{sgn}(r_{Mi})$ ensures that the deviance residuals have the same sign as the martingale residuals.

The original motivation for these residuals is that they are components of the *deviance*. The deviance is a statistic that is used to summarize the extent to which the fit of a model of current interest deviates from that of a model which is a perfect fit to the data. This latter model is called the *saturated* or *full* model, and is a model in which the β-coefficients are allowed to be different for each individual. The statistic is given by

$$D = -2 \left\{ \log \hat{L}_c - \log \hat{L}_f \right\},$$

where $\hat{L}_c$ is the maximized partial likelihood under the current model and $\hat{L}_f$ is the maximized partial likelihood under the full model. The smaller the value of the deviance, the better the model. The deviance can be regarded as a generalization of the residual sum of squares used in modelling normal data to the analysis of non-normal data and features prominently in generalized linear modelling. Note that differences in deviance between two alternative models are the same as differences in the values of the statistic $-2 \log \hat{L}$ introduced in Chapter 3. The deviance residuals are then such that $D = \sum r_{Di}^2$, so that observations that correspond to relatively large deviance residuals are those that are not well fitted by the model.

Another way of viewing the deviance residuals is that they are martingale residuals that have been transformed to produce values that are symmetric about zero when the fitted model is appropriate. To see this, first recall that the martingale residuals r_{Mi} can take any value in the interval $(-\infty, 1)$. For large negative values of r_{Mi}, the term in square brackets in equation (5.7) is dominated by r_{Mi}. Taking the square root of this quantity has the effect of bringing the residual closer to zero. Thus martingale residuals in the range $(-\infty, 0)$ are shrunk towards zero. Now consider martingale residuals in the interval $(0, 1)$. The term $\delta_i \log(\delta_i - r_{Mi})$ in equation (5.7) will only be non-zero for uncensored observations, and will then have the value $\log(1 - r_{Mi})$. As r_{Mi} gets closer to unity, $1 - r_{Mi}$ gets closer to zero and $\log(1 - r_{Mi})$ takes large negative values. The quantity in square brackets in equation (5.7) is then dominated by this logarithmic term, and so the deviance residuals are expanded towards $+\infty$ as the martingale residual reaches its upper limit of unity.

One final point to note is that although these residuals can be expected to be symmetrically distributed about zero when an appropriate model has been fitted, they do not necessarily sum to zero.

5.1.5 Score residuals

Two disadvantages of the residuals described in Sections 5.1.1–5.1.4 is that they depend heavily on the observed survival time and require an estimate of the cumulative hazard function. Both of these disadvantages are overcome in a residual proposed by Schoenfeld (1982), which for the reason given in the sequel will be called a *score residual*. This residual differs from those considered previously in one other important respect. This is that there is not a single value of the residual for each individual, but a set of values, one for each explanatory variable included in the fitted Cox regression model.

The i'th score residual for X_j, the j'th explanatory variable in the model, is given by

$$r_{Sji} = \delta_i \{x_{ji} - a_{ji}\}, \tag{5.8}$$

where x_{ji} is the value of the j'th explanatory variable, $j = 1, 2, \ldots, p$, for the i'th individual in the study,

$$a_{ji} = \frac{\sum_{l \in R(t_i)} x_{jl} \exp(\hat{\beta}' \mathbf{x}_l)}{\sum_{l \in R(t_i)} \exp(\hat{\beta}' \mathbf{x}_l)}, \tag{5.9}$$

and $R(t_i)$ is the set of all individuals at risk at time t_i. Note that non-zero values of these residuals only arise for uncensored observations.

There are a number of ways in which this residual can be derived. The score residual is an estimate of the i'th component of the efficient score for the j'th parameter in the model. From Appendix A, the efficient score is the first derivative of the logarithm of the partial likelihood function with respect to β_j, which is given by

$$\frac{\partial \log L(\beta)}{\partial \beta_j} = \sum_{i=1}^{n} \delta_i \left\{ x_{ji} - \frac{\sum_l x_{jl} \exp(\beta' \mathbf{x}_l)}{\sum_l \exp(\beta' \mathbf{x}_l)} \right\}.$$

The i'th term in this summation, evaluated at $\hat{\beta}$, is then the score residual for x_j.

In Schoenfeld's derivation, the residual is the difference between the j'th explanatory variable and a weighted average of the values of the explanatory variable over individuals at risk at the death time of the i'th individual. The weight used for the l'th individual in the risk set at t_i is $\exp(\hat{\beta}' \mathbf{x}_l)$, and so the coefficient of x_{jl} in equation (5.9) is

$$\exp(\hat{\beta}' \mathbf{x}_l) / \sum_{l \in R(t_i)} \exp(\hat{\beta}' \mathbf{x}_l),$$

which is the component of the maximized partial likelihood for this individual.

Since the estimates of the β's are such that

$$\frac{\partial \log L(\beta)}{\partial \beta_j} \bigg|_{\hat{\beta}} = 0,$$

the score residuals must sum to zero. These residuals also have the property that in large samples the expected value of r_{Sji} is zero, and they are uncorrelated with one another.

In using these residuals, we note that individuals who are unlikely to die at time t_i, relative to those who are at risk of death at t_i, will have small values of the residuals. On the other hand, those individuals who are more likely to die, relative to those at risk, will have relatively large residuals.

In this section a number of residuals have been defined. We conclude with an example that illustrates the calculation of these different types of residual and which shows similarities and differences between them. This example will be used in many illustrations in this chapter, mainly because the relatively small number of observations allows the values of the residuals and other diagnostics to be readily tabulated. However, the methods of this chapter are generally more informative in larger data sets.

Example 5.1 Infection in patients on dialysis
In the treatment of certain disorders of the kidney, dialysis may be used to remove waste materials from the blood. One problem that can occur in patients on dialysis is the occurrence of an infection at the site at which the catheter is inserted. If any such infection occurs, the catheter must be removed, and the infection cleared up. In a study to investigate the incidence of infection, the time from insertion of the catheter until infection was recorded for a group of kidney patients. Sometimes, the catheter has to be removed for reasons other than infection, giving rise to right-censored observations. The data in Table 5.1 give the number of days from insertion of the catheter until its removal following the first occurrence of an infection. The data set includes the values of a variable that indicates the infection status of an individual, which takes the value zero if the catheter was removed for a reason other than the occurrence of an infection, and unity otherwise. Also given are the age of each patient in years and a variable that denotes the sex of each patient (1 = male, 2 = female). These data are taken from McGilchrist and Aisbett (1991), and relate to the 13 patients suffering from diseases of the kidney coded as type 3 in their paper.

When a Cox regression model is fitted to these data, the fitted hazard function for the i'th patient, $i = 1, 2, \ldots, 13$, is found to be

$$\hat{h}_i(t) = \exp \{0.030 \, AGE_i - 2.711 \, SEX_i\} \, \hat{h}_0(t), \qquad (5.10)$$

where AGE_i and SEX_i refer to the age and sex of the i'th patient.

The variable SEX is certainly important, since when SEX is added to the model that contains AGE alone, the decrease in the value of the statistic $-2 \log \hat{L}$ is 6.445 on 1 d.f. This change is highly significant (P = 0.011). On the other hand, there is no statistical evidence for including the variable AGE in the model, since the change in the value of the statistic $-2 \log \hat{L}$ on adding AGE to the model that contains SEX is 1.320 on 1 d.f. (P = 0.251). However, it can be argued that from the clinical viewpoint, the hazard of infection may well depend on age. Consequently, both variables will be retained in the model.

The values of different types of residual for the model in equation (5.10) are displayed in Table 5.2. In this table, r_{Ci}, r_{Mi} and r_{Di} are the Cox–Snell residuals, martingale residuals and deviance residuals, respectively. Also r_{S1i} and r_{S2i} are

Table 5.1 Times to removal of a catheter following a kidney infection.

Patient	Time	Status	Age	Sex
1	8	1	28	1
2	15	1	44	2
3	22	1	32	1
4	24	1	16	2
5	30	1	10	1
6	54	0	42	2
7	119	1	22	2
8	141	1	34	2
9	185	1	60	2
10	292	1	43	2
11	402	1	30	2
12	447	1	31	2
13	536	1	17	2

Table 5.2 Different types of residual after fitting a Cox regression model.

Patient	r_{Ci}	r_{Mi}	r_{Di}	r_{S1i}	r_{S2i}
1	0.3286	0.6714	0.9398	−1.0850	−0.2416
2	0.0785	0.9215	1.8020	14.4930	0.6644
3	1.4331	−0.4331	−0.3828	3.1291	−0.3065
4	0.0938	0.9062	1.7087	−10.2215	0.4341
5	1.7736	−0.7736	−0.6334	−16.5882	−0.5504
6	0.3116	−0.3116	−0.7895	0.0000	0.0000
7	0.2655	0.7345	1.0877	−17.8286	0.0000
8	0.5386	0.4614	0.5611	−7.6201	0.0000
9	1.6523	−0.6523	−0.5480	17.0910	0.0000
10	1.4234	−0.4234	−0.3751	10.2390	0.0000
11	1.4207	−0.4207	−0.3730	2.8575	0.0000
12	2.3927	−1.3927	−1.0201	5.5338	0.0000
13	−	−	−	0.0000	0.0000

the values of score residuals for the variables AGE and SEX respectively.

In this data set, there is just one censored observation, which is for patient number 6. Therefore, the modified Cox–Snell residuals will be the same as the Cox–Snell residuals for all patients except number 6. For this patient, the values of the two forms of modified residuals are $r'_{C6} = 1.3116$ and $r''_{C6} = 1.0048$.

From Table 5.2, we see that the values of the Cox–Snell, martingale and deviance residuals are not defined for patient number 13. This is because patient 13 has the longest time to removal of the catheter, and this time is uncensored. The estimated survivor function is therefore zero at this time, and the Cox–Snell residual is infinite. Since the martingale and deviance residuals are obtained from the values r_{Ci}, they are similarly undefined.

The skewness of the Cox–Snell and martingale residuals is clearly shown in Table 5.2, as is the fact that the Cox–Snell residuals are centred on unity while the martingale and deviance residuals are centred on zero. One unusual feature about the residuals in Table 5.2 is the large number of zero's for the values of the score residual corresponding to SEX. The reason for this is that for infection times greater than 30 days, the value of the variable SEX is always equal to 2. This means that the value of the term a_{ji} for this variable, given in equation (5.9), is equal to 2 for a survival time greater than 30 days, and so the corresponding score residual defined in equation (5.8) is zero.

We must now consider how residuals obtained after fitting a Cox regression model can be used to throw light on the extent to which the fitted model provides an appropriate description of the observed data. We will then be in a position to study the residuals obtained in Example 5.1 in greater detail.

5.2 Plots based on residuals

A number of plots based on residuals can be used in the graphical assessment of assumptions made in fitting the Cox regression model. In particular, residual plots lend themselves to an assessment of the adequacy of the linear predictor and the validity of the proportional hazards assumption. Techniques designed specifically for testing the proportional hazards assumption will be considered later in Section 5.9.

5.2.1 Plots of the Cox–Snell residuals

Since the Cox–Snell residuals are assumed to behave like observations from a unit exponential distribution, when the fitted model is correct they have a mean and variance of unity. This means that simple plots, such as plots of the residuals against the observation number, known as *index plots*, will not lead to a symmetric display. For this reason, the plots are not easy to interpret, and so something a little more sophisticated is needed.

In Section 4.2, we saw how a plot based on the estimated survivor function of a set of survival data could be used to provide information about the distribution of the survival times. Let $\hat{S}(t)$ denote the Kaplan–Meier estimate of the survivor function for the data. If the observed survival times have a Weibull distribution with scale parameter λ and shape parameter γ, then a plot of $\log\{-\log \hat{S}(t)\}$ against $\log t$ will be a straight line with intercept $\log \lambda$ and slope γ. The exponential distribution with mean $1/\lambda$ is a special case of the Weibull distribution, in which the parameter γ is set equal to unity, and so a unit exponential distribution is obtained by setting both λ and γ to unity. Accordingly, if a plot of $\log\{-\log \hat{S}(t)\}$ against $\log t$ gives a straight line with unit slope and zero intercept, then a unit exponential distribution for the survival times is indicated.

The result outlined above can be used to test if the residuals have a unit exponential distribution. After computing the Cox–Snell residuals, r_{Ci}, the Kaplan–Meier estimate of the survivor function of these values is found. This estimate is computed in a similar manner to the Kaplan–Meier estimate of the survivor function of survival times, except that the data on which the estimate is based are now the residuals r_{Ci}. Residuals obtained from censored survival times are themselves taken to be censored. Denoting the estimate by $\hat{S}(r_{Ci})$, the values of $\log\{-\log \hat{S}(r_{Ci})\}$ are plotted against $\log r_{Ci}$. This gives a log-cumulative hazard plot of the residuals. A straight line plot with unit slope and zero intercept will then indicate that the fitted survival model is correct. On the other hand, a plot that displays a systematic departure from the straight line, or yields a line that does not have approximately unit slope or zero intercept, indicates that the model needs to be modified in some way. Models that include additional explanatory variables might then be considered.

When a model includes factors, log-cumulative hazard plots of the Cox–Snell residuals may be formed for each level of any factor in the model. If the fitted model is satisfactory, the points on the plot corresponding to the different levels should overlap. On the other hand, if the residuals for one level of the factor are generally separated from those at the other levels, this suggests that the factor has not been properly taken account of in the model.

Example 5.2 Infection in patients on dialysis
Consider again the data on the time to the occurrence of an infection in kidney patients, described in Example 5.1. In this example, plots based on the residuals in Table 5.2 are used in an assessment of the adequacy of the model fitted to these data.

We first examine whether the Cox–Snell residuals are a plausible sample of observations from a unit exponential distribution. For this, the Kaplan–Meier estimate of the survivor function of the Cox–Snell residuals, $\hat{S}_i(r_{Ci})$, is obtained. The complementary log-log transformation of this estimate, which is the log-cumulative hazard function, $\log \hat{H}_i(r_{Ci})$, is then plotted against the logarithm of the corresponding residual to give a log-cumulative hazard plot of the residuals. The details of this calculation are summarized in Table 5.3, and the log-cumulative hazard plot is shown in Figure 5.1

The relatively small number of observations in this data set makes it difficult to interpret plots of residuals. However, the plotted points in Figure 5.1 are fairly close to a straight line which has a zero intercept and approximately unit slope. This suggests that the model fitted to the data given in Table 5.1 is satisfactory.

Table 5.3 Calculation of the log-cumulative
hazard function of the Cox–Snell residuals.

r_{Ci}	$\log r_{Ci}$	$\hat{S}_i(r_{Ci})$	$\log \hat{H}_i(r_{Ci})$
0.0785	−2.5450	0.9167	−2.4421
0.0938	−2.3661	0.8333	−1.7018
0.2655	−1.3260	0.7500	−1.2459
0.3286	−1.1130	0.6562	−0.8644
0.5386	−0.6189	0.5625	−0.5528
1.4207	0.3512	0.4687	−0.2774
1.4234	0.3531	0.3750	−0.0194
1.4331	0.3599	0.2812	0.2380
1.6523	0.5022	0.1875	0.5152
1.7736	0.5730	0.0937	0.8619
2.3927	0.8724	0.0000	−

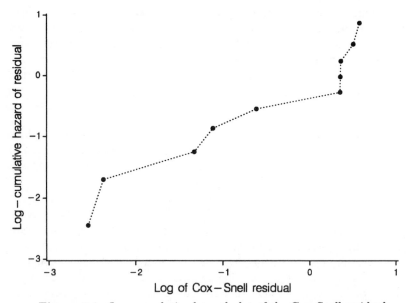

Figure 5.1   Log-cumulative hazard plot of the Cox–Snell residuals.

5.2.2 Plots based on other types of residual

When Cox–Snell residuals have been adjusted for the occurrence of censoring, a number of plots based on these modified residuals can be obtained. For example, modified Cox–Snell residuals can be plotted against the observation number to give an index plot. Residuals with relatively large negative values can indicate that the corresponding observation is not well fitted by the model.

We saw in Chapter 3 that it is the rank order of the survival times, rather than the survival times themselves, that determine the values of the estimated β-parameters in the model. Accordingly, plots of the residuals against the rank order of the survival times can be used to examine model adequacy. A lack of pattern in such a plot would suggest that the fitted model was plausible.

Modified Cox–Snell residuals may also be plotted against explanatory variables that are not in the fitted Cox regression model. If there is no obvious relationship between the residuals and any such variable, this suggests that the variable is not needed in the model. By the same token, plots of residuals against variables included in the model can be used in an informal assessment of the necessity of including those variables. Plots of residuals against explanatory variables can also be used to indicate whether any particular variable needs to be transformed prior to incorporating it in the model. In particular, if the plot of residuals against the values of a particular variable X is non-linear, that variable might be replaced by $\sqrt{X}$ or $\log X$, or polynomial terms such as X^2 or X^3 might need to be included.

Even though censoring has been allowed for in the construction of the modified Cox–Snell residuals, it is a good idea to use separate plotting symbols for censored and uncensored observations in plots derived from them. This enables unusual features associated with censoring to be identified.

Martingale residuals can be used in the construction of the plots that were described for the modified Cox–Snell residuals. In particular, index plots of the residuals, plots against the survival time or the rank of the survival time, and plots against explanatory variables in or out of the model can all be useful in model validation. Index plots can reveal observations not well fitted by the model, and plots against explanatory variables indicate whether the variable needs to be included or whether it is necessary to transform a variable that has already been included in the model. Plots of the residuals against the rank of the survival time can be used to detect departures from proportional hazards.

The type of plots described for modified Cox–Snell residuals and martingale residuals can also be used with deviance residuals. In view of the symmetry that these residuals have when the correct model has been fitted, plots constructed from them are easier to interpret. For example, an index plot will highlight observations with relatively large negative or large positive residuals, making it easier to identify observations for which the model is not a good fit.

When a model includes one or more factors, the points on an index plot of the residuals can be labelled according to the levels of a particular factor. The same pattern should be seen across censored and uncensored observations at each level of the factor.

Plots of the score residuals against survival time, or the rank order of the survival times, should show a random scatter of points, centred on zero, if the fitted model is adequate. If the score residual is calculated for an indicator variable used in connection with fitting a factor in the model, a plot of the residuals against the rank order of the survival times will have a deterministic pattern. Typically, the plot will show two horizontal bands of points either side of the line corresponding to a zero residual, each band corresponding to one of the two values of the indicator variable. This feature is indicated in Example 5.3 below.

Example 5.3 Infection in patients on dialysis
Consider again the data on times to infection in kidney patients. Plots of the martingale and deviance residuals against time to infection, the rank order of the infection time, and the two explanatory variables in the model, AGE and SEX, are shown in Figures 5.2 and 5.3.

A plot of the residuals against the indicator variable SEX is included for completeness, although the patterns in such plots make them difficult to interpret. The other plots display no discernible pattern, confirming that the fitted model is satisfactory. In particular, there is no one residual that stands out from the rest, and no pattern in the residuals over time.

Plots of the score residuals for AGE and SEX against the rank order of the times to infection are shown in Figure 5.4. There is no pattern shown in the plot of the score residuals for AGE. That for SEX provides an indication of the type of pattern that is typical in plots of score residuals for indicator variables. In Figure 5.4(b), the two points above zero are the residuals for $SEX = 2$ (female), while those below zero correspond to $SEX = 1$ (male).

5.3 Some comments and recommendations

In an earlier section it was argued that since the values $-\log S(t_i)$ have a unit exponential distribution, the Cox–Snell residuals, which are estimates of these quantities, should have an approximate unit exponential distribution when the fitted model is correct. This result is then used when interpreting a log-cumulative hazard plot of the residuals. Unfortunately this approximation is not very reliable, particularly in small samples. This is because estimates of the β's, and also of the baseline cumulative hazard function, $H_0(t)$, are needed in the computation of r_{Ci}. The substitution of estimates means that the actual distribution of the residuals is not necessarily unit exponential, but their exact distribution is not known. The distribution of Cox–Snell residuals for $n = 3$ was shown by Lagakos (1980) to be quite dissimilar to a unit exponential sample.

Crowley and Storer (1983) showed empirically that a log-cumulative hazard plot of the residuals is not particularly good at identifying inadequacies in the fitted model. Moreover, in the particular case of the null model, that is, the model that contains no explanatory variates, the log-cumulative hazard plot will be a straight

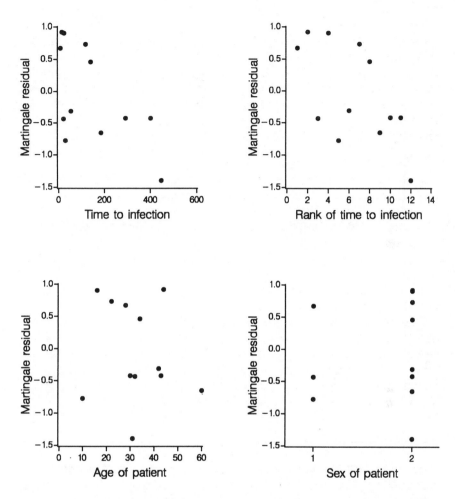

Figure 5.2 Plots of the martingale residuals against (a) time to infection, (b) rank order of time to infection, (c) AGE and (d) SEX.

line with unit slope and zero intercept, even if some explanatory variables should actually be included in the model. The reason for this is that when no covariates are included, the Cox–Snell residual for the i'th individual reduces to $-\log \hat{S}_0(t_i)$. From equation (3.21) in Chapter 3, in the absence of ties this is approximately $\sum_{j=1}^{k} 1/n_j$ at the k'th uncensored survival time, $k = 1, 2, \ldots, r - 1$, where n_j is the number at risk at time t_j. This summation is simply $\sum_{j=1}^{k} 1/(n - j + 1)$, which is the expected value of the k'th order statistic in a sample of size n from a unit exponential distribution.

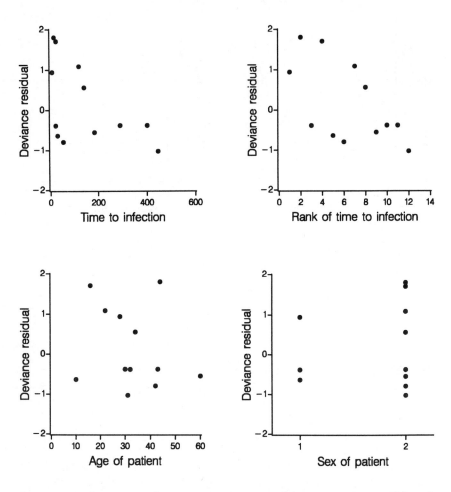

Figure 5.3 Plots of the deviance residuals against (a) time to infection, (b) rank order of time to infection, (c) *AGE* and (d) *SEX*.

A major weakness of plots based on residuals is that there is no objective means of assessing whether there are any deficiencies in the fitted model. Instead, informal judgment is needed on whether, for example, the points in a log-cumulative hazard plot deviate from a straight line, or on whether there is pattern in a plot of residuals against the rank order of the survival times. Furthermore, there is often uncertainty about how plots based on residuals should appear when an appropriate model has been fitted. For example, although martingale residuals are generally preferred to the Cox–Snell residuals, plots of the martingale residuals against survival time can be expected to have a pattern, even when the correct model has

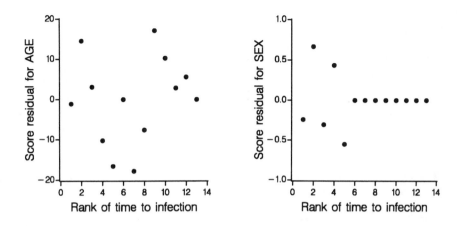

Figure 5.4 Plots of the score residuals for (a) AGE and (b) SEX against rank order of time to infection.

been fitted. This was shown by Henderson and Milner (1991), who go on to propose that estimates of the expected mean of the residuals at each survival time be superimposed on the plot. In this way, patterns that are close to the superimposed line would not be interpreted as a model deficiency.

The score residuals have the disadvantage that they cannot be directly calculated from standard output given by computer packages for survival analysis. However, they do form part of a statistic that is used in the detection of influential values (see Section 5.4.1). In summary, either martingale or deviance residuals are commended for general use. Plots which are likely to be useful include index plots, plots against explanatory variables and plots against the rank order of the survival times. In a more thorough analysis, the score residuals may be computed for each explanatory variable in the model and plotted against the rank order of the survival times. However, note that plots with systematic pattern result from score residuals associated with indicator variables used to introduce factors into the Cox regression model.

Example 5.4 Survival of multiple myeloma patients
In this example we return to the data on the survival times of 48 patients with multiple myeloma, described in Example 1.3. In Example 3.5, a Cox regression model that contained the explanatory variables HB (serum haemoglobin) and $LBUN$ (logarithm of blood urea nitrogen) was found to be a suitable model for the haz-

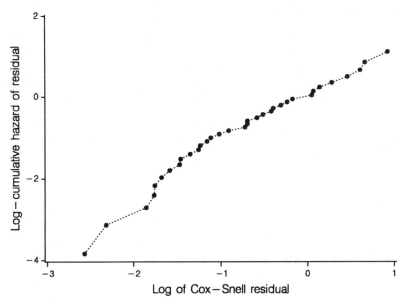

Figure 5.5   Log-cumulative hazard plot of the Cox–Snell residuals.

ard function. We now perform an analysis of the residuals in order to study the adequacy of this fitted model.

First, a log-cumulative hazard plot of the Cox–Snell residuals is shown in Figure 5.5. The line made by the plotted points in Figure 5.5 is not particularly straight, although it does have a zero intercept. If the points that correspond to the five smallest residuals, which appear in the bottom left-hand corner of the plot, are ignored, the remaining points fall on a line of unit slope. The five smallest Cox–Snell residuals correspond to patients 41, 38, 48, 14 and 32, respectively, in Table 1.3 and these patients have survival times of 1, 1, 3, 5 and 1 months respectively. These are some of the shortest survival times in this group of patients, and leads to the conclusion that the model is not a good fit to the data from patients who die in the first few months following entry to the study.

Plots of the deviance and score residuals are now used in a fuller assessment of the adequacy of the fitted Cox regression model. In the three plots that make up Figure 5.6, the martingale residuals are plotted against the rank order of the survival times, and the two explanatory variables in the model, HB and $LBUN$.

In Figure 5.6(a) we see a cluster of observations with martingale residuals close to their upper limit of unity at the shortest survival times. These are the martingale residuals corresponding to the observations with the smallest Cox–Snell residuals. The plot of the martingale residuals against the rank order of the survival times therefore gives a further confirmation of the inadequacy of the model at the shortest survival times. Plots of the martingale residuals against the explanatory variables

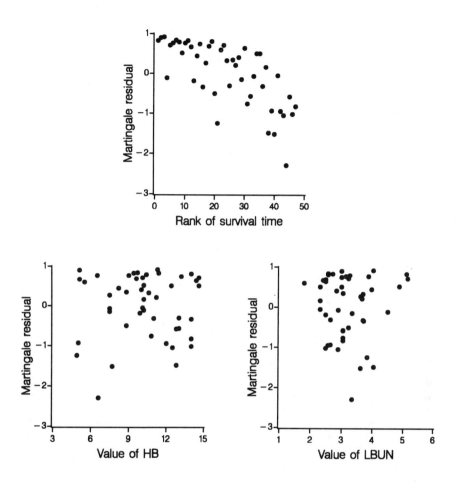

Figure 5.6 Martingale residuals plotted against (a) rank order of survival time, (b) HB and (c) $LBUN$.

HB and $LBUN$ in Figures 5.6(b) and 5.6(c) have no systematic pattern, and the observations with the largest martingale residuals do not seem to be associated with particular values of the explanatory variables.

Much the same conclusions can be drawn from the plots of the deviance residuals in Figure 5.7. In the plot of the deviance residuals against the rank order of the survival times, the large positive residuals in the top left-hand corner of the plot again stand out, but there are no other patterns.

The values of the score residuals for the variables HB and $LBUN$ are plotted against rank order of the survival times in Figure 5.8. There are no anomalies

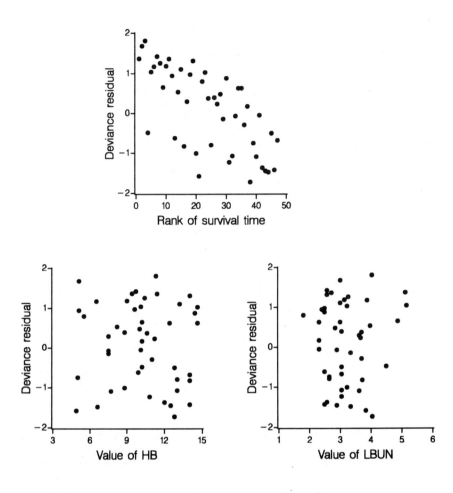

Figure 5.7 Deviance residuals plotted against (a) rank order of survival time, (b) HB and (c) $LBUN$.

shown in the plot of the score residuals for HB against the rank order of the survival times. However, the observations with the three largest score residuals for $LBUN$ are from patients 12, 28 and 32. These patients have survival times of 4, 5 and 1 respectively, and are the only patients to have values of blood urea nitrogen in excess of 100.

The overall conclusion from this analysis is that the fitted model is satisfactory for most of the patients in this study, although there is doubt about its ability to model the hazard function of the patients with the shortest survival times. The next step in the analysis of these data is to investigate whether the data from any of the

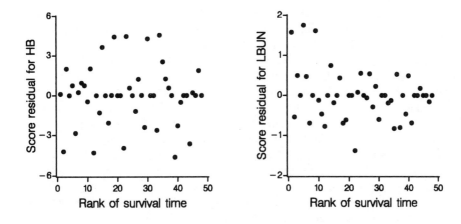

Figure 5.8 Score residuals plotted against rank order of survival time for (a) *HB* and (b) *LBUN*.

patients, and particularly those with smallest survival times, has an undue impact on the form of the fitted hazard function. We will therefore return to these data in Example 5.7.

5.4 Identification of influential observations

In the assessment of model adequacy it is important to determine whether any particular observation has an undue impact on inferences made on the basis of a model fitted to an observed set of survival data. Observations which do have an effect on model-based inferences are said to be *influential*.

As an example, consider a survival study in which a new treatment is to be compared with a standard. In such a comparison, it would be important to determine if the hazard of death on the new treatment, relative to that on the standard, was substantially affected by any one individual. In particular, it might be that when the data record for one individual is removed from the data base, the relative hazard is increased or reduced by a substantial amount. If this happens, the data from such an individual would need to be subject to particular scrutiny.

Conclusions from a survival analysis are often framed in terms of estimates of quantities such as the relative hazard and median survival time, which depend on the estimated values of the β-parameters in the fitted Cox regression model. It is therefore of particular interest to examine the influence of each observation on these estimates. We can do this by examining the extent to which the estimated parameters in the fitted model are affected by omitting in turn the data record for each individual in the study. In some circumstances, the estimates of a subset of the parameters may be of special importance, such as parameters associated with treatment effects. The study of influence may then be limited to just these parameters. On many occasions, the influence that each observation has on the estimated hazard function will be of interest, and it would then be important to identify observations which influence the complete set of parameter estimates under the model. These two aspects of influence are discussed in the following sections.

In contrast to models encountered in the analysis of other types of data, such as the general linear model, the effect of removing one observation from a set of survival data is not easy to study. This is mainly because the log-likelihood function for the Cox regression model cannot be expressed as the sum of a number of terms, in which each term is the contribution to the log-likelihood made by each observation. Instead, the removal of one observation affects the risk sets over which quantities of the form $e^{\beta' x}$ are summed. This means that influence diagnostics are quite difficult to derive and so the following sections of this chapter simply give the relevant results. References to the articles that contain derivations of the quoted formulae are included in the final section of this chapter.

5.4.1 Influence of observations on a parameter estimate

Suppose that we wish to determine whether any particular observation has an untoward effect on $\hat{\beta}_j$, the j'th parameter estimate, $j = 1, 2, \ldots, p$, in a fitted Cox regression model. One way of doing this would be to fit the model to all n observations in the data set, and to then fit the same model to the sets of $n - 1$ observations obtained by omitting each of the n observations in turn. The actual effect that omitting each observation has on the parameter estimate could then be determined. This procedure is computationally expensive, unless the number of observations is very small, and so we use instead an approximation to the amount by which $\hat{\beta}_j$ changes when the i'th observation is omitted, for $i = 1, 2, \ldots, n$. Denote the value of the j'th parameter estimate when the i'th observation is omitted by $\hat{\beta}_{j(i)}$. A number of approximations to the quantity $\hat{\beta}_j - \hat{\beta}_{j(i)}$ have been proposed, but the one to be considered here was presented by Cain and Lange (1984), and is also described in Pettitt and Bin Daud (1989).

As usual, t_i will be used to denote the survival time of the i'th of n individuals, and δ_i will be a censoring indicator that is zero if that time is censored and unity otherwise. We will also take $\hat{\beta}' x_i$ to be the linear predictor for the i'th individual, and define

$$a_{ji} = \frac{\sum_{l \in R(t_i)} x_{jl} \exp(\hat{\beta}' x_l)}{\sum_{l \in R(t_i)} \exp(\hat{\beta}' x_l)},$$

where x_{jl} is the value of the j'th explanatory variable for the l'th individual in $R(t_i)$, the risk set at t_i. This quantity was used in the definition of a score residual, and given previously in equation (5.9).

Now let $\mathbf{d}_i$, $i = 1, 2, \ldots, n$, be a $p \times 1$ vector whose j'th component is

$$d_{ij} = \delta_i(x_{ji} - a_{ji}) + \exp(\hat{\beta}'\mathbf{x}_i) \sum_{t_j \leq t_i} \delta_j \frac{(a_{ji} - x_{ji})}{\sum_{l \in R_{(t_j)}} \exp(\hat{\beta}'\mathbf{x}_l)}, \quad (5.11)$$

for $j = 1, 2, \ldots, p$. There is therefore a separate vector $\mathbf{d}_i$ corresponding to each of the n observations in the data set. Notice that the first term in the expression for d_{ij} is the score residual in equation (5.8) associated with the j'th explanatory variable in the fitted model. It can then be shown that an approximation to $\hat{\beta}_j - \hat{\beta}_{j(i)}$, the change in $\hat{\beta}_j$ on omitting the i'th observation, is the (i, j)'th element of the $n \times p$ matrix

$$\Delta'\mathbf{V}(\hat{\beta}), \quad (5.12)$$

where $\mathbf{V}(\hat{\beta})$ is the $p \times p$ variance–covariance matrix of the vector of parameter estimates in the fitted Cox regression model, and Δ' is the $n \times p$ matrix whose i'th row consists of the components of the vector $\mathbf{d}_i$. This quantity, which is called a *delta-beta*, will be denoted by $\Delta_i\hat{\beta}_j$, so that $\Delta_i\hat{\beta}_j \approx \hat{\beta}_j - \hat{\beta}_{j(i)}$. The expression in equation (5.12) is quite complicated, but it has the merit that it can be computed from quantities available after fitting the model to the full data set.

Observations which influence a particular parameter estimate, the j'th say, will be such that the values of $\Delta_i\hat{\beta}_j$, the delta-beta's for these observations, are larger in absolute value than for other observations in the data set. Index plots of the delta-beta's for each explanatory variable in the model will then reveal whether there are observations that have an undue impact on the parameter estimate for any particular explanatory variable. In addition, a plot of the values of $\Delta_i\hat{\beta}_j$ against the rank order of the survival times yields information about the relation between survival time and influence.

The delta-beta's may be standardized by dividing $\Delta_i\hat{\beta}_j$ by the standard error of $\hat{\beta}_j$ to give a *standardized delta-beta*. The standardized delta-beta can be interpreted as the change in the value of the statistic $\hat{\beta}/\text{s.e.}(\hat{\beta})$, on omitting the i'th observation. Since this statistic can be used in assessing whether a particular parameter has a value significantly different from zero (see Section 3.4 of Chapter 3), the standardized delta-beta can be used to provide information on how the significance of the parameter estimate is affected by the removal of the i'th observation from the data base. Again, an index plot is the most useful way of displaying the standardized delta-beta's. A SAS macro which can be used to compute the values of the unstandardized or standardized delta-beta's is described in Chapter 11.

The statistic in equation (5.12) is an approximation to the actual change in the parameter estimate when the i'th observation is omitted from the fit. The approximation is generally adequate in the sense that observations that have an influence on a parameter estimate will be highlighted. However, the actual effect of omitting any particular observation on model-based inferences will need to be studied. The agreement between the actual and approximate delta-beta's in a particular situation is illustrated in Example 5.5.

Table 5.4 Approximate delta-beta's for AGE ($\hat{\beta}_1$), and SEX ($\hat{\beta}_2$).

Observation	$\Delta_i\hat{\beta}_1$	$\Delta_i\hat{\beta}_2$
1	0.0020	−0.1977
2	0.0004	0.5433
3	−0.0011	0.0741
4	−0.0119	0.5943
5	0.0049	0.0139
6	−0.0005	−0.1192
7	−0.0095	0.1270
8	−0.0032	−0.0346
9	−0.0073	−0.0734
10	0.0032	−0.2023
11	0.0060	−0.2158
12	0.0048	−0.1939
13	0.0122	−0.3157

Example 5.5 Infection in patients on dialysis

In this example, we return to the data on the times to infection following commencement of dialysis. To investigate the influence that the data from each of the 13 patients in the study has on the estimated values of the coefficients of the variables AGE and SEX in the linear component of the fitted Cox regression model, the approximate unstandardized delta-beta's, $\Delta_i\hat{\beta}_1$ and $\Delta_i\hat{\beta}_2$, are obtained. These are given in Table 5.4.

The largest delta-beta for AGE occurs for patient number 13, but there are other delta-beta's with similar values. The actual change in the parameter estimate on omitting the data for this patient is 0.0195, and so omission of this observation increases the hazard of infection relative to the baseline hazard. The standard error of the parameter estimate for AGE in the full data set is 0.026, and so the maximum amount by which this estimate is changed when one observation is deleted is about three-quarters of a standard error. When the data from patient 13 is omitted, the age effect becomes slightly more significant, but the difference is unlikely to be of practical importance.

There are two large delta-beta's for SEX that are quite close to one another. These correspond to the observations from patients 2 and 4. The actual change in the parameter estimate when each observation is omitted in turn is 0.820 and 0.818, and so the approximate delta-beta's underestimate the actual change. The standard error of the estimated coefficient of SEX in the full data set is 1.096, and so again the change in the estimate on deleting an observation is less than one standard error. The effect of deleting either of these two observations is to increase the relative hazard, but again this increase is not great.

To compare the approximate delta-beta's calculated from equation (5.12) with

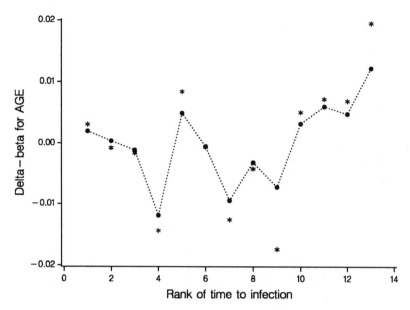

Figure 5.9 Plot of the exact (∗) and approximate (•) delta-beta's for *AGE* against rank order of time to infection.

the actual values, a plot of their values against the rank of the time to infection is given in Figures 5.9 and 5.10. These figures show that the agreement is generally quite good, although there is a tendency for the actual changes in the parameter estimates to be underestimated by the approximation in equation (5.12). The largest difference between the actual and approximate delta-beta for *AGE* is 0.010, which occurs for patient number 8. That for *SEX* is 0.276, which occurs for patient number 2. These differences are about a quarter of the value of the standard error of each parameter estimate.

5.4.2 Influence of observations on the set of parameter estimates

It may happen that the structure of the fitted model is particularly sensitive to one or more observations in the data set. Such observations can be detected using diagnostics that are designed to highlight observations that influence the complete set of parameter estimates in the linear predictor. These diagnostics provide information that is additional to information provided by the delta-beta's. In particular, excluding a given observation from the data set may not have a great influence on any particular parameter estimate, and so will not be revealed from a study of the delta-beta statistics. However, the change in the set of parameter estimates might be such that the form of the estimated hazard function, or values of summary statistics based on the fitted model, change markedly when that observation

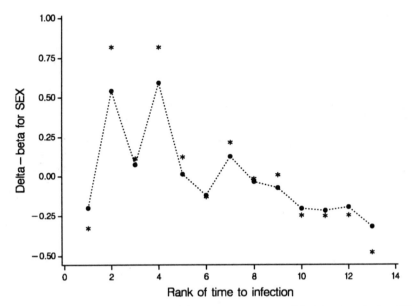

Figure 5.10 Plot of the exact (*) and approximate (•) delta-beta's for *SEX* against rank order of time to infection.

is removed. Statistics for assessing the influence of observations on the set of parameter estimates also have the advantage that there is a single value of the diagnostic for each observation. This makes them easier to use than diagnostics such as the score residuals and delta-beta's.

Over the past few years a number of diagnostics for assessing the influence of each observation on the set of parameter estimates have been proposed. In this section, two will be described, but references to others will be given in the concluding section of this chapter.

One way of assessing the influence of each observation on the overall fit of the model is to examine the amount by which the value of minus twice the logarithm of the maximized partial likelihood, $-2 \log \hat{L}$, under a fitted model, changes when each observation in turn is left out. Write $-2 \log L(\hat{\beta})$ for the value of the maximized log-likelihood when the model is fitted to all n observations, and $-2 \log L(\hat{\beta}_{(i)})$ for the value of the maximized log-likelihood of the n observations when the parameter estimates are computed after omitting the i'th observation from the fit. The diagnostic

$$2 \left\{ \log L(\hat{\beta}) - \log L(\hat{\beta}_{(i)}) \right\}$$

can then be useful in the study of influence.

Pettitt and Bin Daud (1989) show that an approximation to this *likelihood dis-*

placement is

$$LD_i = \mathbf{d}'_i \mathbf{V}(\hat{\beta}) \mathbf{d}_i, \tag{5.13}$$

where $\mathbf{d}_i$ is the $p \times 1$ vector whose j'th component is given in equation (5.11) and $\mathbf{V}(\hat{\beta})$ is the variance–covariance matrix of $\hat{\beta}$, the vector of parameter estimates. The values of this statistic may therefore be straightforwardly obtained from terms used in computing the delta-beta's for each explanatory variable in the model. An index plot, or a plot of the likelihood displacements against the rank order of the survival times, provides an informative visual summary of the values of the diagnostic. Observations that have relatively large values of the diagnostic are influential. Plots against explanatory variables are not recommended, since as demonstrated by Pettitt and Bin Daud (1989), these plots can have a deterministic pattern, even when the fitted model is correct.

Another diagnostic that can be used to assess the impact of each observation on the set of parameter estimates is based on the $n \times n$ symmetric matrix

$$\mathbf{B} = \mathbf{\Delta}' \mathbf{V}(\hat{\beta}) \mathbf{\Delta},$$

where $\mathbf{\Delta}'$ is the $n \times p$ matrix formed from the vectors $\mathbf{d}_i$, and used in equation (5.12). An argument from linear algebra shows that the absolute values of the elements of the $n \times 1$ eigenvector associated with the largest eigenvalue of the matrix $\mathbf{B}$, standardized to have unit length by dividing each component by the square root of the sum of squares of all the components of the eigenvector, is a measure of the sensitivity of the fit of the model to each of the n observations in the data set. Denoting this eigenvector by $\mathbf{l}_{\max}$, the i'th element of $\mathbf{l}_{\max}$ is a measure of the influence of the i'th observation on the set of parameter estimates. The sign of this diagnostic is immaterial, and so plots based on the absolute values, $|\mathbf{l}_{\max}|$ are commended for general use. Index plots of these values, plots against the rank order of the survival times, and against explanatory variables in the model, can all be useful in the assessment of influence.

The standardization to unit length means that the squares of the elements of $\mathbf{l}_{\max}$ must sum to 1.0. Observations for which the squares of the elements of the eigenvector account for a substantial proportion of the total sum of squares of unity will then be those which are most influential. Large elements of this eigenvector will therefore correspond to observations which have most effect on the value of the likelihood function. A final point to note is that unlike other diagnostics, a plot of the elements of $\mathbf{l}_{\max}$ against explanatory variables will not have a deterministic pattern if the fitted model is correct. This means that plots of the absolute values of the elements of $\mathbf{l}_{\max}$ against explanatory variables can be useful in assessing whether there are particular ranges of values of the variates over which the model does not fit well.

Example 5.6 Infection in patients on dialysis
The data first given in Example 5.1 will again be used to illustrate the use of diagnostics designed to reveal observations that influence the complete set of parameter estimates. In Table 5.5, the approximate likelihood displacements from

Table 5.5 Values of the approximate likelihood displacement, LD_i, and the elements of $\mathbf{l}_{max}$.

Observation	LD_i	$|\mathbf{l}_{max}|$
1	0.033	0.161
2	0.339	0.309
3	0.005	0.068
4	0.338	0.621
5	0.050	0.104
6	0.019	0.058
7	0.136	0.291
8	0.027	0.054
9	0.133	0.124
10	0.035	0.193
11	0.061	0.264
12	0.043	0.224
13	0.219	0.464

equation (5.13), and the elements of the vector $|\mathbf{l}_{max}|$, are given. The observations that most affect the value of the maximized log-likelihood when they are omitted are those corresponding to patients 2 and 4. The value of the likelihood displacement diagnostic is also quite large for patient number 13. This means that the set of parameter estimates are most affected by the removal of either of these patients from the data base.

The fourth element of $\mathbf{l}_{max}$ is the largest in absolute value and indicates that omitting the data from patient number 4 has the greatest effect on the pair of parameter estimates. The elements corresponding to patients 2 and 13 are also large relative to the other values, suggesting that the data for these patients are also influential. The sum of the squares of elements 2, 4 and 13 of $\mathbf{l}_{max}$ is 0.70. The total of the sums of squares of the elements is 1.00, and so cases 2, 4 and 13 account for nearly three-quarters of the variability in the elements of $\mathbf{l}_{max}$. Note that the analysis of the delta-beta's in Example 5.5 showed that the observations from patients 2 and 4 most influence the parameter estimate for SEX, while the observation for patient 13 has a greater effect on the estimate for AGE.

In summary, the observations from patients 2, 4 and 13 affect the form of the hazard function to the greatest extent. Omitting each of these in turn gives the following estimates of the linear component in the hazard functions for the i'th individual:

Omitting patient number 2: $0.031\,AGE_i - 3.530\,SEX_i,$

Omitting patient number 4: $0.045\,AGE_i - 3.529\,SEX_i,$

Omitting patient number 13: $0.011\,AGE_i - 2.234\,SEX_i.$

For comparison, the linear component for the full data set is

$$0.030\,AGE_i - 2.711\,SEX_i.$$

To illustrate the magnitude of the change in estimated hazard ratios, consider the relative hazard of infection at time t for a patient aged 50 years relative to one aged 40 years. For the full data set, this is $e^{0.304} = 1.355$. This is increased to 1.365 and 1.564 when patients 2 and 4 respectively are omitted, and decreased to 1.114 when patient 13 is omitted. The effect on the hazard function of removing these patients from the data base is therefore not particularly marked.

In the same way, the hazard of infection at time t for a male patient ($SEX = 1$) relative to a female ($SEX = 2$) is $e^{2.711}$, that is, 5.041 for the full data set. When observations 2, 4, and 13 are omitted in turn, the hazard for males relative to females is 4.138, 4.097 and 9.334 respectively. Omission of the data from patient number 13 appears to have a great effect on the estimated hazard ratio. However, some caution is needed in interpreting this result. Since there are very few males in the data set, the estimated hazard ratio is imprecisely estimated. In fact, a 95% confidence interval for the hazard ratio, when the data from patient 13 are omitted, ranges from 0.012 to 82.96!

5.4.3 Recommendations

In this section a range of influence diagnostics have been presented. Which should be used on a routine basis and which are needed when a more thorough assessment of model adequacy is required? When the inference to be drawn from a model centres on one or two particular parameters, the delta-beta statistic for those parameters, described in Section 5.4.1, will be the most relevant. Plots of these values against rank order of survival times will then be useful. To investigate whether there are observations that have an influence on the set of parameter estimates, the diagnostic based on the absolute values of the elements of l_{max} is probably the most suitable. Plots against rank order of survival times will be informative, but plots against particular explanatory variables might also be revealing.

Example 5.7 Survival of multiple myeloma patients
The effect of individual observations on the estimated values of the parameters of a Cox regression model fitted to the data from Example 1.3 will now be investigated. Plots of the approximate unstandardized delta-beta's for HB and $LBUN$ against the rank order of the survival times are shown in Figures 5.11 and 5.12.

From Figure 5.11, no one observation stands out as having a delta-beta that is different from the rest. However, Figure 5.12 shows that there are three observa-

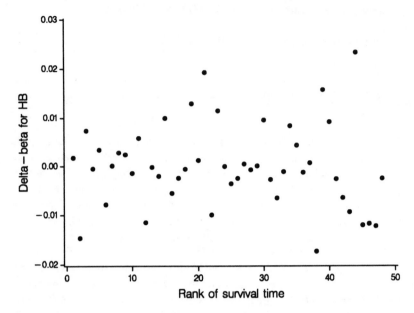

Figure 5.11 Plot of the delta-beta's for HB against rank order of survival time.

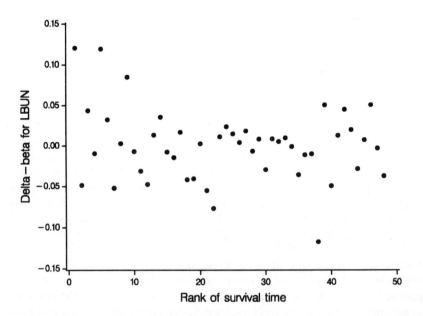

Figure 5.12 Plot of the delta-beta's for $LBUN$ against rank order of survival time.

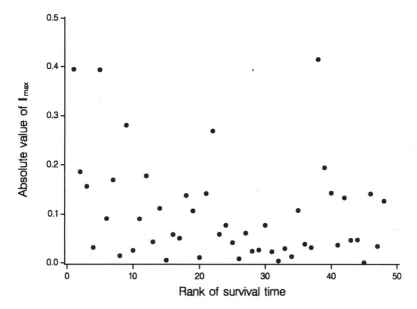

Figure 5.13 Plot of the absolute values of the elements of l_{max} against rank order of survival time.

tions with relatively large positive or large negative delta-beta's. These correspond to patients 12, 25 and 32 in the data given in Table 1.3. Patients 12 and 32 have very short survival times and the two largest values of BUN. Their deletion from the data base decreases the parameter estimate for $LBUN$. Patient number 25 lived for 40 months after trial entry, but has an unusually large value of BUN for someone surviving for such a long period. If the data from this patient are omitted, the coefficient of $LBUN$ in the model is increased.

To identify observations that influence the set of parameter estimates, a plot of the absolute values of the elements of the diagnostic l_{max} against the rank order of the survival times is shown in Figure 5.13. Three observations have relatively large values of $|l_{max}|$, and they again correspond to patients 12, 25 and 32. We may therefore conclude that the influence of these observations can be attributed to their effect on the parameter estimate associated with $LBUN$. Notice that this analysis has shown that the observations that appear unusual in a residual plot are not necessarily influential. In Example 5.4, large score residuals for $LBUN$ were associated with patients 12, 28 and 32, and we now find that only the data for patients 12 and 32 are influential. On the other hand, we also find that patient 25 is now influential. The survival time for this individual was in fact censored and so the corresponding score residual was zero. This observation did not stand out in plots of the martingale or deviance residuals.

The elements of l_{max} are plotted against the two explanatory variables in the

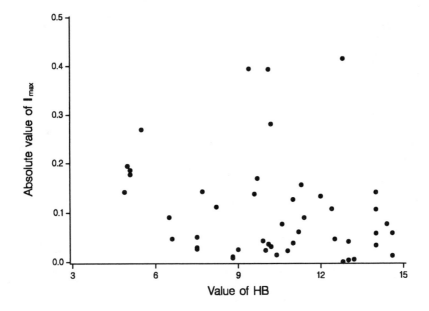

Figure 5.14 Plot of the absolute values of the elements of l_{max} against HB.

model in Figures 5.14 and 5.15. From Figures 5.14 and 5.15, the three patients with influential observations do not have unusual values of HB, but patients 12 and 32 are those with the two largest values of $LBUN$.

5.5 Treatment of influential observations

Once observations have been found to be unduly influential, it is difficult to offer any firm advice on what should be done about them. So much depends on the scientific background to the study.

When possible, the origin of influential observations should be checked. Errors in transcribing and recording categorical and numerical data frequently occur. If any mistakes are found, the data need to be corrected and the analysis repeated. If the observed value of a survival time or other explanatory variables is impossible, and correction is not possible, the corresponding observation should be omitted from the data base before repeating the analysis.

In many situations it will not be possible to confirm that the data corresponding to an influential observation are valid. Certainly, influential observations should not then be rejected outright. In these circumstances, the most appropriate course of action will be to establish the actual effect on the inferences to be drawn from the analysis. For example, if a relative hazard or median survival time is being used in quantifying the size of a treatment effect, the values of these statistics with

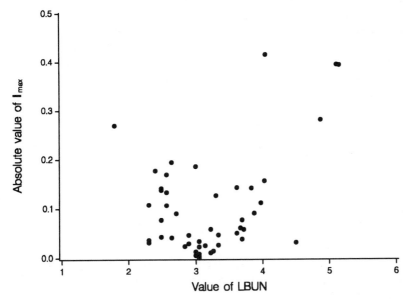

Figure 5.15 Plot of the absolute values of the elements of $\mathbf{l}_{max}$ against $LBUN$.

and without the influential values can be contrasted. If the difference between the results is so small as to not be of practical importance, the queried observations can be retained. On the other hand, if the effect of removing the influential observations is large enough to be of practical importance, analyses based on both the full and reduced data sets will need to be reported. The outcome of consultations with the scientists involved in the study will then be a vital ingredient in the process of deciding on the course of future action.

5.6 * Residuals for the Weibull proportional hazards model

When survival times are assumed to have a Weibull distribution, the parametric form of the proportional hazards model described in Section 4.5 is often appropriate. According to this model, the estimated hazard of death at time t for the i'th individual, $i = 1, 2, \ldots, n$, is

$$\hat{h}_i(t) = \exp(\hat{\beta}'\mathbf{x}_i)\hat{h}_0(t),$$

where $\hat{\beta}'\mathbf{x}_i = \hat{\beta}_1 x_{1i} + \hat{\beta}_2 x_{2i} + \cdots + \hat{\beta}_p x_{pi}$ is the linear predictor, and x_{ji} is the value of the j'th explanatory variable, $j = 1, 2, \ldots, p$, for the i'th individual. The baseline hazard function, $h_0(t)$, is a parametric function whose estimate is

$$\hat{h}_0(t) = \hat{\lambda}\hat{\gamma}t^{\hat{\gamma}-1},$$

where $\hat{\lambda}$ and $\hat{\gamma}$ are respectively the estimated values of the scale and shape parameters of the Weibull distribution.

Residuals that are similar to those used in connection with the Cox regression model can be defined for the Weibull proportional hazards model. Essentially, the only modification that has to be made is that the cumulative hazard function, $H_0(t)$, is now estimated by a parametric function of the survival time, that is,

$$\hat{H}_0(t) = \hat{\lambda} t^{\hat{\gamma}}, \tag{5.14}$$

which depends on two parameters.

Cox–Snell residuals

The Cox–Snell residuals for the Weibull proportional hazards model are as given in equation (5.1), where now $\hat{H}_0(t_i)$ in the expression for r_{Ci} is given in equation (5.14). Thus the Cox–Snell residuals become

$$r_{Ci} = \exp(\hat{\beta}' \mathbf{x}_i) \hat{\lambda} t^{\hat{\gamma}}. \tag{5.15}$$

As before, these residuals can be taken to have a unit exponential distribution when the correct model has been fitted, with censored observations leading to censored residuals.

Cox–Snell residuals for censored observations can be modified by the addition of either 1.0 or 0.693 to give the modified Cox–Snell residuals in equations (5.3) and (5.5). This leads to residuals that can be plotted against observation number, rank order of survival time or explanatory variables in the model.

Martingale residuals

The modified residuals in equation (5.3), or equivalently in equation (5.4), can be adjusted to have zero mean. This leads to the residuals given by

$$r_{Mi} = \delta_i - r_{Ci}, \tag{5.16}$$

where δ_i is the censoring indicator for the i'th observation and r_{Ci} is the Cox–Snell residual given in equation (5.15). These residuals were first given in equation (5.6). Strictly speaking, it is no longer appropriate to refer to these residuals as martingale residuals since the derivation of them based on martingale methods does not carry over to the Weibull model. However, for semantic convenience, we will continue to refer to the quantities in equation (5.16) as martingale residuals.

Deviance residuals

The residuals

$$r_{Di} = \text{sgn}(r_{Mi}) \left[-2 \left\{ r_{Mi} + \delta_i \log(\delta_i - r_{Mi}) \right\} \right]^{1/2},$$

first presented in equation (5.7) can be regarded as an attempt to symmetrize the martingale residuals, making plots based on them easier to interpret. They are not components of the deviance for the fitted Weibull model, but nonetheless it will be convenient to continue to refer to them as deviance residuals.

Score residuals

Score residuals which parallel the score residuals used in connection with the Cox regression model can be defined for the Weibull proportional hazards model. When the model is parametrized as in Section 4.6, the likelihood of a sample of n observations is, from equation (4.44),

$$\prod_{i=1}^{n} \sigma^{-\delta_i} \exp\left[\delta_i \left(\frac{\log t_i - \eta_i}{\sigma}\right) - \exp\left(\frac{\log t_i - \eta_i}{\sigma}\right)\right],$$

where $\eta_i = \mu + \alpha_1 x_{1i} + \alpha_2 x_{2i} + \cdots + \alpha_p x_{pi}$, x_{ji} is the value of the jth covariate, $j = 1, 2, \ldots, p$, for the i'th individual in the study, $i = 1, 2, \ldots, n$, and δ_i is the censoring indicator for the i'th individual. The correspondence between the parameters in this specification of the model, and the β's, λ and γ used elsewhere, was shown in Section 4.6 to be such that

$$\lambda = \exp(-\mu/\sigma), \qquad \gamma = \sigma^{-1}, \qquad \beta_j = -\alpha_j/\sigma,$$

for $j = 1, 2, \ldots, p$.

Differentiating the logarithm of the above likelihood function with respect to the parameters $\mu, \alpha_1, \alpha_2, \ldots, \alpha_p$ and σ gives the following derivatives:

$$\frac{\partial \log L}{\partial \mu} = \sigma^{-1} \sum_{i=1}^{n} \{e^{z_i} - \delta_i\}, \tag{5.17}$$

$$\frac{\partial \log L}{\partial \alpha_j} = \sigma^{-1} \sum_{i=1}^{n} (e^{z_i} - \delta_i) x_{ji}, \tag{5.18}$$

$$\frac{\partial \log L}{\partial \sigma} = \sigma^{-1} \sum_{i=1}^{n} \{z_i e^{z_i} - \delta_i z_i - \delta_i\}, \tag{5.19}$$

where $z_i = (\log t_i - \eta_i)/\sigma$. The i'th component of each derivative, evaluated at the maximum likelihood estimates of the unknown parameters, is then the score residual for the corresponding term.

Now let $\hat{\eta}_i = \hat{\mu} + \hat{\alpha}_1 x_{1i} + \hat{\alpha}_2 x_{2i} + \cdots + \hat{\alpha}_p x_{pi}$, and set $\hat{z}_i = (\log t_i - \hat{\eta}_i)/\hat{\sigma}$. The i'th score residual for the intercept is then

$$\hat{\sigma}^{-1} \left\{e^{\hat{z}_i} - \delta_i\right\},$$

that for the j'th explanatory variable in the model, x_j, is

$$\hat{\sigma}^{-1} \left\{e^{\hat{z}_i} x_{ji} - \delta_i x_{ji}\right\},$$

and that for the scale parameter is

$$\hat{\sigma}^{-1} \left\{\hat{z}_i e^{\hat{z}_i} - \delta_i \hat{z}_i - \delta_i\right\}.$$

These score residuals can be used in a similar manner to the score residuals for the Cox regression model. In particular, index plots or plots of the residuals against the rank order of the survival times, may be useful in assessing model adequacy.

Standardized residuals

There is one further type of residual that is provided by some computer packages that have facilities for fitting the Weibull proportional hazards model. These tend to be known as the *standardized residuals*. Like the score residuals, they can be derived from the representation of the proportional hazards model detailed in Section 4.6. From that section, we have the result that the random variable associated with the survival time of the i'th individual, T_i, can be modelled using

$$\log T_i = \mu + \alpha_1 x_{1i} + \alpha_2 x_{2i} + \cdots + \alpha_p x_{pi} + \sigma \epsilon_i, \qquad (5.20)$$

where ϵ_i has a Gumbel distribution.

Fitting this form of the Weibull model to survival data leads to estimates $\hat{\mu}$, $\hat{\sigma}$, $\hat{\alpha}_1, \hat{\alpha}_2, \ldots, \hat{\alpha}_p$ of the parameters in equation (5.20). The standardized residual is then equal to

$$\{\log t_i - \hat{\mu} - \hat{\alpha}_1 x_{1i} - \hat{\alpha}_2 x_{2i} - \cdots - \hat{\alpha}_p x_{pi}\} / \hat{\sigma}. \qquad (5.21)$$

This has the appearance of a quantity of the form 'observation−fitted value', and would have a Gumbel distribution if the fitted model is correct. This property can be examined using a result given in Section 4.6, according to which the quantity e^ϵ has a unit exponential distribution if ϵ has a Gumbel distribution. This means that the exponential of the standardized residuals should behave as a possibly censored sample from a unit exponential distribution. When we substitute for the parameters μ, σ and $\alpha_1, \alpha_2, \ldots, \alpha_p$, in terms of λ, γ and $\beta_1, \beta_2, \ldots, \beta_p$, the exponentiated standardized residuals,

$$\exp\left\{(\log t_i - \hat{\mu} - \hat{\alpha}_1 x_{1i} - \hat{\alpha}_2 x_{2i} - \cdots - \hat{\alpha}_p x_{pi}) / \hat{\sigma}\right\},$$

reduce to the Cox–Snell residuals in equation (5.15). This means that the standardized residuals have no advantages over the Cox–Snell residuals and so need not be considered further.

5.6.1 Analysis of residuals

The residual plots described in connection with the Cox regression model in Section 5.2 can all be used when the residuals have been obtained from fitting a Weibull model to a set of data. In addition, the comments and recommendations made in Section 5.3 carry over to residuals for the Weibull model with one exception. This is that the caution against comparing the Cox–Snell residuals with the unit exponential distribution does not have as much force for Weibull residuals. The reason for this is that the non-parametric estimate of the baseline cumulative hazard function used in the Cox regression model is replaced by an estimate of a

parametric function in the Weibull model. This function depends on just two parameters, λ and γ, and so fewer parameters are being estimated when a Weibull
model is fitted to survival data. The Cox–Snell residuals for a Weibull model are
therefore more likely to be approximated by a unit exponential distribution, when
the correct model has been fitted. Furthermore, the log-cumulative hazard plot of
these residuals will not necessarily be a straight line when a model with no explanatory variables is fitted.

In summary, a log-cumulative hazard plot of the Cox–Snell residuals, index
plots of martingale or deviance residuals, and plots of martingale or deviance
residuals against the rank order of the survival times or explanatory variables can
all be useful in identifying observations that are not well fitted by the model. Plots
based on deviance residuals are commended for general use, but plots of score
residuals against observation number or rank order of survival time might be examined in a more comprehensive assessment of model adequacy.

Example 5.8 Chemotherapy in ovarian cancer patients
In Example 4.9 of Chapter 4, data on the survival times of patients with ovarian
cancer were presented. The data were analysed using a Weibull proportional hazards model, and the model chosen contained variables corresponding to the age
of the woman, AGE, and the treatment group to which the woman was assigned,
$TREAT$.

A log-cumulative hazard plot of the Cox–Snell residuals is given in Figure 5.16.
In this plot, the plotted points lie on a line that has an intercept and slope close
to zero and unity, respectively. However, there is some evidence of a systematic
deviation from the straight line, giving some cause for concern about the adequacy
of the fitted model.

Plots of the martingale and deviance residuals against the rank order of the survival times are shown in Figures 5.17 and 5.18 respectively. Both of these plots
show a slight tendency for observations with longer survival times, which are actually censored, to have negative residuals.

The graphs in Figure 5.19 show the score residuals for the two variables in the
model, AGE and $TREAT$, plotted against the rank order of the survival times.
The plot of the score residuals for AGE shows that there are three observations
with relatively large residuals. These correspond to patients 14, 4 and 26 in the
original data set given in Table 4.6. However, there does not appear to be anything
unusual about these observations. The score residual for $TREAT$ for patient 26
is also somewhat larger than the others. This points to the fact that the model is
not a good fit to the data from patients 14, 4 and 26.

5.7 * Identification of influential observations

As when fitting the Cox regression model, it will be important to identify observations that exert an undue influence on particular parameter estimates, or on the
complete set of parameter estimates. These two aspects of influence are considered

Figure 5.16   Log-cumulative hazard plot of the Cox–Snell residuals.

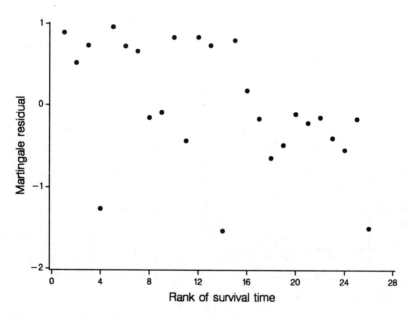

Figure 5.17 Plot of the martingale residuals against rank order of survival time.

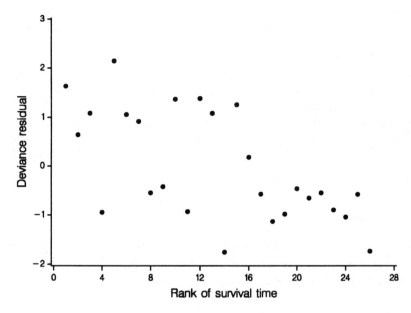

Figure 5.18 Plot of the deviance residuals against rank order of survival time.

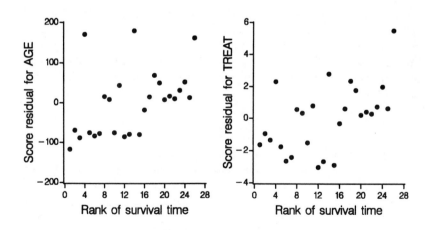

Figure 5.19 Score residuals plotted against rank order of survival time for (a) *AGE* and (b) *TREAT*.

in turn in this section.

In a Technical Report of the Rand Corporation, Hall, Rogers and Pregibon (1982) propose and illustrate a number of influence diagnostics for the Weibull proportional hazards model. These are derived from the parametrization of the model in terms of μ, σ and α described in Section 4.6. Naturally, an observation that is influential in this setting will also be influential under the model that is parametrized in terms of λ, γ and β. These diagnostics are computed from the estimates of all $p + 2$ parameters in the model, and their variance–covariance matrix. For convenience, the vector of $p + 2$ parameters will be denoted by θ, so that $\theta' = (\mu, \alpha_1, \alpha_2, \ldots, \alpha_p, \sigma)$. The vector $\hat{\theta}'$ will be used to denote the corresponding vector of estimates of the parameters.

5.7.1 Influence of observations on a parameter estimate

An approximation to the change in the estimated value of θ_j, the j'th component of the vector θ, on omitting the i'th observation, $\Delta_i \hat{\theta}_j$, has been shown by Hall *et al.* to be the j'th component of the $(p + 2) \times 1$ vector

$$\mathbf{V}(\hat{\theta})\mathbf{s}_i. \tag{5.22}$$

In expression (5.22), $\mathbf{V}(\hat{\theta})$ is the estimated variance–covariance matrix of the parameters in θ, and $\mathbf{s}_i$ is the $(p+2) \times 1$ vector of values of the first partial derivatives of the log-likelihood for the i'th observation, with respect to the $p + 2$ parameters in θ, evaluated at $\hat{\theta}$. The vector $\mathbf{s}_i$ is known as the *score vector* for the i'th observation.

The quantities $\Delta_i \hat{\alpha}_j$ are components 2 to $p-1$ of the vector in expression (5.22), which we will continue to refer to as delta-beta's rather than as delta-alpha's. These values may be standardized by dividing them by the standard error of $\hat{\alpha}_j$, leading to standardized delta-beta's. Index plots or plots of the standardized or unstandardized values of $\Delta_i \hat{\alpha}_j$ provide informative summaries of this aspect of influence.

5.7.2 Influence of observations on the set of parameter estimates

Two summary measures of the influence of the i'th observation on the set of parameters that make up the vector θ have been proposed by Hall, Rogers and Pregibon (1982). These are the statistics F_i and C_i. The quantity F_i is given by

$$F_i = \frac{\mathbf{s}_i' \mathbf{R}^{-1} \mathbf{s}_i}{(p + 2)\{1 - \mathbf{s}_i' \mathbf{R}^{-1} \mathbf{s}_i\}}, \tag{5.23}$$

where the $(p + 2) \times (p + 2)$ matrix $\mathbf{R}$ is the *cross-product matrix of scores*, that is, $\mathbf{R} = \sum_{i=1}^{n} \mathbf{s}_i \mathbf{s}_i'$. Equivalently, $\mathbf{R} = \mathbf{S}'\mathbf{S}$, where $\mathbf{S}$ is the $n \times (p + 2)$ matrix whose i'th row is the transpose of the score vector, $\mathbf{s}_i'$. An alternative measure of

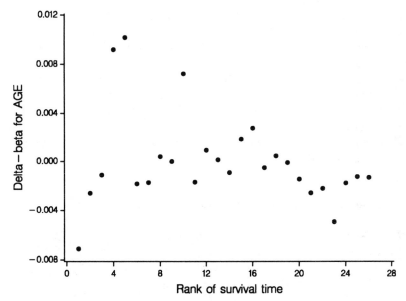

Figure 5.20 Plot of the delta-beta's for AGE against rank order of survival time.

the influence of the i'th observation on the set of parameter estimates is the statistic

$$C_i = \frac{\mathbf{s}_i'\mathbf{V}(\hat{\theta})\mathbf{s}_i}{\{1 - \mathbf{s}_i'\mathbf{V}(\hat{\theta})\mathbf{s}_i\}^2}. \tag{5.24}$$

The statistics F_i and C_i will typically have values that are quite different from each other. However, in each case a relatively large value of the statistic will indicate that the corresponding observation is influential. Exactly how such observations influence the estimates would need to be investigated by omitting that observation from the data set and re-fitting the model.

A SAS macro that can be used to calculate the values of the influence diagnostics presented in this section is described in Chapter 11.

Example 5.9 Chemotherapy in ovarian cancer patients
We now go on to investigate whether there are any influential observations in the data on the survival times following chemotherapy treatment for ovarian cancer.

The unstandardized delta-beta's for AGE and $TREAT$, plotted against the rank order of the survival times, are shown in Figures 5.20 and 5.21. In Figure 5.20, two observations have relatively large values of the delta-beta for AGE. These occur for patients 4 and 5 in the original data set. Both women have short survival times, and in addition one is relatively old at 74 years and the other relatively young at 43 years. The delta-beta's for $TREAT$ displayed in Figure 5.21

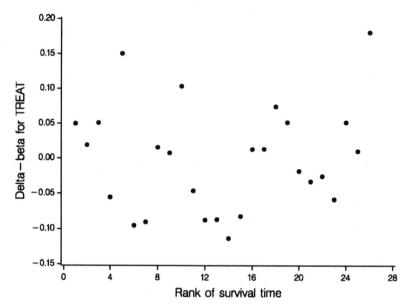

Figure 5.21 Plot of the delta-beta's for $TREAT$ against rank order of survival time.

show no unusual features.

We next investigate the influence of each observation on the set of parameter estimates. Plots of the values of the F-statistic and the C-statistic, defined in equations (5.23) and (5.24), against the rank order of the survival times, are shown in Figures 5.22 and 5.23. Figure 5.22 clearly shows that the observation corresponding to patient 5 is influential, and that the influence of patients 1, 4, 14 and 26 should be investigated in greater detail. Figure 5.23 strongly suggests that the data from patients 5 and 26 is influential.

The linear component of the fitted hazard function for the model fitted to all 26 patients is

$$0.144\, AGE_i - 1.023\, TREAT_i,$$

while that on omitting each of observations 1, 4, 5, 14 and 26 in turn is shown below:

Omitting patient number 1: $0.142\, AGE_i - 1.016\, TREAT_i,$

Omitting patient number 4: $0.175\, AGE_i - 1.190\, TREAT_i,$

Omitting patient number 5: $0.177\, AGE_i - 0.710\, TREAT_i,$

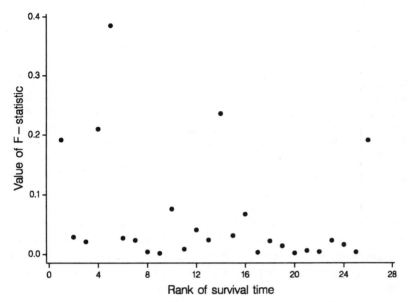

Figure 5.22   Plot of the F-statistic against rank order of survival time.

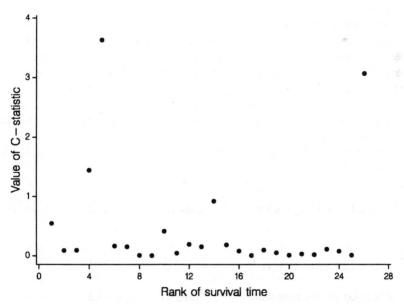

Figure 5.23   Plot of the C-statistic against rank order of survival time.

Omitting patient number 14: $0.149\,AGE_i - 1.318\,TREAT_i$,

Omitting patient number 26: $0.159\,AGE_i - 0.697\,TREAT_i$.

These results show that the effect of omitting the data from patient 1 on the parameter estimates is small. When the data from patient 4 are omitted, the estimated coefficient of AGE is most affected, whereas when the data from patient 14 are omitted, the coefficient of $TREAT$ is changed the most. On leaving out the data from patients 5 and 26, both estimates are considerably affected.

The estimated hazard ratio for a patient on the combined treatment ($TREAT = 2$) relative to one on the single treatment ($TREAT = 1$) is $e^{-1.023} = 0.36$ when the model is fitted to all 26 patients. When the observations from patients 1, 4, 5, 14 and 26 are omitted in turn, the estimated age-adjusted hazard ratios are 0.36, 0.30, 0.49, 0.27 and 0.50, respectively. The data from patients 5 and 26 clearly has the most effect on the estimated hazard ratio; in each case the estimate is increased, and the magnitude of the treatment effect is diminished. Omission of the data from patient 14 decreases the estimated hazard ratio, making the treatment difference greater.

5.8 Testing the assumption of proportional hazards

So far in this chapter we have concentrated on how the adequacy of the linear component of a survival model can be examined. A crucial assumption made when using either the Cox model or the Weibull model is that of proportional hazards. If hazards are not proportional, this means that the linear component of the model varies with time. We must therefore consider how this assumption can be examined critically.

One method of doing this, prior to fitting a model, has already been described in Sections 4.4.1 and 4.5.1 of Chapter 4. Suppose that the survival data are grouped according to the levels of one or more factors, and the Kaplan–Meier estimate of the survivor function of the data in each group is found. Then, a log-cumulative hazard plot, that is, a plot of the negative logarithm of the estimated survivor function against the logarithm of the survival time, will yield parallel curves if the hazards are proportional across the different groups. This method is informative and simple to operate when there are a very small number of groups. But in more highly structured data sets, a different approach needs to be taken.

The time dependency of the linear component of a fitted proportional hazards model is usually a consequence of there being a subset of the explanatory variables in the model which vary with time. In such cases, we require a method that can be used to detect any time dependency in particular covariates, after allowing for the effects of explanatory variables that are known, or expected to be, independent of time. This suggests using a modelling approach to examine the validity of the proportional hazards assumption.

5.8.1 Testing proportional hazards in the Cox model

To examine the assumption of proportional hazards in the Cox regression model, a *time-dependent variable* can be added to the model. The procedure will be described in detail for a particular situation.

Consider a survival study in which each patient has been allocated to one of two groups, corresponding to a standard treatment and a new treatment. Interest may then centre on whether the ratio of the hazard of death at time t in one treatment group relative to the other is independent of survival time. A proportional hazards model for the hazard function of the i'th individual in the study is then

$$h_i(t) = \exp(\beta_1 x_{1i}) h_0(t), \tag{5.25}$$

where x_{1i} is the value of an indicator variable X_1 that is zero for the standard treatment and unity for the new treatment. The relative hazard of death at any time for a patient on the new treatment relative to one on the standard is then e^{β_1}, which is independent of the survival time.

Now define a new explanatory variable X_2, where $X_2 = X_1 t$. If this variable is added to the model in equation (5.25), the hazard of death at time t for the i'th individual becomes

$$h_i(t) = \exp(\beta_1 x_{1i} + \beta_2 x_{2i}) h_0(t). \tag{5.26}$$

The relative hazard at time t is now

$$\exp(\beta_1 + \beta_2 t),$$

since $X_2 = t$ under the new treatment, and zero otherwise. This hazard function depends on t and the model in equation (5.26) is no longer a proportional hazards model. In particular, if $\beta_2 < 0$, the relative hazard decreases with time. This means that the hazard of death on the new treatment, relative to that on the standard, decreases with time. If $\beta_1 < 0$, the interpretation of this would be that the superiority of the new treatment becomes more apparent as time goes on. On the other hand, if $\beta_2 > 0$, the relative hazard of death on the new treatment increases with time, reflecting an increasing risk of death on the new treatment relative to the standard. In the particular case where $\beta_2 = 0$, the relative hazard is constant at e^{β_1}. This means that a test of the hypothesis that $\beta_2 = 0$ is a test of the assumption of proportional hazards. The situation is illustrated in Figure 5.24.

In order to aid in both the computation and interpretation of the parameters in the model of equation (5.26), the variable X_2 can be defined in terms of the deviation from some time, t_0. The estimated values of β_1 and β_2 will then tend to be less highly correlated, and maximization of the appropriate likelihood function will be less difficult. If now X_2 is taken to be such that $X_2 = X_1(t - t_0)$, the value of X_2 is $t - t_0$ for the new treatment and zero for the standard. The relative hazard now becomes

$$\exp\{\beta_1 + \beta_2(t - t_0)\}.$$

In the model of equation (5.26), the quantity e^{β_1} is the hazard of death at time t_0

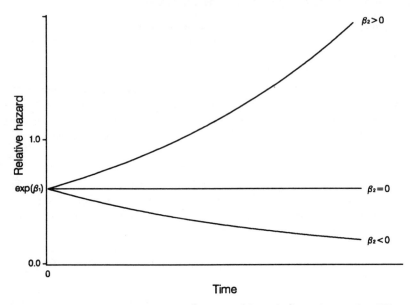

Figure 5.24 Plot of the relative hazard, $\exp\{\beta_1 + \beta_2 t\}$, against t, for different values of β_2.

for an individual on the new treatment relative to one on the standard. In practical applications, t_0 will generally be chosen to provide a convenient interpretation for the time at which this relative hazard is applicable. For example, taking t_0 to be the mean or median survival time means that $\exp(\hat{\beta}_1)$ is the estimated relative hazard of death at this time.

Models that include the time-dependent variable X_2 cannot be fitted by treating X_2 in the same manner as other explanatory variables in the model. The reason for this is that this variable will have different values at different death times, complicating the calculation of the denominator of the partial likelihood function in equation (3.4). Further details on the fitting process will be deferred to Chapter 7.

In many applications, the numerical values of the survival times will be large, and there will be computational problems associated with calculating the value of $\exp(\beta_2 x_{2i})$ in equation (5.26). In such situations, $\log t$ can be used in place of t in the definition of the time-dependent variable X_2. In this version of the model, a test of the hypothesis that $\beta_2 = 0$ is still a test of the assumption of proportional hazards.

On some occasions, the proportionality of hazards with respect to a continuous variable may be in doubt. For example, the relative hazard may depend on how the value of a prognostic variable, such as the weight of an individual, changes with increasing survival time, rather than the weight of the patient at the start of the study. Full details on how time-dependent variables such as this can be incorporated in a Cox regression model, and how such a model can be fitted, are given in Chapter 7.

5.8.2 Testing proportional hazards in the Weibull model

In the Weibull model, the assumption of proportional hazards across a number of groups, g, say, corresponds to the assumption that the shape parameter γ in the baseline hazard function is the same in each group. One way of testing this assumption is to fit a separate Weibull model to each of the g groups, where the linear component of the model is the same in each case. The models fitted to the data from each group will then have different shape parameters as well as different scale parameters. The values of the statistic $-2 \log \hat{L}$ for each of these g separate models are then summed to give a value of $-2 \log \hat{L}$ for a model that has different shape parameter for each group. Denote this by $-2 \log \hat{L}_1$. We then combine the g sets of data and fit a Weibull proportional hazards model that includes parameters associated with the group effect. This model then corresponds to there being a common shape parameter for each group. The inclusion of group effects in the model leads to there being different scale parameters for each group. The value of $-2 \log \hat{L}$ for this model, $-2 \log \hat{L}_0$, say, is then compared with $-2 \log \hat{L}_1$. The difference between the values of these two statistics is the change in $-2 \log \hat{L}$ due to constraining the Weibull shape parameters to be equal, and can be compared with a chi-squared distribution on $g - 1$ degrees of freedom. If the difference is not significant, the assumption of proportional hazards is justified.

Some alternatives to the proportional hazards model are described in Chapter 6 and further comments on how to deal with situations in which the hazards are not proportional are given in Section 10.1 of Chapter 10. An example on the use of this method of testing the proportional hazards assumption is given below.

Example 5.10 Chemotherapy in ovarian cancer patients

We now consider the study of survival following chemotherapy treatment for ovarian cancer, and examine the assumption of proportional hazards in a Cox regression model. To do this, we determine whether the treatment effect is independent of survival time. If so, the relative hazard of death for a woman in one treatment group relative to another is constant over time.

Let the indicator variable C take the value zero if a woman is on the single chemotherapy treatment and unity if a woman is on the combined treatment. Then, $C = TREAT - 1$, where $TREAT$ is the variable associated with treatment group, used in previous analyses of this data set. Also define the variable $CTIME$ to be such that $CTIME = C \times (t - 470)$, where t is the survival time of a patient and the value of 470 days is close to the median survival time of the group of 26 women in the study.

On fitting a Cox regression model that contains the variables AGE, C and the time-dependent variable $CTIME$, the value of the statistic $-2 \log \hat{L}$ is 53.639. Under a proportional hazards model that contains AGE and C alone, the value

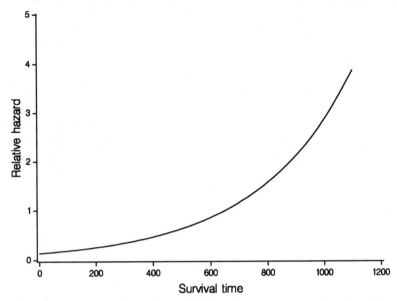

Figure 5.25   The relative hazard of death at time t, as a function of t.

of $-2 \log \hat{L}$ is 54.148, and so the difference in the values of the $-2 \log \hat{L}$ statistic is 0.51 on 1 d.f. This is not significant, and suggests that the hazards for the two treatment groups can be regarded as proportional.

To illustrate how the model that includes the time-dependent variable $CTIME$ can be interpreted, consider the estimated hazard function for the i'th woman, given by

$$\hat{h}_i(t) = \exp\{0.136\,AGE_i - 0.532\,C_i + 0.003\,CTIME_i\}\hat{h}_0(t),$$

where $\hat{h}_0(t)$ is the estimated baseline hazard function. The hazard of death at time t for a woman on the combined treatment, relative to one on the single treatment, is therefore

$$\exp\{-0.532 + 0.003(t - 470)\}.$$

The relative hazard at a time of 470 days is thus $e^{-0.532} = 0.59$. A plot of the relative hazard against time is shown in Figure 5.25, which shows that as the survival time increases, the relative hazard increases. For survival times greater than 626 days, the hazard of the combined treatment relative to the single is greater than unity. The combined chemotherapy treatment is therefore superior in the short term, but for patients surviving beyond 626 days, there is a greater hazard of death than if they had received the single treatment. However, in this example, the relative hazard is not significantly different from a constant value.

Finally, the data of this example are used to illustrate the procedure for testing the assumption that the Weibull shape parameter is the same for the patients in each of the two treatment groups. The first step is to fit a Weibull proportional hazards model that contains AGE alone to the data from the women in each treatment group. When such a model is fitted to the data from those on the single chemotherapy treatment, the value of the statistic $-2 \log \hat{L}$ is 22.851, while that for the women on the combined treatment is 16.757. The sum of the two values of $-2 \log \hat{L}$ is 39.608, which is the value of the statistic for a Weibull model with different shape parameters for the two treatment groups. The value of $-2 \log \hat{L}$ for the model that constrains the shape parameters to be equal is 41.126. The change in $-2 \log \hat{L}$ on constraining the shape parameters to be equal is therefore 1.52, which is not significant when compared with a chi-squared distribution on one degree of freedom. The two shape parameters may therefore be taken to be equal.

5.9 Further reading

General introductions to the ideas of model checking in linear models are included in Draper and Smith (1981) and Montgomery and Peck (1982). Cook and Weisberg (1982) give a more detailed account of the theory underlying residuals and influence diagnostics in a number of situations. Atkinson (1985) describes model checking in linear models from a practical viewpoint, and McCullagh and Nelder (1989) and Aitkin et al. (1989) discuss this topic in the context of generalized linear models.

Many textbooks devoted to the analysis of survival data, and particularly those of Lawless (1982), Kalbfleisch and Prentice (1980) and Cox and Oakes (1984), include sections on the use of residuals. Hinkley, Reid and Snell (1991) and Hastie and Tibshirani (1990) also include brief discussions on methods for assessing the adequacy of models fitted to survival data.

Early articles on the use of residuals in checking the adequacy of survival models include Kay (1977) and Crowley and Hu (1977). These papers include a discussion on the Cox–Snell residuals, which are based on the general definition of residuals given by Cox and Snell (1968). See also Crowley and Storer (1983) for a practical application of the methods. Reviews of diagnostic procedures in survival analysis were given in the mid-1980s by Kay (1984) and Day (1985).

Martingale residuals were proposed by Barlow and Prentice (1988). Essentially the same residuals were proposed by Lagakos (1980) and their use is discussed by Therneau, Grambsch and Fleming (1990) and Henderson and Milner (1991). Deviance residuals were also introduced in Therneau, Grambsch and Fleming (1990). In a recent account of survival analysis based on the theory of counting processes, Fleming and Harrington (1991) show how martingale residuals can be used and gives detailed practical examples. The score residuals for the Cox model were proposed by Schoenfeld (1982).

Influence diagnostics for the Cox regression model have been considered by many authors, but the major papers are those of Cain and Lange (1984), Reid and Crepeau (1985), Storer and Crowley (1985), Pettitt and Bin Daud (1989) and

Weissfeld (1990). Pettitt and Bin Daud (1990) show how time-dependence in the Cox proportional hazards model can be detected by smoothing the score residuals.

Residuals and influence measures for the Weibull proportional hazards model are described by Hall, Rogers and Pregibon (1982). More recent discussions of methods are given by Weissfeld and Schneider (1990) and Escobar and Meeker (1992). A SAS macro for evaluating influence diagnostics for the Weibull proportional hazards model is described by Escobar and Meeker (1988). These papers involve arguments based on local influence, a topic which is explored in general terms by Cook (1986).

Some other graphical methods for evaluating survival models not mentioned in this chapter have been proposed by Cox (1979) and Arjas (1988). Gray (1990) describes the use of smoothed estimates of cumulative hazard functions in evaluating the fit of a Cox model.

Most of the diagnostic procedures presented in this chapter rely on an informal evaluation of tabular or graphical presentations of particular statistics. In addition to these procedures, a variety of significance tests have recently been proposed that can be used to asses the goodness of fit of the model. Examples include the methods of Schoenfeld (1980), Andersen (1982), Nagelkerke, Oosting and Hart (1984), Ciampi and Etezadi-Amoli (1985), Moreau, O'Quigley and Mesbah (1985), Gill and Schumacher (1987) and O'Quigley and Pessione (1989). A review of goodness of fit tests for the Cox regression model is included in Lin and Wei (1991). These procedures are not yet generally available in computer software for survival analysis and further work is needed to evaluate their performance in the face of particular types of departure from the fitted model. For these reasons, fuller details on these methods have not been given here.

- -

6

Some other parametric models for survival data

Although the proportional hazards model finds widespread applicability in the analysis of survival data, there is another general model that is of some importance. This is the *accelerated failure time model*. In circumstances where the proportional hazards assumption is not tenable, models based on this general class may prove to be fruitful. Again, the Weibull distribution turns out to be a natural choice for the distribution of survival times in the accelerated failure time model, but there are other probability distributions which might be adopted. This chapter therefore begins with a brief survey of alternative distributions for survival data. The accelerated failure time model is then considered in detail in Sections 6.3–6.5.

One other general family of survival models, known as the *proportional odds model* may be useful in some circumstances. However, software for fitting this model is not widely available, and so this model is only briefly described in Section 6.6.

6.1 Probability distributions for survival data

The Weibull distribution will not necessarily provide a satisfactory model for survival times in all circumstances, and so alternatives to this distribution need to be considered. Although any continuous distribution for non-negative random variables might be used, the properties of the *log-logistic distribution* make it a particularly attractive alternative to the Weibull distribution.

6.1.1 The log-logistic distribution

One limitation of the Weibull hazard function is that it is a monotonic function of time. However, situations in which the hazard function changes direction can arise. For example, following a heart transplantation, a patient faces an increasing hazard of death over the first ten days or so after the transplant, while the body adapts to the new organ. The hazard then decreases with time as the patient recovers. In situations such as this, a unimodal hazard function may be appropriate.

A particular form of unimodal hazard is the function

$$h(t) = \frac{e^\theta \kappa t^{\kappa-1}}{1 + e^\theta t^\kappa}. \tag{6.1}$$

This hazard function decreases monotonically if $\kappa \le 1$, but if $\kappa > 1$, the hazard has a single mode. The survivor function corresponding to the hazard function in equation (6.1) is given by

$$S(t) = [1 + e^\theta t^\kappa]^{-1}, \tag{6.2}$$

and the probability density function is

$$f(t) = \frac{e^\theta \kappa t^{\kappa-1}}{(1 + e^\theta t^\kappa)^2}.$$

This is the density of a random variable T that has a *log-logistic distribution*, with parameters θ, κ. The distribution is so called because the variable $\log T$ has a *logistic distribution*, a symmetric distribution whose probability density function is very similar to that of the normal distribution.

The p'th percentile of the log-logistic distribution is

$$t(p) = \left(\frac{100 - p}{p e^\theta}\right)^{1/\kappa},$$

and so the median of the distribution is

$$t(50) = e^{-\theta/\kappa}. \tag{6.3}$$

The hazard functions for log-logistic distributions with a median of 20 and $\kappa = 0.5$, 2.0 and 5.0 are shown in Figure 6.1. The corresponding values of θ for these distributions are -1.5, -6.0 and -15.0, respectively.

6.1.2 * Some other distributions for survival data

Other distributions for survival data that might be considered are the gamma and lognormal distributions, both of which are defined for random variables that take positive values. However these two distributions suffer from the considerable disadvantage that their survivor and hazard functions can only be expressed in terms of integrals. In particular, the hazard function for the gamma distribution with mean ρ/λ and variance ρ/λ^2 is

$$h(t) = \frac{\lambda^\rho t^{\rho-1} e^{-\lambda t}}{\Gamma(\rho)\{1 - \Gamma_{\lambda t}(\rho)\}},$$

where $\Gamma(\rho)$ is a gamma function and $\Gamma_{\lambda t}(\rho)$ is the incomplete gamma function given by

$$\Gamma_{\lambda t}(\rho) = \frac{1}{\Gamma(\rho)} \int_0^{\lambda t} u^{\rho-1} e^{-u} \, du.$$

This integral has to be evaluated numerically.

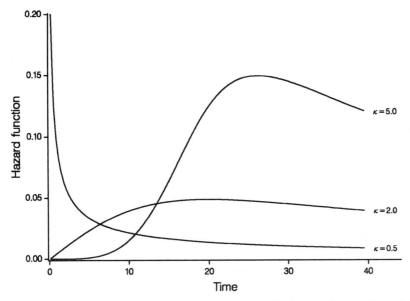

Figure 6.1 Hazard functions for a log-logistic distribution with a median of 20 and $\kappa = 0.5$, 2.0 and 5.0.

A random variable T is said to have a lognormal distribution with parameters μ and σ if $\log T$ has a normal distribution with mean μ and variance σ^2. The probability density function of T is given by

$$f(t) = \frac{1}{\sigma\sqrt{(2\pi)}}t^{-1}\exp\left\{-(\log t - \mu)^2/2\sigma^2\right\},$$

for $0 \leq t < \infty$, from which the survivor and hazard functions can be derived. Again, these functions can only be expressed in terms of integrals.

The hazard functions for the gamma and lognormal distributions are clearly unsuitable for general use. Moreover, in view of the similarity of the normal and logistic distributions, the lognormal model will tend to be very similar to the log-logistic model. Also, the Weibull and gamma distributions will generally lead to very similar results.

When the hazard of death is expected to increase or decrease with time in the short term, and to then become constant, a hazard function that follows a *general exponential curve* or *Mitscherlich curve* may be appropriate. We would then take the hazard function to be

$$h(t) = \theta - \beta e^{-\gamma t},$$

where $\theta > 0, \beta > 0$ and $\gamma > 0$. The general shape of this function is depicted in Figure 6.2. This function has a value of $\theta - \beta$ when $t = 0$ and increases to a

Figure 6.2 An asymptotic hazard function, where $h(t) = \theta - \beta e^{-\gamma t}$.

horizontal asymptote at a hazard of θ. Similarly the function

$$h(t) = \theta + \beta e^{-\gamma t},$$

where $\theta > 0, \beta > 0$ and $\gamma > 0$, could be used to model a hazard which decreases from $\theta + \beta$ to a horizontal asymptote at θ.

Using equation (1.5), the corresponding survivor function can be found, from which the probability density function can be obtained. In fact, the probability distribution corresponding to this specification of the hazard function is known as the *Gomperz–Makeham distribution*.

To model a hazard function which decreases and then increases symmetrically about the minimum value, a quadratic hazard function might be suitable. Thus if

$$h(t) = \theta + \beta t + \gamma t^2,$$

for values of θ, β and γ which give the required shape of hazard and ensure that $h(t) \geq 0$, explicit forms for the survivor function and probability density function can be obtained.

Another form of hazard function which decreases to a single minimum and increases thereafter is the 'bathtub' hazard. The model with

$$h(t) = \alpha t + \frac{\beta}{1 + \gamma t}$$

provides a straightforward representation of this form of hazard.

6.2 Exploratory analyses

When the number of observations in a sample is reasonably large, an empirical estimate of the hazard function could be obtained using the method described in Section 2.2.1. A plot of the estimated hazard function may then suggest a suitable parametric form for the hazard function. For example, if the hazard plot is found to be unimodal, a log-logistic distribution could be used for the survival times.

The adequacy of the log-logistic model in describing a single sample of survival times can be explored using a procedure similar to that described in Section 4.2. From equation (6.2), the survivor function for this distribution is

$$S(t) = [1 + e^{\theta} t^{\kappa}]^{-1}.$$

Consequently, the odds of surviving beyond time t are

$$\frac{S(t)}{1 - S(t)} = e^{-\theta} t^{-\kappa},$$

and so the log-odds of survival beyond t can be expressed as

$$\log \left\{ \frac{S(t)}{1 - S(t)} \right\} = -\theta - \kappa \log t.$$

If the survivor function is estimated using the Kaplan–Meier estimate and the estimated log-odds of survival beyond t are plotted against $\log t$, a straight line plot will be obtained if a log-logistic model for the survival times is suitable. Estimates of the parameters of the log-logistic distribution, θ and κ, can be obtained from the intercept and slope of the straight line plot.

The suitability of other parametric models can be investigated along similar lines.

Example 6.1 Time to discontinuation of the use of an IUD
A plot of $\log\{\hat{S}(t)/[1 - \hat{S}(t)]\}$ against $\log t$ for the data on the times to discontinuation of an IUD is shown in Figure 6.3. From this plot, it appears that the relationship between the estimated log-odds of discontinuing use of the contraceptive after time t and $\log t$ is reasonably straight, suggesting that a log-logistic model could be used to model the observed data.

Notice that there is very little difference in the extent of departures from linearity in the plots in Figures 4.6 and 6.3. This means that either the Weibull distribution or the log-logistic distribution is likely to be satisfactory, even though the estimated hazard function under these two distributions may be quite different. Indeed, when survival data are obtained for a relatively small number of individuals, as in this example, there will often be little to choose between alternative distributional models for the data. The model that is the most convenient for the purpose in hand will then be adopted.

Figure 6.3 A plot of the estimated log-odds of discontinuation after t against $\log t$ for the data from Example 1.1.

6.3 The accelerated failure time model

The accelerated failure time model is a general model for survival data, in which explanatory variables measured on an individual are assumed to act multiplicatively on the time-scale, and so affect the rate at which an individual proceeds along the time axis. This means that the models can be interpreted in terms of the speed of progression of a disease, an interpretation that has immediate intuitive appeal.

Before the general form of the model is presented, the model for comparing the survival times of two groups of patients is described. Suppose that patients are randomized to receive one of two treatments, a standard treatment S or a new treatment N. Under an accelerated failure time model, the survival time of an individual on the new treatment is taken to be a multiple of the survival time for an individual on the standard treatment. Thus the effect of the new treatment is to 'speed up' or 'slow down' the passage of time. Under this assumption, the probability that an individual on the new treatment survives beyond time t is the probability that an individual on the standard treatment survives beyond time ϕt, where ϕ is an unknown positive constant.

Now let $S_S(t)$ and $S_N(t)$ be the survivor functions for individuals in the two treatment groups. Then, the accelerated failure time model specifies that

$$S_N(t) = S_S(\phi t),$$

for any value of the survival time t. One interpretation of this model is that the lifetime of an individual on the new treatment is ϕ times the lifetime that the individual would have experienced under the standard treatment. When the end-point of concern is the death of a patient, values of ϕ less than unity correspond to an acceleration in the time to death of an individual assigned to the new treatment. The standard treatment would then be the more suitable in terms of promoting longevity. On the other hand, if the end-point is the recovery from some disease state, values of ϕ greater than unity occur when the effect of the new treatment is to speed up the recovery time. In these circumstances the new treatment would be superior to the standard. The parameter ϕ is therefore termed the *acceleration factor*.

The acceleration factor can also be interpreted in terms of the median survival times of patients on the new and standard treatments, $t_N(50)$ and $t_S(50)$, say. These values are such that $S_N\{t_N(50)\} = S_S\{t_S(50)\} = 0.5$. Now, under the accelerated failure time model, $S_N\{t_N(50)\} = S_S\{\phi t_N(50)\}$ and so it follows that $t_S(50) = \phi t_N(50)$. In other words, under the accelerated failure time model, the median survival time of a patient on the new treatment is ϕ times that of a patient on the standard treatment. This interpretation of the acceleration factor is particularly appealing to clinicians.

From the relationship between the survivor function, probability density function and hazard function given in equation (1.3), the relationship between the density and hazard functions for individuals in the two treatment groups is

$$f_N(t) = \phi f_S(\phi t),$$

and

$$h_N(t) = \phi h_S(\phi t).$$

Now let X be an indicator variable which takes the value zero for an individual in the group receiving the standard treatment, and unity for one who receives the new treatment. The hazard function for the i'th individual can then be expressed as

$$h_i(t) = \phi^{x_i} h_0(\phi^{x_i} t), \tag{6.4}$$

where x_i is the value of X for the i'th individual in the study. Putting $x_i = 0$ in this expression shows that the function $h_0(t)$ is the hazard function for an individual on the standard treatment. This is again referred to as the baseline hazard function. The hazard function for an individual on the new treatment is then $\phi h_0(\phi t)$.

The parameter ϕ must be non-negative and so it is convenient to set $\phi = e^\beta$. The accelerated failure time model in equation (6.4) then becomes

$$h_i(t) = e^{\beta x_i} h_0(e^{\beta x_i} t). \tag{6.5}$$

so that the hazard function for an individual on the new treatment is $e^\beta h_0(e^\beta t)$.

6.3.1 Comparison with the proportional hazards model

To illustrate the difference between a proportional hazards model and the accelerated failure time model, again suppose that the survival times of individuals in two groups, Group I and Group II, say, are to be modelled. Further suppose that for the individuals in Group I, the hazard function is given by

$$h_0(t) = \begin{cases} 0.5 & \text{if } t \leq 1, \\ 1.0 & \text{if } t > 1, \end{cases}$$

where the time-scale is measured in months. This type of hazard function arises from a *piecewise exponential model*, since a constant hazard in each time interval implies exponentially distributed survival times, with different means, in each interval. This model provides a simple way of representing a variable hazard function, and may be appropriate in situations where there is a constant short-term risk which increases abruptly after a threshold time.

Now let $h_P(t)$ and $h_A(t)$ denote the hazard function for individuals in Group II under a proportional hazards model and an accelerated failure time model respectively. Consequently, we may write

$$h_P(t) = \psi h_0(t),$$

and

$$h_A(t) = \phi h_0(\phi t),$$

for the two hazard functions. Using the result $S(t) = \exp\{-\int_0^t h(u)\,du\}$, the baseline survivor function is

$$S_0(t) = \begin{cases} e^{-0.5t} & \text{if } t \leq 1, \\ e^{-0.5-(t-1)} & \text{if } t > 1. \end{cases}$$

Since $S_0(t) > 0.61$ if $t < 1$, the median occurs in the second part of the survivor function and is when $\exp\{-0.5 - (t-1)\} = 0.5$. The median survival time for those in Group I is therefore 1.39 months.

The survivor functions for the individuals in Group II under the two models are

$$S_P(t) = [S_0(t)]^{\psi},$$

and

$$S_A(t) = S_0(\phi t),$$

respectively.

To illustrate the difference between the hazard functions under proportional hazards and accelerated failure time models, consider the particular case where $\psi = \phi = 2.0$. The median survival time for individuals in Group II is the same under each model, and is equal to 0.69 months. The hazard functions for the two groups under both models are shown in Figure 6.4 and the corresponding survivor functions are shown in Figure 6.5. Under the accelerated failure time model, the increase in the hazard for Group II from 1.0 to 2.0 occurs sooner than under the

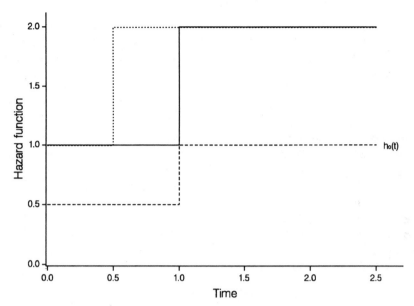

Figure 6.4 The hazard functions for individuals in Group I, $h_0(t)$, and in Group II under (a) a proportional hazards model (—) and (b) an accelerated failure time model ($\cdots$).

proportional hazards model. The 'kink' in the survivor function also occurs earlier under the accelerated failure time model.

6.3.2 The percentile-percentile plot

The *percentile-percentile plot*, also known as the *quantile-quantile plot* or the *Q-Q plot*, provides an exploratory method for assessing the validity of an accelerated failure time model for two groups of survival data. Recall that the p'th percentile of a distribution is the value $t(p)$ which is such that the estimated survivor function at time $t(p)$ is $1 - (p/100)$, for any value of p in the interval $(0, 100)$. The p'th percentile is therefore such that

$$t(p) = S^{-1}\left(\frac{100 - p}{100}\right).$$

Now let $t_0(p)$ and $t_1(p)$ be the p'th percentiles estimated from the survivor functions of the two groups of survival data. The values of p might be taken to be $10, 20, \ldots, 90$, so long as the number of observations in each of the two groups is not too small. The percentiles of the two groups may therefore be expressed as

$$t_0(p) = S_0^{-1}\left(\frac{100 - p}{100}\right), \qquad t_1(p) = S_1^{-1}\left(\frac{100 - p}{100}\right),$$

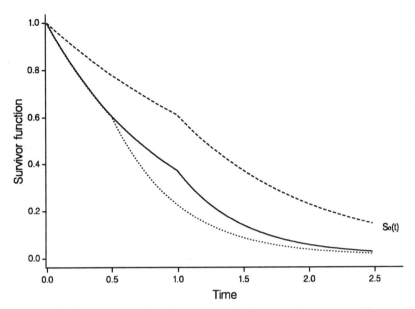

Figure 6.5 The survivor functions for individuals in Group I, $S_0(t)$, and in Group II under (a) a proportional hazards model (—) and (b) an accelerated failure time model $(\cdots)$.

where $S_0(t)$ and $S_1(t)$ are the survivor functions for the two groups. It then follows that

$$S_1\{t_1(p)\} = S_0\{t_0(p)\}, \qquad (6.6)$$

for any given value of p.

Under the accelerated failure time model, $S_1(t) = S_0(\phi t)$, and so the p'th percentile for the second group, $t_1(p)$, is such that

$$S_1\{t_1(p)\} = S_0\{\phi t_1(p)\}.$$

Using equation (6.6),

$$S_0\{t_0(p)\} = S_0\{\phi t_1(p)\},$$

and hence

$$t_0(p) = \phi t_1(p).$$

Now let $\hat{t}_0(p)$, $\hat{t}_1(p)$ be the estimated percentiles in the two groups, so that

$$\hat{t}_0(p) = \hat{S}_0^{-1}\left(\frac{100-p}{100}\right), \qquad \hat{t}_1(p) = \hat{S}_1^{-1}\left(\frac{100-p}{100}\right).$$

A plot of the quantity $\hat{t}_0(p)$ against $\hat{t}_1(p)$, for suitably chosen values of p, should give a straight line through the origin if the accelerated failure time model is appropriate. The slope of this line will be an estimate of the acceleration factor ϕ. This

Table 6.1 Estimated percentiles of the distributions of survival times for women with tumours which were negatively or positively stained.

Percentile	Negative staining	Positive staining
10	47	13
20	69	26
30	148	35
40	181	48
50	–	61
60	–	113
70	–	143
80	–	–
90	–	–

plot may therefore be used in an exploratory assessment of the adequacy of the accelerated failure time model. In this sense, it is an analogue of the log-cumulative hazard plot, used in Section 4.4.1 to examine the validity of the proportional hazards model.

Example 6.2 Prognosis for women with breast cancer
In this example, the data on the survival times of women with breast tumours that were negatively or positively stained, originally given as Example 1.2 in Chapter 1, is used to illustrate the percentile-percentile plot. The percentiles of the distribution of the survival times in each of the two groups can be estimated from the Kaplan–Meier estimate of the respective survivor functions. These are given in Table 6.1.

The relatively small numbers of death times and the censoring pattern in the data from the two groups of women mean that not all of the percentiles can be estimated. The percentile-percentile plot will therefore have just four pairs of points. For illustration, this is shown in Figure 6.6. The points fall on a line that is reasonably straight, suggesting that the accelerated failure time model would not be inappropriate. However, this conclusion must be regarded with some caution in view of the limited number of points in the plot.

The slope of a straight line drawn through the points in Figure 6.6 is approximately equal to three, which is a rough estimate of the acceleration factor. The interpretation of this is that for women whose tumours were positively stained the disease process is speeded up by a factor of three, relative to those whose tumours were negatively stained. We can also say that the median survival time for women with negatively stained tumours is estimated to be three times that of women with positively stained tumours.

Figure 6.6 Percentile-percentile plot for the data on the survival times of breast cancer patients.

6.3.3 The general accelerated failure time model

Equation (6.5) shows how the accelerated failure time model can be generalized to the situation where the values of p explanatory variables have been recorded for each individual in a study. According to the general accelerated failure time model, the hazard function of the i'th individual at time t, $h_i(t)$, is then such that

$$h_i(t) = e^{\eta_i} h_0(e^{\eta_i} t), \qquad (6.7)$$

where

$$\eta_i = \beta_1 x_{1i} + \beta_2 x_{2i} + \cdots + \beta_p x_{pi}$$

is the linear component of the model, in which x_{ji} is the value of the j'th explanatory variable, X_j, $j = 1, 2, \ldots, p$, for the i'th individual, $i = 1, 2, \ldots, n$. As in the proportional hazards model, the baseline hazard function $h_0(t)$ is the hazard of death at time t for an individual for whom the values of the p explanatory variables are all equal to zero.

6.3.4 The Weibull accelerated failure time model

The next step in the specification of the accelerated failure time model is to impose a probability distribution on the survival times. If this is chosen to be the Weibull distribution with scale parameter λ and shape parameter γ, written $W(\lambda, \gamma)$, the baseline hazard function is

$$h_0(t) = \lambda \gamma t^{\gamma - 1}.$$

The hazard function for the i'th individual is then, from equation (6.7), given by

$$
\begin{aligned}
h_i(t) &= e^{\eta_i} \lambda \gamma (e^{\eta_i} t)^{\gamma - 1} \\
&= (e^{\eta_i})^\gamma \lambda \gamma t^{\gamma - 1},
\end{aligned}
$$

so that the survival time of this individual has a $W(\lambda e^{\gamma \eta_i}, \gamma)$ distribution. The Weibull distribution is therefore said to possess the *accelerated failure time property*. Indeed, this is the only distribution that has both the proportional hazards and accelerated failure time properties.

Because the Weibull distribution has both the proportional hazards property and the accelerated failure time property, there is a direct correspondence between the parameters under the two models. If the baseline hazard function is the hazard function of a $W(\lambda, \gamma)$ distribution, the survival times under the proportional hazards model have a $W(\lambda e^{\eta_i}, \gamma)$ distribution, while those under the accelerated failure time model have a $W(\lambda e^{\gamma \eta_i}, \gamma)$ distribution. It then follows that when the coefficients of the explanatory variables in the linear component of the proportional hazards model are divided by γ, we get the corresponding β-coefficients in the accelerated failure time model. In the particular case of comparing two groups, an acceleration factor of $\phi = e^\beta$ under the accelerated failure time model corresponds to a hazard ratio of $\phi^\gamma = e^{\gamma \beta}$ in a proportional hazards model.

6.3.5 The log-logistic accelerated failure time model

Now suppose that the survival times have a log-logistic distribution. If the baseline hazard function in the general accelerated failure time model in equation (6.7) is derived from a log-logistic distribution with parameters θ, κ, this function is given by

$$h_0(t) = \frac{e^\theta \kappa t^{\kappa - 1}}{1 + e^\theta t^\kappa}.$$

Under the accelerated failure time model, the hazard of death at time t for the i'th individual is

$$h_i(t) = e^{\eta_i} h_0(e^{\eta_i} t),$$

and so

$$h_i(t) = \frac{e^{\eta_i} e^\theta \kappa (e^{\eta_i} t)^{\kappa - 1}}{1 + e^\theta (e^{\eta_i} t)^\kappa},$$

that is,

$$h_i(t) = \frac{e^{\theta + \kappa \eta_i} \kappa t^{\kappa - 1}}{1 + e^{\theta + \kappa \eta_i} t^\kappa}.$$

It then follows that the survival time for the i'th individual also has a log-logistic distribution with parameters $\theta + \kappa \eta_i$ and κ. The log-logistic distribution therefore has the accelerated failure time property. However, this distribution does not have the proportional hazards property.

6.4 * Log-linear form of the accelerated failure time model

In Section 4.6, an alternative representation of the proportional hazards model was presented. In this formulation, a log-linear model is adopted for the random variable associated with the survival time of the i'th individual, T_i, and we write

$$\log T_i = \mu + \alpha_1 x_{1i} + \alpha_2 x_{2i} + \cdots + \alpha_p x_{pi} + \sigma \epsilon_i.$$

Using equation (4.40), the survivor function of T_i is then given by

$$
\begin{aligned}
S_i(t) &= P(T_i \geq t) \\
&= P(\log T_i \geq \log t) \\
&= P\left(\epsilon_i \geq \frac{\log t - \mu - \alpha_1 x_{1i} - \alpha_2 x_{2i} - \cdots - \alpha_p x_{pi}}{\sigma} \right).
\end{aligned}
$$

In Section 4.6, we saw that if the random variable ϵ_i in this model is taken to have a Gumbel distribution, we obtain the Weibull proportional hazards model. In view of the correspondence between the parameters of a Weibull proportional hazards model and those of a Weibull accelerated failure time model, the log-linear model also provides a representation of the Weibull accelerated failure time model. Now suppose that ϵ has a logistic distribution so that the probability density and survivor function of ϵ is

$$ f(\epsilon) = \frac{e^\epsilon}{(1 + e^\epsilon)^2}, \qquad S(\epsilon) = \frac{1}{1 + e^\epsilon}. $$

The survivor function of T_i is then

$$ S_i(t) = \left[1 + \exp\left\{ \frac{\log t - \mu - \alpha_1 x_{1i} - \alpha_2 x_{2i} - \cdots - \alpha_p x_{pi}}{\sigma} \right\} \right]^{-1}. \tag{6.8} $$

Now consider the general accelerated failure time model, in which a log-logistic distribution with parameters θ, κ is used to model the survival time of an individual for whom the values of the p explanatory variables in the model are all equal to zero. The survival time of the i'th individual will also have a log-logistic distribution with parameters $\theta + \kappa \eta_i$, κ, where $\eta_i = \beta_1 x_{1i} + \beta_2 x_{2i} + \cdots + \beta_p x_{pi}$. From equation (6.2), the survivor function of T_i is

$$ S_i(t) = \frac{1}{1 + e^{\theta + \kappa \eta_i} t^\kappa}, $$

and comparing this expression with that for the survivor function in equation (6.8), we see that the parameters θ, κ and β_i can be expressed in terms of μ, σ and α_i. Specifically,

$$ \theta = -\mu/\sigma, \qquad \kappa = \sigma^{-1}, \qquad \beta_i = -\alpha_i. $$

This shows that a log-linear model can be used to represent both the Weibull and log-logistic accelerated failure time models. In fact, the log-linear representation of the accelerated failure time model can be used in conjunction with a wide range of survival time distributions, including the lognormal and gamma distributions referred to in Section 6.1.

As noted earlier, the log-linear representation of the accelerated failure time model is important because most computer software packages for survival analysis, including SAS and BMDP, use this version of the model. The parameter estimates given by these packages on fitting an accelerated failure time model with log-logistic survival times are therefore estimates of μ, σ and the α_i's, rather than θ, κ and the β_i's.

In order to interpret the parameter estimates given in computer output it is vital to know how the model is parametrized in the computer package being used for the analysis. Further details on the parametrization of the model used by different packages are given in Chapter 11.

When a log-logistic accelerated failure time model has been parametrized as above, estimates of quantities such as the acceleration factor or median survival time can be obtained directly from the estimates of μ, σ and the α_i's. For example, the acceleration factor for the i'th individual is $\exp\{-(\alpha_1 x_{1i} + \alpha_2 x_{2i} + \cdots + \alpha_p x_{pi})\}$ and the median survival time is simply

$$t(50) = \exp(\mu + \alpha_1 x_{1i} + \alpha_2 x_{2i} + \cdots + +\alpha_p x_{pi}).$$

The standard error of the estimated value of $t(50)$ may be found from the variance–covariance matrix of the parameter estimates using the general method outlined in Appendix B.

6.5 Fitting and comparing accelerated failure time models

In principle, accelerated failure time models can be fitted using the method of maximum likelihood. The likelihood function is best derived from the log-linear representation of the model, after which iterative methods are used to obtain the estimates. Since the method is very similar to that used in fitting the proportional hazards model, which was outlined in Section 4.5, details will not be included here.

Once a model has been fitted, the value of the statistic $-2 \log \hat{L}$ can be computed and used in making comparisons between nested models, just as for the proportional hazards model. Specifically, to compare two nested models, the difference in the values of the statistic $-2 \log \hat{L}$ for the two models is calculated and compared with percentage points of the chi-squared distribution.

The adequacy of a fitted model can be assessed using plots based on residuals. The Cox–Snell residuals can be obtained directly from the fitted survivor function, and from these residuals, analogues of the martingale residuals can be derived.

Once an appropriate model has been found, the model can be interpreted in terms of the estimated value of the acceleration factor for particular individuals, or in terms of the median and other percentiles of the distribution of survival times.

Example 6.3 Prognosis for women with breast cancer

In this example, accelerated failure time models are fitted to the data on the survival times of women with breast cancer. The Weibull accelerated failure time model is first considered. Under this model, the hazard of death at time t for the i'th woman is

$$h_i(t) = e^{\beta x_i} h_0(e^{\beta x_i} t),$$

where x_i takes the value zero if the i'th woman had a negatively stained tumour, and unity if the tumour was positively stained. For the Weibull distribution, the baseline hazard function is

$$h_0(t) = \lambda \gamma t^{\gamma - 1},$$

which is the hazard function for women with negatively stained tumours. Hence,

$$h_i(t) = e^{\gamma \beta x_i} \lambda \gamma t^{\gamma - 1}.$$

On fitting this model, the estimated values of the parameters are given by $\hat{\lambda} = 0.00414$, $\hat{\gamma} = 0.937$ and $\hat{\beta} = 0.997$. The acceleration factor, e^{β}, is estimated by $e^{0.997} = 2.71$. The survival time of a woman with a positively stained tumour is therefore accelerated by a factor of about 2.7 under this model. This is in broad agreement with the estimated slope of the percentile-percentile plot for this data set, found in Example 6.2.

The median survival time under the Weibull accelerated failure time model is

$$t(50) = \left[\frac{\log 2}{\lambda e^{\gamma \beta x_i}} \right]^{1/\gamma}.$$

The estimated median survival time for a woman with negative staining ($x_i = 0$) is 236 days, while that for women with positive staining ($x_i = 1$) is 87 days. The median survival time for women with positively stained tumours is therefore about one third that of those whose tumours were negatively stained.

The estimated hazard functions for the two groups of women under the Weibull accelerated failure time model are shown in Figure 6.7. The correspondence between the Weibull accelerated failure time model and the Weibull proportional hazards model means that the hazard ratio under the latter model is $e^{\gamma \beta}$ which is estimated to be 2.55.

We now fit the log-logistic accelerated failure time model to the same data set. Now, the baseline hazard function is

$$h_0(t) = \frac{e^{\theta} \kappa t^{\kappa - 1}}{1 + e^{\theta} t^{\kappa}},$$

and so the hazard function for the i'th woman in the study is

$$h_i(t) = \frac{e^{\theta + \kappa \beta x_i} \kappa t^{\kappa - 1}}{1 + e^{\theta + \kappa \beta x_i} t^{\kappa}}.$$

The estimated values of the parameters in this model are found to be given by $\hat{\theta} = -6.786$, $\hat{\kappa} = 1.243$ and $\hat{\beta} = 1.149$. The acceleration factor is e^{β}, which

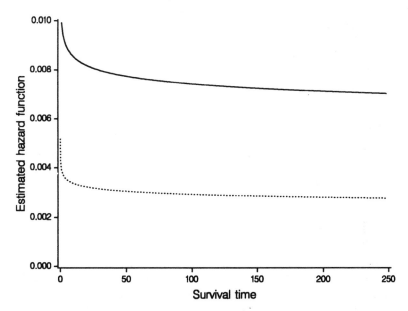

Figure 6.7 Estimated hazard functions under the Weibull accelerated failure time model for women with positively stained (—) and negatively stained ($\cdots$) tumours.

is estimated by 3.16. This is slightly greater than that found under the Weibull accelerated failure time model.

The median survival time under this model is given by

$$t(50) = \exp\left\{ -\frac{\theta + \kappa\beta x_i}{\kappa} \right\},$$

from which the estimated median survival time for a women with negative staining is 235 days, while that for women with positive staining is 75 days. These values are very close to those obtained under the Weibull accelerated failure time model.

The estimated hazard functions for the two groups of women under this model are shown in Figure 6.8. This can be compared with the graph in Figure 6.7. The hazard functions for those in Group I are very similar under the two models. However, the hazard function for those in Group II under the log-logistic model is different from that under the Weibull model. An analysis of residuals may help in choosing between these two models, although with this small data set, such an analysis is unlikely to be very informative.

Example 6.4 Comparison of two treatments for prostatic cancer
In a further illustration of modelling survival data using the log-logistic accelerated failure time model, the data from a clinical trial to compare two treatments for

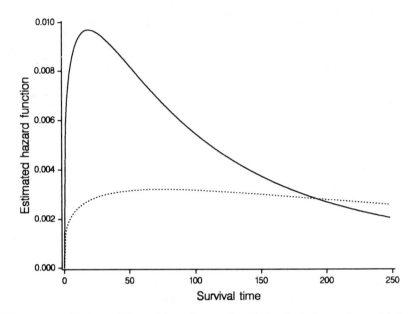

Figure 6.8 Estimated hazard functions under the log-logistic accelerated failure time model for women with positively stained (—) and negatively stained ($\cdots$) tumours.

prostatic cancer are considered. These data were first given in Example 1.4, and analysed using a Cox regression model in Examples 3.6 and 3.10.

To identify the terms which should be in the linear component of the log-logistic accelerated failure time model, the procedure described in Example 3.6 can again be followed. The values of the statistic $-2 \log \hat{L}$ on fitting models with all combinations of the four prognostic variables, AGE, SHB, $SIZE$ and $INDEX$, are shown in Table 6.2. As in Example 3.6, the variables $SIZE$ and $INDEX$ are the ones that are needed in the model. When either of these variables is omitted, the corresponding increase in the value of $-2 \log \hat{L}$ is significant, and neither AGE nor SHB reduce $-2 \log \hat{L}$ by a significant amount when they are added to the model.

When the term corresponding to the treatment effect, $TREAT$, is added to the model that contains $SIZE$ and $INDEX$, $-2 \log \hat{L}$ decreases to 21.245. When this reduction of 1.867 is compared with a chi-squared distribution on 1 d.f., the reduction is not significant at the 10% level (P = 0.172). There is no evidence of any interaction between $TREAT$ and the prognostic variables $SIZE$ and $INDEX$, and so the conclusion is that there is no statistically significant treatment effect.

The magnitude of the treatment effect can be assessed by calculating the acceleration factor. The fitted hazard function for the i'th patient, $i = 1, 2, \ldots, 38$, is

Table 6.2 Values of $-2 \log \hat{L}$ for models fitted to the data from Example 1.4.

Terms in model	$-2 \log \hat{L}$
null model	35.806
AGE	35.752
SHB	35.700
$SIZE$	27.754
$INDEX$	27.965
$AGE + SHB$	35.657
$AGE + SIZE$	27.652
$AGE + INDEX$	27.859
$SHB + SIZE$	27.722
$SHB + INDEX$	26.873
$SIZE + INDEX$	23.112
$AGE + SHB + SIZE$	27.631
$AGE + SHB + INDEX$	26.870
$AGE + SIZE + INDEX$	23.002
$SHB + SIZE + INDEX$	22.895
$AGE + SHB + SIZE + INDEX$	22.727

given by

$$\hat{h}_i(t) = e^{\hat{\eta}_i} \hat{h}_0(e^{\hat{\eta}_i} t),$$

where

$$\hat{\eta}_i = 0.029 \, SIZE_i + 0.293 \, INDEX_i - 0.573 \, TREAT_i,$$

and $SIZE_i$ and $INDEX_i$ are the values of tumour size and Gleason index for the i'th individual and $TREAT_i$ is zero if the ith individual is in the placebo group and unity if in the treated group. The estimated parameters in the estimated baseline hazard function, $\hat{h}_0(t)$, are given by $\hat{\theta} = 22.644$ and $\hat{\kappa} = 2.956$.

The estimated acceleration factor for an individual in the treated group, relative to one in the control group, is $e^{-0.573} = 0.56$. The interpretation of this result is that after allowing for the size and index of the tumour, the effect of the treatment with DES is to slow down the progression of the cancer by a factor of about 2. This effect might be of clinical importance, even though it is not statistically significant. However, before accepting this interpretation, the adequacy of the fitted model should be checked using an appropriate analysis of residuals.

A confidence interval for the acceleration factor is found by exponentiating the confidence limits for the logarithm of the acceleration factor. In this example, the logarithm of the acceleration factor is the estimated coefficient of $TREAT$ in the model for the hazard function, which is -0.573, and the standard error of this estimate is 0.473. Thus a 95% confidence interval for the acceleration factor has limits of $\exp\{-0.573 \pm 1.96 \times 0.473\}$, and the required interval is from 0.70 to 4.48. Notice that this interval estimate includes unity, which is consistent with the earlier finding of a non-significant treatment difference.

6.6 The proportional odds model

In this general model, the odds of an individual surviving beyond some time t are expressed as

$$\frac{S_i(t)}{1 - S_i(t)} = e^{\eta_i} \frac{S_0(t)}{1 - S_0(t)}, \tag{6.9}$$

where

$$\eta_i = \beta_1 x_{1i} + \beta_2 x_{2i} + \cdots + \beta_p x_{pi}$$

is a linear combination of the values of p explanatory variables, $X_1, X_2, \ldots, X_p$, measured on the i'th individual, and $S_0(t)$, the baseline survivor function, is the survivor function for an individual whose explanatory variables all take the value zero.

In this model, the explanatory variates act multiplicatively on the odds of survival beyond t. The logarithm of the ratio of the odds of survival beyond t for the i'th individual, relative to an individual for whom the explanatory variables are all equal to zero, is therefore just η_i. The model is therefore a linear model for the log-odds ratio.

Now consider the particular case of a two-group study, in which individuals receive either a standard treatment or new treatment. Let the single indicator variable X take the value zero if an individual is on the standard treatment and unity if on the new. The odds of the i'th individual surviving beyond time t is then

$$\frac{S_i(t)}{1 - S_i(t)} = e^{\beta x_i} \frac{S_0(t)}{1 - S_0(t)},$$

where x_i is the value of X for the i'th individual, $i = 1, 2, \ldots, n$. Thus if $S_N(t)$ and $S_S(t)$ are the survivor functions for individuals on the new and standard treatments respectively,

$$\frac{S_N(t)}{1 - S_N(t)} = e^{\beta} \frac{S_S(t)}{1 - S_S(t)},$$

and the log-odds ratio is simply β. The parameters in the linear component of the model therefore have an immediate interpretation.

As for the proportional hazards model, a non-parametric estimate of the baseline hazard function can be obtained. The model is then fitted by estimating the β-parameters in the linear component of the model and the baseline survivor function from the data. A method for accomplishing this has been described by Bennett (1983a), but details will not be included here. Fully parametric versions of the proportional odds model can be derived by using a specific probability distribution for the survival times. One such model is described below in Section 6.6.1.

One particularly important property of the proportional odds model concerns the ratio of the hazard function for the i'th individual to the baseline hazard, $h_i(t)/h_0(t)$. It can be shown that this ratio converges from the value $e^{-\eta_i}$ at time

$t = 0$, to unity at $t = \infty$. To show this, the model in equation (6.9) can be rearranged to give

$$S_i(t) = S_0(t) \left\{ e^{-\eta_i} + (1 - e^{-\eta_i}) S_0(t) \right\}^{-1},$$

and taking logarithms, we get

$$\log S_i(t) = \log S_0(t) - \log \left\{ e^{-\eta_i} + (1 - e^{-\eta_i}) S_0(t) \right\}. \tag{6.10}$$

Using the general result from equation (1.4) that the hazard function is

$$h_i(t) = -\frac{d}{dt} \log S_i(t),$$

we obtain

$$h_i(t) = h_0(t) - \frac{(1 - e^{-\eta_i}) f_0(t)}{e^{-\eta_i} + (1 + e^{-\eta_i}) S_0(t)}$$

after differentiating both sides of equation (6.10) with respect to t, where $f_0(t)$ is the baseline probability density function. After some rearrangement, this equation becomes

$$h_i(t) = h_0(t) - \frac{f_0(t)}{(e^{\eta_i} - 1)^{-1} + S_0(t)}. \tag{6.11}$$

From equation (1.3), we also have that $h_0(t) = f_0(t)/S_0(t)$ and substituting for $f_0(t)$ in equation (6.11) gives

$$h_i(t) = h_0(t) \left\{ 1 - \frac{S_0(t)}{(e^{\eta_i} - 1)^{-1} + S_0(t)} \right\}.$$

Finally, after further rearrangement, the hazard ratio is given by

$$\frac{h_i(t)}{h_0(t)} = [1 + (e^{\eta_i} - 1) S_0(t)]^{-1}.$$

As t increases from 0 to ∞, the baseline survivor function decreases monotonically from 1 to 0. When $S_0(t) = 1$, the hazard ratio is $e^{-\eta_i}$ and as t increases to ∞, the hazard ratio converges to unity.

In practical applications, it is common for the hazard functions obtained for patients in two or more groups to converge with time. For example, in a follow-up study of patients in a clinical trial, the effect on survival of the treatment, or the initial stage of disease, may wear off. Similarly, in studies where a group of patients with some disease are being compared with a control group of disease-free individuals, an effective cure of the disease would lead to the survival experience of each group becoming more similar over time. This suggests that the proportional odds model with its property of convergent hazard functions would be of considerable value. However, there are two reasons why this general model has not been widely used in practice. The first of these is that computer software for fitting the model is not generally available. The second is that the model is likely to give similar results to a Cox regression model that includes a time-dependent

variable to produce non-proportional hazards. This particular approach to modelling survival data with non-proportional hazards was described in Section 5.8,
and is considered more fully in the next chapter.

6.6.1 The log-logistic proportional odds model

If survival times for individuals are assumed to have a log-logistic distribution,
the baseline survivor function is

$$S_0(t) = [1 + e^\theta t^\kappa]^{-1},$$

where θ and κ are unknown parameters. The baseline odds of survival beyond
time t are then given by

$$\frac{S_0(t)}{1 - S_0(t)} = e^{-\theta} t^{-\kappa}.$$

The odds of the i'th individual surviving beyond time t are therefore

$$\frac{S_i(t)}{1 - S_i(t)} = e^{\eta_i - \theta} t^{-\kappa},$$

and so the survival time of the i'th individual has a log-logistic distribution with
parameters $\theta - \eta_i$ and κ. The log-logistic distribution therefore has the *proportional
odds property*, and the distribution is the natural one to use in conjunction with
the proportional odds model. In fact, it is the only distribution to share both the
accelerated failure time property and the proportional odds property.

This result means that the β-parameters under the proportional odds model can
be obtained from those under the accelerated failure time model, and vice versa.
In particular, the estimated coefficients of the explanatory variable in the linear
component of the proportional odds model are obtained by multiplying those in
the accelerated failure time model by $-\hat{\kappa}$, where $\hat{\kappa}$ is the estimated value of κ
under the accelerated failure time model. This enables the result of the survival
analysis to be interpreted in terms of an acceleration factor or the ratio of the odds
of survival beyond some time, whichever is the more convenient.

As for other models for survival data, the proportional odds model can be fitted using the method of maximum likelihood. Alternative models may then be
compared on the basis of the statistic $-2 \log \hat{L}$.

In a two-group study, a preliminary examination of the likely suitability of the
model can easily be undertaken. The log-odds of the i'th individual surviving beyond time t are

$$\log \left\{ \frac{S_i(t)}{1 - S_i(t)} \right\} = \beta x_i - \theta - \kappa \log t,$$

where x_i is the value of an indicator variable that takes the value zero if an individual is in one group and unity if in the other. The Kaplan–Meier estimate of
the survivor function is then obtained for the individuals in each group and the es-
timated log-odds of survival beyond time t, $\log \left\{ \hat{S}_i(t) / [1 - \hat{S}_i(t)] \right\}$, are plotted

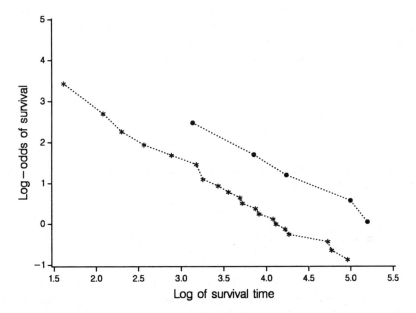

Figure 6.9 Estimated values of the log-odds of survival beyond t plotted against $\log t$ for women with positively stained (∗) and negatively stained (•) tumours.

against $\log t$. If the plot shows two parallel straight lines, this would indicate that the log-logistic model was appropriate. If the lines were straight but not parallel, this would suggest that the parameter κ in the model was not the same for each treatment group. Parallel curves in this plot suggest that although the proportional odds assumption is valid, the survival times cannot be taken to have a log-logistic distribution.

Example 6.5 Prognosis for women with breast cancer
In this example to illustrate the use of the proportional odds model, the model is fitted to the data on the survival times of breast cancer patients. In order to assess the likely suitability of the proportional odds model, the Kaplan–Meier estimate of the survivor function for the negatively and positively stained women is computed. For the two groups of women, the log-odds of survival beyond time t are estimated and plotted against $\log t$. The resulting graph is shown in Figure 6.9. The lines are reasonably straight and parallel, and so we go on to use the log-logistic proportional odds model to summarize these data.

The model can be fitted using software for fitting the log-logistic accelerated failure time model. In Example 6.3, this latter model was fitted to the data on the survival of breast cancer patients. The estimated value of κ and θ in the proportional odds model are 1.243 and -6.786, the same as those in the accelerated failure time model. However, the estimated value of β in the linear component of the proportional odds model is $\hat{\beta} = -1.149 \times 1.243 = -1.428$. This is an estimate of the logarithm of the ratio of the odds of a positively stained woman surviving beyond time t, relatively to one who is negatively stained. The corresponding odds ratio is $e^{-1.428} = 0.24$, so that the odds of a woman surviving beyond t are about four times greater if that woman has a negatively stained tumour.

6.7 Further reading

The properties of random variables that have probability distributions such as the logistic, lognormal and gamma, are presented in Johnson and Kotz (1970). Use of the quadratic hazard function was discussed by Gaver and Acar (1979) and the bathtub hazard function was proposed by Hjorth (1980).

Cox and Oakes (1984) show that the Weibull distribution is the only one to have both the proportional hazards property and the accelerated failure time property. They also demonstrate that the log-logistic distribution is the only one that shares the accelerated failure time property and the proportional odds property.

A non-parametric version of the accelerated failure time model, which does not require the specification of a probability distribution for the survival data, has been introduced by Wei (1992). This paper, and the published discussion, Fisher (1992), includes comments on whether the accelerated failure time should be used more widely in the analysis of survival data.

The application of the accelerated failure time and proportional odds models to the analysis of reliability data has recently been described by Crowder *et al.* (1991). The general proportional odds model for survival data was introduced by Bennett (1983a). A fuller account of the topic is contained in Bennett (1983b), which includes details on a FORTRAN program for fitting the model. Bennett (1983c) describes the log-logistic proportional odds model. GLIM macros for fitting the model are described in Bennett and Whitehead (1981).

The piecewise exponential model, in which hazards are constant over particular time intervals, was introduced by Breslow (1974). Breslow also points out that the Cox regression model is equivalent to a piecewise exponential model with constant hazards between each death time. GLIM macros for fitting the piecewise exponential model were given by Whitehead (1983). The piecewise exponential model and the use of the normal, lognormal, logistic and log-logistic distributions for modelling survival times are described in Aitkin *et al.* (1989). This book also includes a listing of GLIM macros for fitting these models.

A more general way of modelling survival data is to use a general family of distributions for survival times which includes the Weibull and log-logistic as special cases. The choice between alternative distributions can then be made within a likelihood framework. In particular, the exponential, Weibull, log-logistic, lognormal and gamma distributions are special cases of the generalized F-distribution described by Kalbfleisch and Prentice (1980). However, this methodology will only tend to be informative in the analysis of data sets in which the number of death times is relatively large.

7

Time-dependent variables

In earlier chapters, we have seen how the dependence of the hazard function for an individual on the values of certain explanatory variables can be modelled. When explanatory variables are incorporated in a model for survival data, the values taken by such variables are those recorded at the time origin of the study. For example, consider the study to compare two treatments for prostatic cancer first described in Example 1.4 of Chapter 1. Here, the age of a patient, serum haemoglobin level, size of the tumour, value of the Gleason index, and of course the treatment group, were all recorded at the time when a patient was entered into the study. The impact of these variables on the hazard of death is then evaluated.

In many studies that generate survival data, individuals are monitored for the duration of the study. During this period, the values of certain explanatory variables may be recorded on a regular basis. Thus in the example on prostatic cancer, the size of the tumour and other variables may be recorded at regular intervals. If account can be taken of the values of explanatory variables as they evolve, it might be possible to provide a better forecast of the median survival time of a patient. Thus again in connection with the prostatic cancer study, more recent values of the size of the tumour may provide a better indication of future life expectancy than the value at the time origin.

Variables whose values change over time are known as *time-dependent variables* and in this chapter we see how such variables can be incorporated in models used in the analysis of survival data. In this process, the most recent value of a time-dependent variable is used at each specific time in the modelling procedure.

7.1 Types of time-dependent variable

It is useful to consider two types of variable that change over time, which may be referred to as *internal variables* and *external variables*.

Internal variables are variables which only relate to a particular individual in a study, and can only be measured while a patient is alive. Examples include measures of lung function such as vital capacity and peak flow rate, white blood cell count, systolic blood pressure and so on. On the other hand, external variables are time-dependent variables which do not necessarily require the survival of a patient for their existence.

One type of external variable is a variable which changes in such a way that its value will be known in advance at any future time. The most obvious example is the age of a patient, in that once the age at the time origin is known, that patient's age at any future time will be known exactly. However there are other examples, such as the dose of a drug that is to be varied in a predetermined manner during the course of a study.

Another type of external variable is one that exists totally independently of any particular individual, such as the level of atmospheric sulphur dioxide, or air temperature. Changes in the values of such quantities may well have an effect on the lifetime of individuals, as in studies concerning the management of patients with certain types of respiratory disease.

These different types of time-dependent variable can be introduced into the Cox proportional hazards model. The resulting model will simply be referred to as the Cox regression model, and is described in the following section.

7.2 A model with time-dependent variables

According to the Cox proportional hazards model described in Chapter 3, the hazard of death at time t for the i'th of n individuals in a study can be written

$$h_i(t) = \exp\left\{\sum_{j=1}^{p} \beta_j x_{ji}\right\} h_0(t),$$

where x_{ji} is the baseline value of the j'th explanatory variable, $X_j, j = 1, 2, \ldots, p$, for the i'th individual, $i = 1, 2, \ldots, n$, and $h_0(t)$ is the baseline hazard function. Generalizing this model to the situation in which some of the explanatory variables are time-dependent, we write $x_{ji}(t)$ for the value of the j'th explanatory variable at time t in the i'th individual. The Cox regression model then becomes

$$h_i(t) = \exp\left\{\sum_{j=1}^{p} \beta_j x_{ji}(t)\right\} h_0(t). \tag{7.1}$$

In this model, the baseline hazard function $h_0(t)$ is interpreted as the hazard function for an individual for whom all the variables are zero at the time origin, and remain at this same value through time.

It is important to note that in the model given in equation (7.1), the values of the variables $x_{ji}(t)$ depend on the time t, and so the relative hazard $h_i(t)/h_0(t)$ is also time dependent. This means that the hazard of death at time t is no longer proportional to the baseline hazard, and the model is no longer a proportional hazards model.

To provide an interpretation of the β-parameters in this model, consider the ratio of the hazard functions at time t for two individuals, the r'th and s'th, say. This is given by

$$\frac{h_r(t)}{h_s(t)} = \exp\{\beta_1[x_{r1}(t)-x_{s1}(t)]+\beta_2[x_{r2}(t)-x_{s2}(t)]+\cdots+\beta_p[x_{rp}(t)-x_{sp}(t)]\}.$$

The coefficient β_j, $j = 1, 2, \ldots, p$, can therefore be interpreted as the log-hazard ratio for two individuals whose value of the j'th explanatory variable at time t differs by one unit, with the two individuals having the same values of all the other $p-1$ variables at that time.

The survivor function for the i'th individual is, from equation (1.6) in Chapter 1, obtained from the integrated hazard function, and is given by

$$S_i(t) = \exp\left\{-\int_0^t \exp\left(\sum_{j=1}^p \beta_j x_{ji}(u)\right) h_0(u)\,\mathrm{d}u\right\}.$$

This survivor function therefore depends not only on the baseline hazard function $h_0(t)$, but also on the values of the time-dependent variables over the interval from 0 to t. Moreover, the result that $S_i(t)$ can be expressed as a power of the baseline survivor function, $S_0(t)$, given in equation (3.19) of Chapter 3, no longer holds. This means that the survivor function is generally difficult to obtain for any individual.

7.2.1* Fitting the Cox model

When the Cox regression model is extended to incorporate time-dependent variables, the partial log-likelihood function, from equation (3.5) in Chapter 3, can be generalized to

$$\sum_{i=1}^n \delta_i \left\{\sum_{j=1}^p \beta_j x_{ji}(t_i) - \log \sum_{l \in R(t_i)} \exp\left(\sum_{j=1}^p \beta_j x_{jl}(t_i)\right)\right\}, \qquad (7.2)$$

in which $R(t_i)$ is the risk set at time t_i, the death time of the i'th individual in the study, $i = 1, 2, \ldots, n$, and δ_i is a censoring indicator which is zero if the survival time of the i'th individual is censored and unity otherwise. This expression can then be maximized to give estimates of the β-parameters.

In order to use equation (7.1) in this maximization process, the values of each of the variables in the model must be known at each death time for all individuals in the risk set at time t_i. This is no problem for external variables whose values are pre-ordained, but it may be a problem for external variables that exist independently of the individuals in a study, and certainly for internal variables.

To illustrate the problem, consider a trial of two maintenance therapies for patients who have suffered a myocardial infarct. The serum cholesterol level, X, of such patients may well be measured at the time when a patient is admitted to the study and at regular intervals of time thereafter. It is then plausible that the hazard

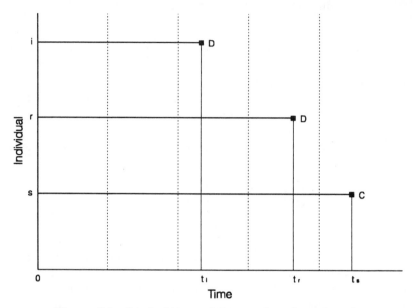

Figure 7.1   Survival times of three patients in patient time.

of death for any particular patient, the i'th, say, at time t, $h_i(t)$, is more likely to be influenced by the value of the explanatory variable X at time t than at the time origin, $t = 0$.

Now suppose that the i'th individual dies at time t_i and that there are two other individuals, labelled r and s, in the risk set at time t_i. We further suppose that individual r dies at time t_r, where $t_r > t_i$, and that the survival time of individual s, t_s, is censored at some time after t_r. The situation is illustrated graphically in Figure 7.1. In this figure, the vertical dotted lines refer to points in patient time when the value of X is measured.

If individuals r and s are the only two in the risk set at time t_i, and X is the only explanatory variable that is measured, the contribution of the i'th individual to the log-likelihood function in expression (7.2) will be

$$\beta x_i(t_i) - \log \sum_l \exp\{\beta x_l(t_i)\},$$

where $x_i(t_i)$ is the value of X for the i'th individual at their death time, t_i, and l in the summation takes the values i, r, and s. This expression is therefore equal to

$$\beta x_i(t_i) - \log \left\{ e^{\beta x_i(t_i)} + e^{\beta x_r(t_i)} + e^{\beta x_s(t_i)} \right\}.$$

This shows that the value of the time-dependent variable X is needed at the death time of the i'th individual and at time t_i for individuals r and s. In addition, the value of the variable X will be needed for individuals r and s at t_r, the death time

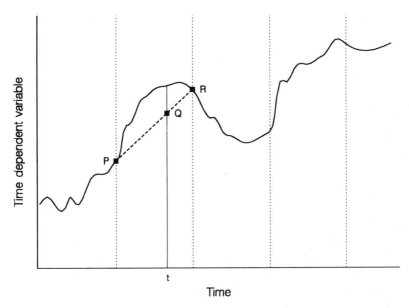

Figure 7.2 Computation of the value of a time-dependent variable at intermediate times.

of individual r.

For terms in a model that are explicit functions of time, such as interactions between a variable or factor measured at baseline and time, there is no difficulty in evaluating the values of the time-dependent variables at any time for any individual. Indeed, it is usually straightforward to incorporate such variables in the Cox model when using statistical software that has facilities for dealing with time-dependent variables. For other variables, such as serum cholesterol level, the values of the time-dependent variable at times other than that at which it was measured has to be approximated. There are then several possibilities.

One option is to use the last recorded value of the variable before the time at which the value of the variable is needed. When the variable has been recorded for an individual before and after the time when the value is required, the value closest to that time might be used. Another possibility is to use linear interpolation between consecutive values of the variable.

Figure 7.2 illustrates these approximations. In this figure, the continuous curve depicts the actual value of a time-dependent variable at any time. The dotted vertical lines signify times when the variable is actually measured. If the value of the variable is required at time t in this figure, we could use either the value at P, the last recorded value of the variable, the value at R, the value closest to t, or the value at Q, the linearly interpolated value between P and R.

Linear interpolation is clearly not an option when a time-dependent variable is

Figure 7.3   Values of a time-dependent categorical variable.

a categorical variable. In addition, some categorical variables may be such that individuals can only progress through the levels of the variable in a particular direction. For example, the performance status of an individual may only be expected to deteriorate, so that the performance status for any individual might only change from 'good' to 'fair' and from 'fair' to 'poor'. As another example, following a biopsy, a variable associated with the occurrence of a tumour will take two values corresponding to absence and presence. It might then be very unlikely for the status to change from 'present' to 'absent' in consecutive biopsies.

Anomalous changes in the values of time-dependent variables can be detected by plotting the values of the variable against time for each patient. This may then lead on to a certain amount of data editing. For example, consider the plot in Figure 7.3 which shows the biopsy result (absent or present) for a particular patient at a number of time points. In this diagram, at least one of the observations made at times t_A and t_B must be incorrect. The observation at t_A might then be changed to 'absent' or that at t_B to 'present'.

7.3 Some applications of time-dependent variables

One application of time-dependent variables is in connection with evaluating the assumption of proportional hazards. This was discussed in detail in Section 5.8 of Chapter 5. In this application, a term βt is added to the linear part of the Cox model, and the null hypothesis that β is zero is tested. If β is found to be significantly different from zero, there is evidence that the assumption of proportional hazards is not valid.

In many circumstances, the waiting time from the occurrence of some catastrophic event until a patient receives treatment may be strongly associated with the patient's survival. For example, in a study of factors affecting the survival of patients who have had a myocardial infarct, the time from the infarct to when the patient arrives in an intensive care unit may be crucial. Some patients may die before receiving treatment in the Unit, while those who arrive at the Unit soon after their infarct will tend to have a more favourable prognosis than those for whom treatment is delayed. It will be important to take account of this aspect of the data when assessing the effects of other explanatory variables on the survival times of these patients.

In a similar example, Crowley and Hu (1977) show how a time-dependent variable can be used in organ transplantation studies. Here, one feature of interest is the effect of a transplant on the patient's survival time. Suppose that in a study on the effectiveness of a particular type of organ transplant, a patient is judged to be suitable for a transplant at some time t_0. They then wait some period of time until a suitable donor organ is found, during which time the patients are unlikely to receive any beneficial therapy. Suppose that if the patient survives this period, they receive a transplant at time t_1.

In studies of this sort, the survival times of patients who have received a transplant cannot be compared with those who have not had a transplant in the usual way. The reason for this is that in order to receive a transplant, a patient must survive the waiting time to transplant. Consequently, the group who survive to transplant is not directly comparable with the group who receive no such transplant. Similarly, it is not possible to compare the times that the patients who receive a transplant survive after the transplant with the survival times of the group not receiving a transplant. Here, the time origin would be different for the two groups, and so the groups are not comparable at the time origin. This means that it is not possible to identify a time origin from which standard methods for survival analysis can be used.

The solution to this problem is to introduce a time-dependent variable $X_1(t)$, which takes the value zero if a patient has not received a transplant at time t, and unity otherwise. Adopting a Cox regression model, the hazard of death for the i'th individual at time t is then

$$h_i(t) = \exp\{\eta_i + \beta_1 x_{1i}(t)\}h_0(t),$$

where η_i is a linear combination of other non-time-dependent explanatory variables whose values have been recorded at the time origin for the i'th individual, and $x_{1i}(t)$ is the value of X_1 for that individual at time t.

Under this model, the hazard function is $\exp(\eta_i)h_0(t)$ for patients who do not receive a transplant before time t, and $\exp\{\eta_i + \beta_1\}h_0(t)$ thereafter. The effect of transplant on the patient's survival experience is then reflected in β_1. In particular, for two patients who have the same values of other explanatory variables in a model, e^{β_1} is the hazard of death at time t for the patient who receives a transplant before time t, relative to the hazard at that time for the patient who does not. Values of $-2\log\hat{L}$ can be compared after fitting the models with and without the time-dependent variable X_1; a significant difference in these values means that the transplant has an effect on survival.

In a refinement to this model, Cox and Oakes (1984) suggested that the term $\beta_1 x_{1i}(t)$ be replaced by $\beta_1 + \beta_2 \exp\{-\beta_3(t - t_1)\}$ for patients receiving a trans-

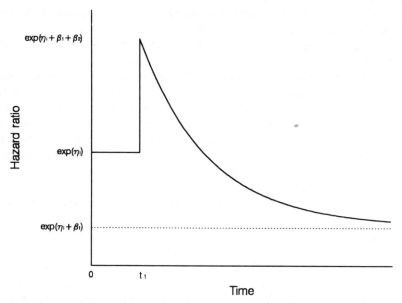

Figure 7.4 The hazard ratio $\exp\{\eta_i + \beta_1 + \beta_2 e^{-\beta_3(t-t_1)}\}$, $t > t_1$, for individual i who receives a transplant at t_1.

plant at time t_1. In this model, the effect of the transplant is to increase the hazard to some value $\exp(\eta_i + \beta_1 + \beta_2)h_0(t)$ immediately after the transplant, when $t = t_1$, and to then decrease exponentially to $\exp(\eta_i + \beta_1)h_0(t)$, which is less than the initial hazard $\exp(\eta_i)h_0(t)$ if $\beta_1 < 0$. See Figure 7.4, which shows graphically the behaviour of the hazard ratio, $h_i(t)/h_0(t)$, for a transplant patient for whom η_i is the linear component of the model. Although this is an attractive model, it does have the disadvantage that specialist software is required to fit it.

In situations where a particular explanatory variable is changing rapidly, new variables which reflect such changes may be defined. The dependence of the hazard on the values of such variables can then be explored. For example, in an oncological study, the percentage increase in the size of a tumour over a period of time might be a more suitable prognostic variable than either the size of the tumour at the time origin, or the time-dependent version of that variable. If this route is followed, the computational burden of fitting time-dependent variables can be avoided.

7.4 Comparison of treatments

In order to examine the magnitude of a treatment effect after taking account of variables that change over time, the values of $-2 \log \hat{L}$ for a model that contains the time-dependent variables and any other prognostic factors is compared with that for the model that contains the treatment term in addition to these other variables. If no treatment effect is revealed, one explanation could be that the time-dependent variable has *masked* the treatment difference.

To fix ideas, consider the example of a study to compare two cytotoxic drugs in the treatment of patients with leukaemia. Here, a patient's survival time may well depend on subsequent values of that patient's white blood cell count. If the effect of the treatment is to increase white blood cell count, no treatment difference will be identified after including white blood cell count as a time-dependent variable in the model. On the other hand, the treatment effect may appear in the absence of this variable. An interpretation of this is that the time-dependent variable has accounted for the treatment difference, and so provides an explanation as to how the treatment has been effective.

In any event, much useful information will be gained from a comparison of the results of an analysis that incorporates time-dependent variables with an analysis that uses baseline values alone.

7.5 Two examples

In this section two examples of survival analyses that involve time-dependent variables are given. In the first, the data from Example 4.9 of Chapter 4 on the comparison of two chemotherapy treatments for ovarian cancer are analysed to explore whether there is an interaction between age and survival time. The second example is designed to illustrate how information on a time-varying explanatory variate recorded during the follow-up period can be incorporated in a model for survival times. Studies in which the values of certain explanatory variables are recorded regularly throughout the follow-up period of each patient generate large sets of data. For this reason, artificial data from a small number of individuals will be used to illustrate the methodology. In both examples, the procedure proc phreg of the statistical package SAS has been used to fit models that contain time-dependent variables.

Example 7.1 Chemotherapy in ovarian cancer patients
When a Cox proportional hazards model that contains the variables AGE, the age of a patient at the time origin, and $TREAT$, the treatment group, is fitted to the data on the survival times of patients with ovarian cancer, the estimated hazard function for the i'th of 26 patients in the study is

$$\hat{h}_i(t) = \exp\{0.147\, AGE_i - 0.796\, TREAT_i\} h_0(t).$$

The value of the statistic $-2 \log \hat{L}$ for this model is 54.148.

We now fit a model that contains AGE and $TREAT$, and a term corresponding to an interaction between age and survival time. This interaction will be modelled by including the time-dependent variable $AGET$, whose values are formed from the product of AGE and the survival time t, that is, $AGET = AGE \times t$. Since the values of $AGET$ are dependent upon t, this time-dependent variable cannot be fitted in the same manner as AGE and $TREAT$. When $AGET$ is added to the model, the fitted hazard function becomes

$$\hat{h}_i(t) = \exp\{0.216\,AGE_i - 0.664\,TREAT_i - 0.0002\,AGE_it\}h_0(t).$$

Under this model, the hazard of death at t for a patient of a given age on the combined treatment ($TREAT = 2$), relative to one of the same age on the single treatment ($TREAT = 1$), is $\exp(-0.664) = 0.52$, which is not very different from the value of 0.45 found using the model that does not contain the variable $AGET$. However, the log-hazard ratio for a patient aged a_2 years, relative to one aged a_1 years, is

$$0.216(a_2 - a_1) - 0.0002(a_2 - a_1)t$$

at time t. This model therefore allows the log-hazard ratio for AGE to be linearly dependent on survival time.

   The value of $-2\log \hat{L}$ for the model that contains AGE, $TREAT$ and $AGET$ is 53.613. The change in $-2\log \hat{L}$ on adding the variable $AGET$ to a model that contains AGE and $TREAT$ is therefore 0.53, which is not significant (P = 0.465). We therefore conclude that the time-dependent variable $AGET$ is not in fact needed in the model.

Example 7.2 Data from a hypothetical cirrhosis study
Although the data to be used in this example are artificial, it is useful to provide a background against which these data can be considered. Suppose therefore that 12 patients have been recruited to a study on the treatment of cirrhosis of the liver. The patients are randomized to receive either a placebo or a new treatment that will be referred to as Liverol. Six patients are allocated to Liverol and six to the placebo. At the time when the patient is entered into the study, the age and baseline value of the patient's bilirubin level are recorded. Patients are supposed to return to the clinic three, six and twelve months after the commencement of treatment, and yearly thereafter. On these occasions, the bilirubin level is again measured and recorded. Data are therefore available on how the bilirubin level changes in each patient throughout the duration of the study. The natural logarithm of the bilirubin value (in μmol/l) will be used in this analysis. The variables measured at baseline are summarized below:

$TIME$:	Survival time of the patient in days
$CENS$:	Censoring indicator (0 = censored, 1 = uncensored)
$TREAT$:	Treatment group (0 = placebo, 1 = Liverol)
AGE :	Age of the patient in years
LBR :	Logarithm of bilirubin level

The values of these variables are given in Table 7.1. Table 7.2 gives the values of the logarithm of the bilirubin value at each time in the follow-up period for each patient.

   In taking log(bilirubin) to be a time-dependent variable, for each patient the value of the variate at any time t is that recorded at the last follow-up visit before t.

Table 7.1 Survival times of 12 patients in a study on cirrhosis of the liver.

Patient	$TIME$	$CENS$	$TREAT$	AGE	LBR
1	281	1	0	46	3.2
2	604	0	0	57	3.1
3	457	1	0	56	2.2
4	384	1	0	65	3.9
5	341	0	0	73	2.8
6	842	1	0	64	2.4
7	1514	1	1	69	2.4
8	182	0	1	62	2.4
9	1121	1	1	71	2.5
10	1411	0	1	69	2.3
11	814	1	1	77	3.8
12	1071	1	1	58	3.1

This is equivalent to the assumption that the values of LBR for a given individual follow a step-function in which the values are assumed constant between any two adjacent time points.

We first analyse these data taking the log(bilirubin) values to be time-dependent. Let $LBRT$ be the time-dependent variate formed from the log(bilirubin) values. The values of $-2\log \hat{L}$ on fitting Cox regression models to the data are then given in Table 7.3. It is clear from this table that the hazard function depends on the time-dependent variable $LBRT$, and that after allowing for this, the effect of AGE is slight. We therefore add the treatment effect $TREAT$ to the model that contains $LBRT$ alone. The effect of this is that $-2\log \hat{L}$ is reduced from 12.053 to 10.678, a reduction of 1.375 on 1 d.f. This reduction is not significant (P = 0.241) leading to the conclusion that after taking account of the dependence of the hazard of death on the evolution of the log(bilirubin) values, no treatment effect is discernible.

The estimated hazard function for the i'th individual is given by

$$\hat{h}_i(t) = \exp\{3.614\,LBR_i(t) - 1.480\,TREAT_i\}h_0(t),$$

where $LBR_i(t)$ is the value of log(bilirubin) for the i'th patient at time t. The estimated ratio of the hazard of death at time t for two individuals on the same treatment who have values of LBR that differ by one unit at t is $e^{3.614} = 37.10$.

For comparison, the data are analysed using the baseline value of log(bilirubin) alone. A Cox proportional hazards model is used, and the values of $-2\log \hat{L}$ on fitting particular models are as shown in Table 7.4.

Now, both AGE and LBR appear to be needed in the model, although the evidence for including AGE as well as LBR is not very strong. When $TREAT$ is added to the model that contains AGE and LBR, the reduction in the value of $-2\log \hat{L}$ is 5.182 on 1 d.f. This is significant at the 5% level (P = 0.023). The

Table 7.2 Follow-up times and log(bilirubin) values for the 12 patients in the cirrhosis study.

Patient	Follow-up time	Log(bilirubin)
1	47	3.8
	184	4.9
	251	5.0
2	94	2.9
	187	3.1
	321	3.2
3	61	2.8
	97	2.9
	142	3.2
	359	3.4
	440	3.8
4	92	4.7
	194	4.9
	372	5.4
5	87	2.6
	192	2.9
	341	3.4
6	94	2.3
	197	2.8
	384	3.5
	795	3.9
7	74	2.9
	202	3.0
	346	3.0
	917	3.9
	1411	5.1
8	90	2.5
	182	2.9
9	101	2.5
	410	2.7
	774	2.8
	1043	3.4
10	182	2.2
	847	2.8
	1051	3.3
	1347	4.9
11	167	3.9
	498	4.3
12	108	2.8
	187	3.4
	362	3.9
	694	3.8

Table 7.3 Values of $-2\log\hat{L}$ for models with a time-dependent variable.

Terms in model	$-2\log\hat{L}$
null model	25.121
AGE	22.135
LBRT	12.053
AGE, LBRT	11.147

Table 7.4 Values of $-2\log\hat{L}$ for models without a time-dependent variable.

Terms in model	$-2\log\hat{L}$
null model	25.121
AGE	22.135
LBR	21.662
AGE, LBR	18.475

coefficient of $TREAT$ is -3.052, indicating that the drug Liverol is effective in reducing the hazard of death. Indeed, other things being equal, Liverol reduces the hazard of death by a factor of 0.047.

One possible explanation for the difference between the results of these two analyses is that the effect of the treatment is to change the values of the bilirubin level, so that after changes in these values over time have been allowed for, no treatment effect is visible.

7.6 Further reading

The possibility of incorporating time-dependent variables in a proportional hazards model was raised by Cox (1972). The appropriate partial likelihood function was given in his paper, and discussed in greater detail in Cox (1975). Kalbfleisch and Prentice (1980) include a detailed account of the construction of the partial likelihood function. The classification of time-dependent variables outlined in Section 7.1 is due to Prentice and Kalbfleisch (1979), who amplify on this in Kalbfleisch and Prentice (1980). Andersen (1992) reviews the uses of time-dependent variables in survival analysis and includes an example on which the hypothetical study of Example 7.2 is loosely based.

The model described in Section 7.3 in connection with organ transplantation was presented by Crowley and Hu (1977) and Cox and Oakes (1984) in an analysis of the 'Stanford heart transplant data'. This famous data set is given in Crowley and Hu (1977) and Kalbfleisch and Prentice (1980); an update is provided by Cox and Oakes (1984). See also Aitkin, Laird and Francis (1983) and the ensuing discussion.

Relatively little work has been done on incorporating time-dependent variables in a fully parametric model for survival data. However, Petersen (1986) shows how a parametric model with time-dependent variables can be fitted.

8

Interval-censored survival data

In many studies where the response variable is a survival time, the exact time of failure will not be known. Instead, the failure will be known to have occurred during a particular interval of time. Data in this form are known as *grouped* or *interval-censored* survival data.

Interval-censored data commonly arise in studies where there is a non-lethal end-point, such as the recurrence of a disease or condition. However, most survival analyses are based on interval-censored data, in the sense that the survival times are often taken as the nearest day, week or month. In this chapter, some methods for analysing interval-censored data will be described and illustrated.

8.1 Modelling interval-censored survival data

Throughout this chapter, methods for the analysis of interval-censored survival data will be discussed in the context of a study on disease recurrence. In the management of patients who have been cured of ulcers, carcinomas or other recurrent conditions, the patients are usually provided with medication to maintain their recovery. These patients are subsequently examined at regular intervals in order to detect whether a recurrence has occurred. Naturally, some patients may experience symptoms of a recurrence and be subsequently diagnosed as having had a recurrence at a time other than one of the scheduled screening times.

Now suppose that the study is designed to compare two maintenance therapies, a new and a standard treatment, say, and that a number of explanatory variables are recorded for each individual when they are recruited to the study. The vector $\mathbf{x}_i$ will be used to denote the set of values of p explanatory variables, $X_1, X_2, \ldots, X_p$, for the i'th individual in the study. The first of these variables, X_1, will be taken to be an indicator variable corresponding to the treatment group, where $X_1 = 0$ if an individual is on the standard treatment and $X_1 = 1$ if on the new treatment.

Clearly, one way of analysing such data is to ignore the interval censoring. A survival analysis is then carried out on the times of a detected recurrence. However, the data set used in this analysis will be based on a mixture of recurrences

detected at scheduled screening times, known as *screen-detected recurrences* and
recurrences diagnosed following the occurrence of symptoms or *interval-detected
recurrences*. This leads to a difficulty in interpreting the results of the analysis.

To illustrate the problem, consider a study to compare two treatments for sup-
pressing the recurrence of an ulcer, a new and a standard treatment, say. Also
suppose that both treatments have exactly the same effect on the recurrence time,
but that the new treatment suppresses symptoms. The recurrence of an ulcer in
a patient on the new treatment will then tend to be detected later than that in a
patient on the standard treatment. Therefore, interval-detected recurrences will be
identified sooner in a patient on the standard treatment. The interval-detected recurrence times will then be shorter for this group of patients, indicating an apparent
advantage of the new treatment over the standard.

If the time interval between successive screenings is short, relative to the average time to recurrence, there will be few interval-detected recurrences. The standard application of methods for survival analysis will not then be inappropriate.

Example 8.1 Recurrence of an ulcer
In a double blind clinical trial to compare treatments for the inhibition of relapse
after primary therapy has healed an ulcer, patients are randomized to receive one
or other of two treatments, labelled A and B. Regular visits to a clinic were arranged for the patients, and endoscopies were performed 6 months and 12 months
after randomization. A positive endoscopy result indicates that an ulcer has recurred in the time since the last negative result. Information is therefore obtained
on whether or not an ulcer has recurred in the interval from 0 to 6 months or in the
interval from 6 to 12 months. Additionally, some patients presented at the clinic
between scheduled visits suffering from symptoms of recurrence. These patients
had endoscopies at these times in order to detect if a recurrence had in fact occurred.

At entry to the trial, the age of each person, in years, and the duration of verified
disease (1 = less than 5 years, 2 = greater than or equal to five years) was recorded,
in addition to the treatment group (A or B). There are two variables associated with
ulcer detection in the data set, namely the time of the last visit, in months, and the
result of the endoscopy (1 = no ulcer detected, 2 = ulcer detected). Those with
times other than 6 or 12 months had presented with symptoms between scheduled
visits.

The study itself was multinational and the full set of data is given in Whitehead
(1989). In this example, only the data from Belgium will be used, and the relevant
data are given in Table 8.1.

Once an ulcer is detected by endoscopy, a patient is treated for this and is then
no longer in the study. There were some patients who either did not have an endoscopy six months after trial entry, or who dropped out after a negative unsched-
uled endoscopy in the first six months. These patient have been omitted from the
data set on the grounds that there is no information about whether an ulcer has
recurred in the first six months of the study. This means that those patients in Ta-

Table 8.1 Data on the recurrence of an ulcer following treatment for the primary disease.

Patient	Age	Duration	Treatment	Time of last visit	Result
1	48	2	B	7	2
2	73	1	B	12	1
3	54	1	B	12	1
4	58	2	B	12	1
5	56	1	A	12	1
6	49	2	A	12	1
7	71	1	B	12	1
8	41	1	A	12	1
9	23	1	B	12	1
10	37	1	B	5	2
11	38	1	B	12	1
12	76	2	B	12	1
13	38	2	A	12	1
14	27	1	A	6	2
15	47	1	B	6	2
16	54	1	A	6	1
17	38	1	B	10	2
18	27	2	B	7	2
19	58	2	A	12	1
20	75	1	B	12	1
21	25	1	A	12	1
22	58	1	A	12	1
23	63	1	B	12	1
24	41	1	A	12	1
25	47	1	B	12	1
26	58	1	A	3	2
27	74	2	A	2	2
28	75	2	A	6	1
29	72	1	A	12	1
30	59	1	B	12	2
31	52	1	B	12	1
32	75	1	B	12	2
33	76	1	A	12	1
34	34	2	A	6	1
35	36	1	B	12	1
36	59	1	B	12	1
37	44	1	A	12	2
38	28	2	B	12	1
39	62	1	B	12	1
40	23	1	A	12	1
41	49	1	B	12	1
42	61	1	A	12	1
43	33	2	B	12	1

Table 8.2 Values of $-2\log\hat{L}$ on fitting a Cox regression model to the data on the times to a recurrence of an ulcer.

Variables in model	$-2\log\hat{L}$
None	79.189
DUR	79.157
AGE	78.885
$AGE + DUR$	78.872
$AGE + DUR + TREAT$	78.747
$TREAT$	79.097

ble 8.1 whose last visit was greater than 6 months after randomization would have had a negative endoscopy at 6 months.

In modelling the data from this study, duration of disease is denoted by an indicator variable DUR which is zero when the duration is less than 5 years and unity otherwise. The treatment effect is denoted by a variable $TREAT$ which takes the value zero if an individual is on treatment A and unity if on treatment B. The patients age is reflected in the continuous variate AGE.

We first analyse the recurrence times in Table 8.1 ignoring the interval censoring. The recurrence times of those patients who have not had a detected recurrence by the time of their last visit are taken to be censored. The data are now analysed using the Cox proportional hazards model, described in Chapter 3.

From the values of the $-2\log\hat{L}$ statistic for different models, given in Table 8.2, it is clear that neither age nor duration of disease are important prognostic factors. Moreover, the reduction in $-2\log\hat{L}$ on adding the treatment effect to the model, adjusted or unadjusted for the prognostic factors, is nowhere near significant.

The estimated coefficient of $TREAT$ in the model that contains $TREAT$ alone is 0.189, and the standard error of this estimate is 0.627. The estimated hazard of a recurrence under treatment B ($TREAT = 1$), relative to that under treatment A ($TREAT = 0$), is therefore $\exp(0.189) = 1.21$. The standard error of the estimated hazard ratio is found using equation (3.10) in Chapter 3, and is 0.758. The fact that the estimated hazard ratio is greater than unity gives a slight indication that treatment A is superior to treatment B, but not significantly so.

8.2 Modelling the recurrence probability in the follow-up period

Suppose that patients are followed up to time t_s, at which time the last scheduled screening test is carried out. Information on whether or not a recurrence was detected at any time up to and including the last screen is then recorded. Let $p_i(t_s)$ be the probability of a recurrence up to time t_s for the i'th patient, $i = 1, 2, \ldots, n$, with explanatory variables x_i. We now adopt a Cox proportional hazards model, according to which the hazard of a recurrence at t_s, for the i'th patient, is given by

$$h_i(t_s) = \exp(\eta_i)h_0(t_s),$$

where $\eta_i = \beta'\mathbf{x}_i = \beta_1 x_{1i} + \beta_2 x_{2i} + \cdots + \beta_p x_{pi}$, and $h_0(t_s)$ is the baseline hazard function at t_s.

The probability that the i'th individual experiences a recurrence after time t_s is the survivor function $S_i(t_s)$, so that $S_i(t_s) = 1 - p_i(t_s)$. Now, from equation (3.19) in Section 3.8 of Chapter 3,

$$S_i(t_s) = [S_0(t_s)]^{\exp(\eta_i)}, \tag{8.1}$$

where $S_0(t_s)$ is the value of the survivor function at t_s for an individual on the standard treatment for whom all the other explanatory variables are zero. The probability of a recurrence up to time t_s under this model is therefore

$$p_i(t_s) = 1 - [S_0(t_s)]^{\exp(\eta_i)},$$

and so

$$\log[-\log\{1 - p_i(t_s)\}] = \eta_i + \log\{-\log S_0(t_s)\}.$$

Writing $\beta_0 = \log\{-\log S_0(t_s)\}$, the model can be expressed as

$$\log[-\log\{1 - p_i(t_s)\}] = \beta_0 + \beta_1 x_{1i} + \beta_2 x_{2i} + \cdots + \beta_p x_{pi}. \tag{8.2}$$

This is a linear model for the complementary log-log transformation of the probability of a recurrence up to time t_s. The model can be fitted to data on a binary response variable that takes the value zero for those individuals in the study who have not experienced a recurrence before t_s, the time of the last screen, and unity otherwise.

As in modelling survival data, models fitted to data on a binary response variable can be compared on the basis of the statistic $-2\log \hat{L}$. Here, $\hat{L}$ is the maximized likelihood of the binary data, and $-2\log \hat{L}$ is generally known as the *deviance*. Differences in the deviance for two nested models have an asymptotic chi-squared distribution, and so models fitted to binary data can be compared in the same manner as the models used in survival analysis.

When the model in equation (8.2) is fitted to the observed data, the estimate of the constant, $\hat{\beta}_0$, is an estimate of $\log\{-\log S_0(t_s)\}$, from which an estimate of the baseline survivor function at t_s can be obtained. Also, the ratio of the hazard of a recurrence for an individual on the new treatment, relative to one on the standard, is $\exp(\beta_1)$. This can be estimated by $\exp(\hat{\beta}_1)$, where $\hat{\beta}_1$ is the parameter estimate corresponding to X_1, the indicator variable that corresponds to the treatment group. Values of the hazard ratio less than unity suggest that the risk of a recurrence at any time is smaller under the new treatment than under the standard. A confidence interval for the hazard ratio may be obtained from the standard error of $\hat{\beta}_1$ in the usual manner.

This method of estimating the hazard ratio from interval-censored survival data is not particularly efficient, since data on the times that a recurrence is detected are

Table 8.3 Deviances on fitting complementary log-log models to data on the recurrence of an ulcer in 12 months.

Variables in model	Deviance	d.f.
Constant	48.902	42
DUR	48.899	41
AGE	48.573	41
$TREAT$	48.531	41
$DUR + AGE$	48.565	40
$DUR + TREAT$	48.531	40
$AGE + TREAT$	48.175	40
$DUR + AGE + TREAT$	48.172	39
$DUR + AGE + TREAT + TREAT \times AGE$	47.944	38 '
$DUR + AGE + TREAT + TREAT \times DUR$	48.062	38

not utilized. However, the method is appropriate when interest simply centres on the risk of a recurrence in a specific time period. It is also the method that would be adopted in modelling quantities such as the probability of a relapse in the first year of treatment, or the probability of no recurrence in a five-year period after trial entry.

Example 8.2 Recurrence of an ulcer
We now model the probability of an ulcer recurring in the 12 months following recruitment to the study described in Example 8.1. Of the 43 patients in the data set, 11 of them had experienced a recurrence in this 12-month period, namely patients 1, 10, 14, 15, 17, 18, 26, 27, 30, 32 and 37. A binary response variable is now defined, which takes the value unity if a patient has experienced a recurrence and zero otherwise. A model in which the complementary log-log transformation of the recurrence probability is related to age, duration of disease and treatment group is then fitted to the binary observations.

Table 8.3 gives the deviances on fitting complementary log-log models with different terms to the binary response variable. All the models fitted include a constant term. The deviances in this table have been obtained using the software package GLIM, but other packages, such as Genstat, SAS and BMDP, also have the necessary facilities.

In this example, the effects of age, duration of disease and treatment group have been modelled using the variates AGE, DUR, and $TREAT$, defined in Example 8.1. However, factors corresponding to duration and treatment could have been used in conjunction with packages that allow factors to be included directly. This would not make any difference to the deviances in Table 8.3, but it may have an effect on the interpretation of the parameter estimates. See Sections 3.2 and 3.7 for fuller details.

It is clear from Table 8.3 that no variable reduces the deviance by a significant amount. For example, the change in the deviance on adding $TREAT$ to the model that only contains a constant is 0.371, which is certainly not significant when compared to percentage points of the chi-squared distribution on 1 d.f. Approximately the same change in deviance results when $TREAT$ is added to the model that contains AGE and DUR, showing that the treatment effect is of a similar magnitude after allowing for these two variables. Moreover, there is no evidence whatsoever of an interaction between treatment and the variables AGE and DUR.

On fitting a model that contains $TREAT$ alone, the estimated coefficient of $TREAT$ is 0.378, with a standard error of 0.629. Thus the ratio of the hazard of a recurrence before 12 months in a patient on treatment B ($TREAT = 1$), relative to that for a patient on treatment A ($TREAT = 0$), is $\exp(0.378) = 1.46$. The risk of a recurrence in the year following randomization is thus greater under treatment B than it is under treatment A, but not significantly so. This hazard ratio is not too different from the value of 1.21 obtained in Example 8.1. The standard error of the estimated hazard ratio, again found using equation (3.10) in Chapter 3, is 0.918, which is also very similar to that found in Example 8.1.

A 95% confidence interval for the log-hazard ratio has limits of $0.378 \pm 1.96 \times 0.629$, and so the corresponding interval estimate for the hazard ratio itself is $(0.43, 5.01)$. Notice that this interval includes unity, a result which was foreshadowed by the non-significant treatment effect.

The estimated constant term in this fitted model is -1.442. This is an estimate of $\log\{-\log S_0(12)\}$, the survivor function at 12 months for a patient on treatment A. The estimated probability of a recurrence after 12 months for a patient on treatment A is therefore $\exp(-e^{-1.442}) = 0.79$. The corresponding value for a patient on treatment B is $0.79^{\exp(0.378)} = 0.71$. The probabilities of a recurrence in the first 12 months are therefore 0.21 for a patient on treatment A, and 0.29 for a patient on treatment B. This again shows that patients on treatment B have a slightly higher probability of the recurrence of an ulcer in the year following randomization.

8.3 * Modelling the recurrence probability at different times

In this method for analysing interval-censored survival data, information about whether or not a recurrence is detected at different examination times is taken into account.

Suppose that patients enter a study at time 0 and are followed up to time t_k. During the course of this follow-up period, the individuals are screened on a regular basis in order to detect a recurrence of the disease or condition under study. Denote the examination times by $t_1, t_2, \ldots, t_k$. Further, let t_0 denote the time origin, so that $t_0 = 0$ and let $t_{k+1} = \infty$.

For each individual, information will be recorded on whether or not a recurrence has occurred at times $t_1, t_2, \ldots, t_k$. It can then be determined whether a given individual has experienced a recurrence in the j'th time interval from t_{j-1} to t_j. Thus a patient who has a recurrence detected at time t_j has an actual recurrence time of t, where $t_{j-1} \leq t < t_j$, $j = 1, 2, \ldots, k$. Note that the study will not provide any information about whether or not a recurrence occurs after the final screening time, t_k.

Now let p_{ij} be the probability of a recurrence being detected in the i'th patient, $i = 1, 2, \ldots, n$, at time t_j, so that p_{ij} is the probability that patient i experiences a recurrence in the j'th time interval, $j = 1, 2, \ldots, k + 1$. Also let π_{ij} be the probability that the i'th of n patients is found to be free of the disease at time t_{j-1} and has a recurrence in the j'th time interval, $j = 1, 2, \ldots, k + 1$. This is therefore the conditional probability of a recurrence in the j'th interval, given that the recurrence occurs after t_{j-1}. Using T_i to denote the random variable associated with the recurrence time of the i'th individual, we therefore have

$$p_{ij} = P(t_{j-1} \leq T_i < t_j),$$

and

$$\pi_{ij} = P(t_{j-1} \leq T_i < t_j \mid T_i \geq t_{j-1}),$$

for $j = 1, 2, \ldots, k + 1$. It then follows that

$$p_{ij} = (1 - \pi_{i1})(1 - \pi_{i2}) \cdots (1 - \pi_{i,j-1})\pi_{ij}, \qquad (8.3)$$

for $j = 2, 3, \ldots, k + 1$, with $p_{i1} = \pi_{i1}$.

Now let r_{ij} be unity if the i'th patient has a recurrence detected in the interval from t_{j-1} to t_j, $j = 1, 2, \ldots, k + 1$, and zero if no recurrence is detected in that interval. Also let s_{ij} be unity if a patient has a detected recurrence after t_j, and zero otherwise. Then

$$s_{ij} = r_{i,j+1} + r_{i,j+2} + \cdots + r_{i,k+1},$$

for $j = 1, 2, \ldots, k$.

The sample likelihood of the $n(k + 1)$ values r_{ij} is

$$\prod_{i=1}^{n} \prod_{j=1}^{k+1} p_{ij}^{r_{ij}},$$

and on substituting for p_{ij} from equation (8.3), the likelihood function becomes

$$\prod_{i=1}^{n} \prod_{j=1}^{k+1} \{(1 - \pi_{i1}) \cdots (1 - \pi_{i,j-1})\pi_{ij}\}^{r_{ij}}.$$

This function can be written as

$$\prod_{i=1}^{n} \pi_{i1}^{r_{i1}} [(1 - \pi_{i1})\pi_{i2}]^{r_{i2}} \cdots [(1 - \pi_{i1}) \cdots (1 - \pi_{ik})\pi_{i,k+1}]^{r_{i,k+1}},$$

which reduces to

$$\prod_{i=1}^{n} \pi_{i,k+1}^{r_{i,k+1}} \prod_{j=1}^{k} \pi_{ij}^{r_{ij}} (1 - \pi_{ij})^{s_{ij}}. \qquad (8.4)$$

When the end-point is the recurrence of a disease or condition, any such recurrence will not be detected after the time of the last examination, so that no recurrence can be detected in the $(k+1)$'th time interval (t_k, ∞). This means that $r_{i,k+1} = 0$ and so the term $\pi_{i,k+1}^{r_{i,k+1}}$ in equation (8.4) will be unity. On the other hand, if the end-point is death, the probability of death in the interval (t_{k+1}, ∞), conditional on being alive at time t_{k+1} is unity. Hence $\pi_{i,,k+1} = 1$, and again $\pi_{i,k+1}^{r_{i,k+1}}$ will be unity. The likelihood function in equation (8.4) therefore becomes

$$\prod_{i=1}^{n}\prod_{j=1}^{k} \pi_{ij}^{r_{ij}}(1 - \pi_{ij})^{s_{ij}}. \tag{8.5}$$

This is the likelihood function for nk observations r_{ij} from a binomial distribution with response probability π_{ij} and where the binomial denominator is $r_{ij} + s_{ij}$. This denominator is equal to unity when a patient is at risk of having a detected recurrence after time t_j, and zero otherwise. In fact, the denominator is zero when both r_{ij} and s_{ij} are equal to zero, and the likelihood function in expression (8.5) is unaffected by observations for which $r_{ij} + s_{ij} = 0$. Data records for which the binomial denominator is zero are therefore uninformative, and so they can be omitted from the data set. If there are m observations remaining after these deletions, so that $m \leq nk$, the likelihood function in expression (8.5) is that of m observations from binomial distributions with parameters 1 and π_{ij}, in other words m Bernoulli observations.

The next step is to note that for the i'th patient

$$1 - \pi_{ij} = P(T_i \geq t_j \mid T_i \geq t_{j-1}),$$

so that

$$1 - \pi_{ij} = \frac{S_i(t_j)}{S_i(t_{j-1})}.$$

Adopting a proportional hazards model for the recurrence times, the hazard of a recurrence being detected at time t_j in the i'th individual can be expressed as

$$h_i(t_j) = \exp(\eta_i)h_0(t_j),$$

where $h_0(t_j)$ is the baseline hazard at t_j, and η_i is the risk score for the i'th individual. Notice that this assumption means that the hazards need only be proportional at the scheduled screening times t_j, and not at intermediate times. This is less restrictive than the usual proportional hazards assumption which requires that hazards be proportional at every time.

Using the result in equation (8.1),

$$1 - \pi_{ij} = \left[\frac{S_0(t_j)}{S_0(t_{j-1})}\right]^{\exp(\eta_i)},$$

and on taking logarithms we find that

$$\log(1 - \pi_{ij}) = \exp(\eta_i)\log\left\{S_0(t_j)/S_0(t_{j-1})\right\}.$$

Consequently,

$$\log\{-\log(1 - \pi_{ij})\} = \eta_i + \log\left[-\log\{S_0(t_j)/S_0(t_{j-1})\}\right]$$
$$= \eta_i + \gamma_j,$$

say. This is a linear model for the complementary log-log transformation of π_{ij}, in which the parameters γ_j, $j = 1, 2, \ldots, k$, are associated with the k time intervals. The model can be fitted using standard methods for modelling binary data.

In modelling the probability of a recurrence in the j'th time interval for the i'th patient, π_{ij}, the data are the values r_{ij}. Data records for which both r_{ij} and s_{ij} are equal to zero are omitted, and so the binomial denominator is unity for each remaining observation. The parameters γ_j are incorporated in the model by fitting terms corresponding to a k-level factor associated with the period of observation, or by including suitable indicator variables as described in Section 3.2. Note that a constant term is not included in the model. The estimates of the β-coefficients in η_i, obtained on fitting this model, can again be interpreted as log-hazard ratios. Also, estimates of the γ_j can be used to obtain estimates of the π_{ij}. This process is illustrated in Example 8.3 below.

Example 8.3 Recurrence of an ulcer
The data on the time to detection of an ulcer recurrence, given in Example 8.1, are now analysed using the method described above. To prepare the data set for analysis using this approach, the two additional variables, $PERIOD$ and R, are introduced. The first of these, $PERIOD$, is used to signify the period, and the variable is given the value unity for each observation. The second variable, R, contains the values r_{i1}, $i = 1, 2, \ldots, 43$, and so R is equal to unity if an ulcer is detected in the first period and zero otherwise. For these data, patients 10, 14, 15, 26 and 27 experienced a recurrence in the interval from 0 to 6 months, and so the value of R is unity for these five individuals and zero for the remaining 38.

We then add a second block of data to this set. This block is a duplication of the records for the patients who have not had a recurrence at the six-month mark and for whom the last visit is made after 6 months. There are 38 patients who have not had a recurrence at six months, but three of these, patients 16, 28 and 34, took no further part in the study. The second block of data therefore contains 35 records. The variable $PERIOD$ now takes the value 2 for these 35 observations, since they correspond to the second time period. The variable R contains the values r_{i2} for this second block of data. Therefore, R takes the value unity for patients 1, 17, 18, 30, 32 and 37 and zero otherwise, since these are the only six patients who have a detectable recurrence at 12 months.

The combined set of data has $43 + 35 = 78$ rows, and includes the variable $PERIOD$ which defines the period in which an endoscopy is performed ($1 = 0$–6 months, $2 = 6$–12 months), and variable R which defines the endoscopy result ($0 = $ negative, $1 = $ positive). The value of s_{ij} is unity for all records except those for which $r_{ij} = 1$ when it is zero. The binomial denominators $r_{ij} + s_{ij}$ are therefore equal to unity for each patient, since every patient in the extended data set is at

Table 8.4 Modified data on the recurrence of an ulcer in two periods for the first 18 patients.

Patient	Age	Duration	Treat-ment	Time of last visit	Result	*PERIOD*	R
1	48	2	B	7	2	1	0
1	48	2	B	7	2	2	1
2	73	1	B	12	1	1	0
2	73	1	B	12	1	2	0
3	54	1	B	12	1	1	0
3	54	1	B	12	1	2	0
4	58	2	B	12	1	1	0
4	58	2	B	12	1	2	0
5	56	1	A	12	1	1	0
5	56	1	A	12	1	2	0
6	49	2	A	12	1	1	0
6	49	2	A	12	1	2	0
7	71	1	B	12	1	1	0
7	71	1	B	12	1	2	0
8	41	1	A	12	1	1	0
8	41	1	A	12	1	2	0
9	23	1	B	12	1	1	0
9	23	1	B	12	1	2	0
10	37	1	B	5	2	1	1
11	38	1	B	12	1	1	0
11	38	1	B	12	1	2	0
12	76	2	B	12	1	1	0
12	76	2	B	12	1	2	0
13	38	2	A	12	1	1	0
13	38	2	A	12	1	2	0
14	27	1	A	6	2	1	1
15	47	1	B	6	2	1	1
16	54	1	A	6	1	1	0
17	38	1	B	10	2	1	0
17	38	1	B	10	2	2	1
18	27	2	B	7	2	1	0
18	27	2	B	7	2	2	1

risk of a detectable recurrence. Instead of giving a full listing of the modified data set, the records for the patient × period combinations for the first 18 patients are shown in Table 8.4.

The dependence of the complementary log-log transformation of the probabilities π_{ij} on certain explanatory variables can now be investigated by fitting models to the binary response variable R. Each model includes a two-level factor, *PERIOD*, associated with the period, but no constant term. The factor will have

Table 8.5 Deviances on fitting complementary log-log models that do not include a constant to the variable R.

Terms fitted in model	Deviance	d.f.
$PERIOD$	62.982	76
$PERIOD + AGE$	62.685	75
$PERIOD + DUR$	62.979	75
$PERIOD + AGE + DUR$	62.685	74
$PERIOD + AGE + DUR + TREAT$	62.566	73
$PERIOD + AGE + DUR + TREAT + TREAT \times AGE$	62.278	72
$PERIOD + AGE + DUR + TREAT + TREAT \times DUR$	62.552	72
$PERIOD + TREAT$	62.884	75

2 d.f., and the corresponding term in the model is γ_j, $j = 1, 2$. The deviances for the models fitted are summarized in Table 8.5.

From this table we see that the effect of adding either AGE or DUR to the model that contains $PERIOD$ alone is to reduce the deviance by less than 0.3. There is therefore no evidence that the age of a person or the duration of disease are associated with a recurrence. Adding $TREAT$ to the model that contains $PERIOD$ alone, the reduction in deviance is 0.10 on 1 d.f. This leads us to conclude that there is no significant difference between the two treatments. The treatment effect, after adjusting for the variables AGE and DUR, is of a similar magnitude.

To check whether there are interactions between treatment and the two prognostic factors, we look at the effect of adding the terms $TREAT \times AGE$ and $TREAT \times DUR$ to that model that contains $PERIOD, AGE$ and DUR. From Table 8.5, the resulting change in deviance is very small, and so there is no evidence of any such interactions.

In summary, the modelling process shows that π_{ij}, the probability that the i'th patient has a recurrence in the j'th period, does not depend on the patient's age or the duration of the disease, and, more importantly, does not dependent on the treatment group.

To further quantify the treatment effect, consider the model that includes both $TREAT$ and $PERIOD$. The equation of the fitted model can be written as

$$\log\{-\log(1 - \hat{\pi}_{ij})\} = \hat{\gamma}_j + \hat{\beta} TREAT_i, \tag{8.6}$$

where γ_j is the effect of the j'th period, $j = 1, 2$, and $TREAT_i$ is the value of the indicator variable $TREAT$, for the i'th individual. This variable is zero if that patient is on treatment A and unity otherwise.

The estimated coefficient of $TREAT$ in this model is 0.195 and the standard error of this estimate is 0.626. The hazard of a recurrence on treatment B at any given time, relative to that on treatment A, is $\exp(0.195) = 1.21$. Since this exceeds unity, there is the suggestion that the risk of recurrence is less on treatment

Table 8.6 Fitted and observed probabilities of an ulcer recurring in the two time periods.

Period	Treatment A		Treatment B	
	fitted	observed	fitted	observed
$(0, 6)$	0.104	0.158	0.125	0.083
$(6, 12)$	0.153	0.077	0.183	0.227

A than on treatment B, but the evidence for this is not statistically significant. The standard error of the estimated hazard ratio is 0.757. For comparison, from Example 8.2, the estimated hazard ratio at 12 months was found to be 1.46, with a standard error of 0.918. These values are very similar to those obtained in this example. Moreover, the results of analyses that accommodate interval censoring are comparable to those found in Example 8.1, in which the Cox proportional hazards model was used without taking account of the fact that the data are interval-censored.

The model in equation (8.6) can be used to provide estimates of the π_{ij}. The estimates of the period effects in this model are $\hat{\gamma}_1 = -2.206$, $\hat{\gamma}_2 = -1.794$, and so the estimated probability of a recurrence in the first period, for a patient on treatment A, denoted $\hat{\pi}_{A1}$, is given by

$$\log\{-\log(1 - \hat{\pi}_{A1})\} = \hat{\gamma}_1 + \hat{\beta} \times 0.$$

Therefore,

$$\log\{-\log(1 - \hat{\pi}_{A1})\} = -2.206,$$

and from this, $\hat{\pi}_{A1} = 0.104$. Other fitted probabilities can be calculated in a similar manner, and the results of these calculations are shown in Table 8.6. The corresponding observed proportions of individuals with a recurrence for each combination of treatment and period are also displayed. The agreement between the observed and fitted probabilities is good, which indicates that the model is a good fit.

If desired, probabilities of a recurrence in either period 1 or period 2 could also be estimated. The probability that a patient on treatment A has a recurrence in either period 1 or period 2 is

P[recurrence in $(0, 6)$]+P[recurrence in $(6, 12)$ and no recurrence in $(0, 6)$].

The joint probability of a recurrence in $(6, 12)$ and no recurrence in $(0, 6)$ can be expressed as

P[recurrence in $(6, 12)$ | no recurrence in $(0, 6)$]$\times$P[no recurrence in $(0, 6)$],

and so the required probability is estimated by

$$\hat{\pi}_{A1} + \hat{\pi}_{A2}(1 - \hat{\pi}_{A1}) = 0.104 + 0.153 \times 0.896 = 0.241.$$

Similarly, that for treatment B is

$$\hat{\pi}_{B1} + \hat{\pi}_{B2}(1 - \hat{\pi}_{B1}) = 0.125 + 0.183 \times 0.875 = 0.285.$$

This again indicates the superiority of treatment A, but there is insufficient data for this effect to be declared significant.

8.4 Discussion

The method for analysing interval-censored data that has been described and illustrated in Section 8.3 will generally be the most appropriate. However, this approach is not optimal since recurrences detected between scheduled examinations, that is, interval-detected recurrences, are only counted at the next examination time. If the intervals between successive examination times are not too large, the difference between the results of an analysis based on the model in Section 8.3, and one which uses the actual times of interval-detected recurrences, will be negligible. In fact, if the number of intervals is not too small, and the time between successive examinations not too large, the results will not be too different from an analysis that assumes the recurrence times to be continuous, outlined in Section 8.1.

As mentioned earlier, the model described in Section 8.3 only requires hazards to be proportional at scheduled screening times. This means that the model is useful when the hazards are not necessarily proportional between screening times. Furthermore, the model could be relevant in situations where although actual survival times are available, the hazards can only be taken to be proportional at specific times. Further comments on methods for analysing survival data where hazards are non-proportional are included in Chapter 10.

One final point concerns the design of studies where events such as disease recurrence can only be detected at an examination. If the model described in Section 8.3 is to be used in the analysis, the examination times must be the same for each patient in the study. Care must therefore be taken to ensure that each patient keeps to the prescribed examination schedule.

8.5 Further reading

Much of this chapter is based on the summary of methods for processing interval-censored survival data given by Whitehead (1989). The approach described in Section 8.3 is based on Prentice and Gloeckler (1978).

Collett (1991) describes a model-based approach to the analysis of binary data. This book includes a description of the facilities available for modelling binary data in the major software packages. Other books that include material on the analysis of binary data include Cox and Snell (1989), Dobson (1990) and Morgan (1992).

For the method of Section 8.3 to be feasible, the schedule of screening times must be the same for each patient in the study. Severe complications arise if this is not the case. The form of the appropriate likelihood function in this situation is described by Finkelstein (1986), but computer software to implement the methodology is not readily available.

The use of the complementary log-log transformation in the analysis of interval-censored data was described by Thompson (1981). Becker and Melbye (1991) show how a log-linear model can be used to obtain an estimate of the survivor function from interval-censored data, assuming a constant hazard in each interval. Whitehead (1992) shows how the log-rank test for comparing two treatment groups can be adapted for use in processing interval-censored survival data. However, since this approach does not allow other explanatory variables to be taken account of, details are not given here.

9

Sample size requirements for a survival study

There are many aspects of the design of a clinical trial that must be considered when the response variable is a survival time. These include all the usual matters, such as patient eligibility, definition of the treatments, the method of randomization to be employed in allocating patients to treatment group, and the use of blinding. In addition, care must be taken to define both the time origin and the end-point of the study in a clear and unambiguous manner. Consideration might also be given to whether the study should be based on a fixed number of patients, or whether a sequential design should be adopted, in which the study continues until there is a sufficient number of events to be able to distinguish between two treatments. The desirability of performing interim analyses might also need to be addressed.

In a book of this nature, there is insufficient space to be able to go into any of these topics in detail. Fortunately there are a number of excellent books that include material on the design of clinical trials, some of which include material on the design of trials for survival analysis, such as Pocock (1983).

However, there is one matter in the design of fixed sample size studies that will be discussed here. This is the crucial issue of the number of patients that are required in such studies. If too few patients are recruited, there may be insufficient information available in the data to enable a treatment difference to be pronounced significant. On the other hand, it is unethical to waste resources in studies that are unnecessarily large.

9.1 Distinguishing between two treatment groups

Many survival studies are concerned with distinguishing between two alternative treatments. For this reason, a study to compare the survival times of patients who receive a new treatment with those who receive a standard will be used as the focus for this chapter.

Suppose that in this study, there are two groups of patients, and that the standard treatment is allocated to the patients in Group I, while the new treatment is allocated to those in Group II. Assuming a proportional hazards model for the survival times, the hazard of death at time t for a patient on the new treatment, $h_N(t)$, can be written as

$$h_N(t) = \psi h_S(t),$$

where $h_S(t)$ is the hazard function at t for a patient on the standard treatment and ψ is the unknown hazard ratio. We will also define $\theta = \log \psi$ to be the log-hazard ratio. If θ is zero, there is no treatment difference. On the other hand, negative values of θ indicate that survival is longer under the new treatment, while positive values of θ indicate that patients survive longer on the standard treatment.

In order to test the null hypothesis that $\theta = 0$, the log-rank test described in Section 2.5 can be used. As was shown in Section 3.9, this is equivalent to using the score test of the null hypothesis of equal hazards in the Cox regression model. In this chapter, sample size requirements will be based on the log-rank test statistic, but the results will naturally apply when an analysis based on the Cox regression model is envisaged.

In a survival study, the occurrence of censoring means that it is not usually possible to measure the actual survival times of all patients in the study. However, it is the number of actual deaths that is important in the analysis, rather than the total number of subjects. Accordingly, the first step in determining the number of patients in a study is to calculate the number of deaths that must be observed. We then go on to determine the required number of subjects.

9.2 Calculating the required number of deaths

To determine the sample size requirement for a study, we calculate the number of subjects needed for there to be a certain chance of declaring θ to be significantly different from zero when the true, but unknown, log-hazard ratio is θ_R. Here, θ_R is the *reference value* of θ. It will be a reflection of the magnitude of the treatment difference that it is important to detect, using the test of significance. In a study to compare a new treatment with a standard there is likely to be a minimum worthwhile improvement and a maximum envisaged improvement. The actual choice of θ_R will then lie between these two values. In practice, θ_R might be chosen on the basis of the increase in the median survival time that is to be detected, or in terms of the probability of survival beyond some time. This is discussed further and illustrated later in Example 9.1.

More formally, the required number of deaths is taken to be such that there is a probability of $1 - \beta$ of declaring the observed log-hazard ratio to be significantly different from zero, using a hypothesis test with a specified significance level of α, when in fact $\theta = \theta_R$. The quantity $1 - \beta$ is the probability of rejecting the null hypothesis when it is in fact false, and is known as the *power* of the test. The term β is the probability of not rejecting the null hypothesis when it is false and is sometimes known as the *type II error*. Both α and β are taken to be small. Typical values will be $\alpha = 0.05$ and $\beta = 0.1$, and with these values there would

Table 9.1   Values of the function $c(\alpha, \beta)$.

Value of α	Value of $1 - \beta$			
	0.80	0.90	0.95	0.99
0.10	6.18	8.56	10.82	15.77
0.05	7.85	10.51	13.00	18.37
0.01	11.68	14.88	17.81	24.03
0.001	17.08	20.90	24.36	31.55

be a 90% chance of declaring the observed difference between two treatments to be significant at the 5% level. The exact specification of α and β will to some extent depend on the circumstances. If it is important to detect a difference as being significant at a lower level of significance, or if there needs to be a higher chance of declaring a result to be significant, α and β will need to be modified accordingly.

The required number of deaths in a survival study, d, can be obtained from the equation

$$d = \frac{4(z_{\alpha/2} + z_\beta)^2}{\theta_R^2}, \tag{9.1}$$

where $z_{\alpha/2}$ and z_β are the upper $\alpha/2$- and upper β-points, respectively, of the standard normal distribution. It is convenient to write $c(\alpha, \beta) = (z_{\alpha/2} + z_\beta)^2$ in equation (9.1), giving

$$d = 4c(\alpha, \beta)/\theta_R^2. \tag{9.2}$$

The values of $c(\alpha, \beta)$ for commonly chosen values of the significance level α and power $1 - \beta$ are given in Table 9.1.

Calculation of the required number of deaths then requires that a value for θ_R be identified, and appropriate values of α and β chosen. Table 9.1 is then used in conjunction with equation (9.2) to give the number of deaths required in a study.

The derivation of the result in equation (9.2) assumes that the same number of individuals is to be assigned to each treatment group. If this is not the case, a modification has to be made. In particular, if the proportions of individuals to be allocated to Groups I and II are π_1 and π_2 respectively, the required total number of deaths becomes

$$d = \frac{c(\alpha, \beta)}{\pi_1 \pi_2 \theta_R^2}.$$

Notice that an imbalance in the number of individuals in the two treatment groups leads to an increase in the total number of deaths required. The derivation also includes an approximation which means that the calculated number of deaths could be an underestimate. Some judicious rounding up of the calculated value is therefore suggested to compensate for this.

The actual derivation of the formula for the required number of deaths is important and so details are given below in Section 9.2.1. This section can be omitted without loss of continuity. It is followed by an example which illustrates the calculations.

9.2.1* Derivation of the required number of deaths

An expression for the required number of deaths is now derived on the basis of a log-rank test to compare two treatment groups. As in Section 2.5, suppose that there are r distinct death times, $t_{(1)} < t_{(2)} < \cdots < t_{(r)}$, amongst the individuals in the study, and that in the i'th group there are d_{ij} deaths at the j'th ordered death time $t_{(j)}$, for $i = 1, 2$ and $j = 1, 2, \ldots, r$. Also suppose that the number at risk at $t_{(j)}$ in the i'th group is n_{ij}, and write $n_j = n_{1j} + n_{2j}$ for the total number at risk at $t_{(j)}$ and $d_j = d_{1j} + d_{2j}$ for the number who die at $t_{(j)}$. The log-rank statistic is then

$$U = \sum_{j=1}^{r}(d_{1j} - e_{1j}),$$

where e_{1j} is the expected number of deaths in Group I at $t_{(j)}$, given by $e_{1j} = n_{1j}d_j/n_j$, and the variance of the log-rank statistic is

$$V = \sum_{j=1}^{r} \frac{n_{1j}n_{2j}d_j(n_j - d_j)}{n_j^2(n_j - 1)}. \tag{9.3}$$

When using the log-rank test, the null hypothesis that $\theta = 0$ is rejected if the absolute value of U is sufficiently large, that is, if $|U| > k$, say, where $k > 0$ is a constant. We therefore require that

$$P(|U| > k; \theta = 0) = \alpha, \tag{9.4}$$

and

$$P(|U| > k; \theta = \theta_R) = 1 - \beta. \tag{9.5}$$

for a two-sided $100\alpha\%$ significance test to have a power $1 - \beta$.

We now quote without proof a result given in Sellke and Siegmund (1983), according to which the log-rank statistic, U, has an approximate normal distribution with mean θV and variance V, for small values of θ. Indeed, the result that $U \sim N(0, V)$ under the null hypothesis $\theta = 0$, is used as a basis for the log-rank test. Then, since

$$P(|U| > k; \theta = 0) = P(U > k; \theta = 0) + P(U < -k; \theta = 0),$$

and U has an $N(0, V)$ distribution when $\theta = 0$, a distribution that is symmetric about zero,

$$P(U > k; \theta = 0) = P(U < -k; \theta = 0).$$

It then follows from equation (9.4) that

$$P(U > k; \theta = 0) = \frac{\alpha}{2}. \tag{9.6}$$

Next, we note that

$$P(|U| > k; \theta = \theta_R) = P(U > k; \theta = \theta_R) + P(U < -k; \theta = \theta_R).$$

For the sort of values of k that are likely to be used in the hypothesis test, either $P(U < -k; \theta = \theta_R)$ or $P(U > k; \theta = \theta_R)$ will be negligible. For example, if the new treatment is expected to increase survival so that θ_R is taken to be less than zero, the probability of U having a value in excess of k, $k > 0$, will be small. So without loss of generality we will take

$$P(|U| > k; \theta = \theta_R) \approx P(U < -k; \theta = \theta_R).$$

We now denote the upper $100p\%$ point of the standard normal distribution by z_p. Then $\Phi(z_p) = 1 - p$, where $\Phi(\ )$ stands for the standard normal distribution function. The quantity $\Phi(z_p)$ therefore represents the area under a standard normal density function to the left of the value z_p. Now, since $U \sim N(0, V)$ when $\theta = 0$,

$$P(U > k; \theta = 0) = 1 - P(U \leq k; \theta = 0) = 1 - \Phi\left(\frac{k}{\sqrt{(V)}}\right),$$

and using equation (9.6) we have that

$$\Phi\left(\frac{k}{\sqrt{(V)}}\right) = 1 - (\alpha/2).$$

Therefore,

$$\frac{k}{\sqrt{(V)}} = z_{\alpha/2},$$

where $z_{\alpha/2}$ is the upper $\alpha/2$-point of the standard normal distribution, and so k can be expressed as

$$k = z_{\alpha/2}\sqrt{(V)}. \tag{9.7}$$

In a similar manner, since $U \sim N(\theta_R V, V)$ when $\theta = \theta_R$,

$$P(U < -k; \theta = \theta_R) = \Phi\left(\frac{-k - \theta_R V}{\sqrt{(V)}}\right) \approx 1 - \beta,$$

and so we take

$$\frac{-k - \theta_R V}{\sqrt{(V)}} = z_\beta,$$

where z_β is the upper β-point of the standard normal distribution. If we now substitute for k from equation (9.7), we get

$$-z_{\alpha/2}\sqrt{(V)} - \theta_R V = z_\beta\sqrt{(V)},$$

and so V needs to be such that

$$V = (z_{\alpha/2} + z_\beta)^2/\theta_R^2, \tag{9.8}$$

to meet the specified requirements.

When the number of deaths is few relative to the number at risk, the expression for V in equation (9.3) is approximately

$$\sum_{j=1}^{r} \frac{n_{1j}n_{2j}d_j}{n_j^2}. \tag{9.9}$$

Moreover, if θ is small, and recruitment to each treatment group proceeds at a similar rate, then $n_{1j} \approx n_{2j}$, for $j = 1, 2, \ldots, r$, and so

$$\frac{n_{1j}n_{2j}}{n_j^2} = \frac{n_{1j}n_{2j}}{(n_{1j} + n_{2j})^2} \approx \frac{n_{1j}^2}{(2n_{1j})^2} = \frac{1}{4}.$$

Then, V is given by

$$V \approx \sum_{j=1}^{r} d_j/4 = d/4,$$

say, where $d = \sum_{j=1}^{r} d_j$ is the total number of deaths amongst the individuals in the study.

Finally, using equation (9.8), we now require d to be such that

$$\frac{d}{4} = \frac{(z_{\alpha/2} + z_\beta)^2}{\theta_R^2},$$

which leads to the required number of deaths being that given in equation (9.1).

At later death times, that is, when the values of j in expression (9.9) are close to r, the numbers of subjects at risk in the two groups will be small. This is likely to mean that n_{1j} and n_{2j} will be quite different at the later death times, and so $n_{1j}n_{2j}/n_j^2$ will be less than 0.25. This in turn means that $V < d/4$ and so the required number of deaths will tend to be underestimated.

Example 9.1 Survival from chronic active hepatitis
Patients suffering from chronic active hepatitis rapidly progress to an early death from liver failure. A new treatment has become available and so a clinical trial is planned to evaluate the effect of this new treatment on the survival times of patients suffering from the disease. As a first step, information is obtained on the survival times in years of patients in a similar age range who have received the standard therapy. The Kaplan–Meier estimate of the survivor function derived from such data is shown in Figure 9.1.

From this estimate of the survivor function, the median survival time is 3.3 years, and the survival rates at two, four and six years can be taken to be given by $S(2) = 0.70$, $S(4) = 0.45$, and $S(6) = 0.25$.

The new treatment is expected to increase the survival rate at five years from 0.35, the value under the standard treatment, to 0.55. This information can be used to calculate a value for θ_R. To do this, we use the result that if the hazard functions are assumed to be proportional, the survivor function for an individual on the new treatment at time t is

$$S_N(t) = [S_S(t)]^\psi, \tag{9.10}$$

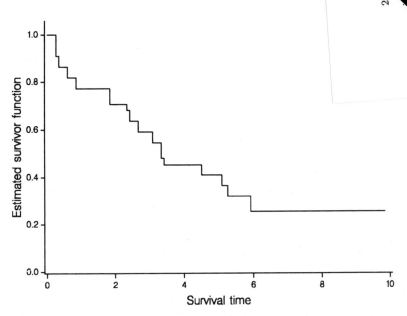

Figure 9.1 Estimated survivor function for patients receiving a standard treatment for hepatitis.

where $S_S(t)$ is the survivor function for an individual on the standard treatment at t and ψ is the hazard ratio. Therefore,

$$\psi = \frac{\log S_N(t)}{\log S_S(t)},$$

and so the value of ψ corresponding to an increase in $S(t)$ from 0.35 to 0.55 is

$$\psi_R = \frac{\log(0.55)}{\log(0.35)} = 0.57.$$

With this information, the survivor function for a patient on the new treatment can be estimated by $[S_S(t)]^{\psi_R}$. In particular, $S_N(2) = 0.82$, $S_N(4) = 0.63$, and $S_N(6) = 0.45$. A plot of the two survivor functions is shown in Figure 9.2.

The median survival time under the new treatment can be estimated from this estimate of the survivor function. Using Figure 9.2, the median survival time under the new treatment is estimated to be about six years.

To calculate the number of deaths that would be required in a study to compare the two treatments, we will take $\alpha = 0.05$ and $1 - \beta = 0.90$. With these values of α and β, the value of the function $c(\alpha, \beta)$ from Table 9.1 is 10.51. Substituting for $c(0.05, 0.1)$ in equation (9.2) and taking $\theta_R = \log \psi_R = \log(0.57) = -0.563$, the number of deaths required to have a 90% chance of detecting a hazard ratio of

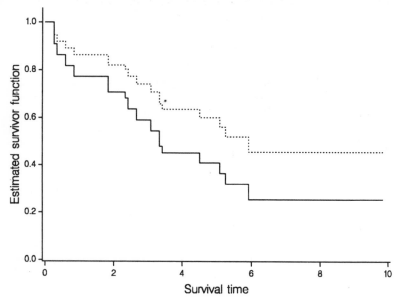

Figure 9.2 Estimated survivor functions for individuals on the standard treatment (—) and the new treatment ($\cdots$).

0.57 to be significant at the 5% level is then given by

$$d = \frac{4 \times 10.51}{0.563^2} = 133.$$

Allowing for possible underestimation, this can be rounded up to 140 deaths in total. This means that approximately 70 deaths would need to be observed in each treatment group.

The calculations described above are only going to be of direct use when a study is to be continued until a given number of those entering the study have died. Most trials will be designed on the basis of the number of patients to be recruited and so we must now examine how this number can be calculated.

9.3 Calculating the required number of patients

In order to calculate the actual number of patients that are required in a survival study, we need to consider the probability of death over the duration of a study. Typically, patients are recruited over an *accrual period* of length a. After recruitment is complete, there is an additional *follow-up period* of length f. The total duration of a study will therefore be of length $a + f$. Notice that if f is small, or even zero, there will need to be correspondingly more patients recruited in order to achieve a specific number of deaths.

Once the probability of a patient dying in the study has been evaluated, the required number of patients will be found from

$$n = \frac{d}{P(\text{death})},\qquad(9.11)$$

where d is the required number of deaths found from equation (9.2). According to a result derived in the next section, the probability of death can be taken as

$$P(\text{death}) = 1 - \frac{1}{6}\{\bar{S}(f) + 4\bar{S}(0.5a + f) + \bar{S}(a + f)\}.\qquad(9.12)$$

where

$$\bar{S}(t) = \frac{S_S(t) + S_N(t)}{2},$$

and $S_S(t)$ and $S_N(t)$ are the estimated values of the survivor functions for individuals on the standard and new treatments, respectively, at time t.

The above result shows how the required number of patients can be calculated for a trial with an accrual period of a and a follow-up period of f. Of course, the duration of the accrual period and follow-up period will depend on the recruitment rate. So suppose that the recruitment rate is expected to be m patients per month and that d deaths are required. If n patients are to be entered into the study over a period of a months, this means that n/a need to be recruited in each month. In practice, information is likely to be available on the accrual rate, m, that can be expected. The number recruited in an accrual period of length a is then ma and so the expected number of deaths in the study is

$$ma \times P(\text{death}).$$

Values of a and f which make this value close to the number of deaths required can then be found numerically, for example, by trying out different values of a and f. This algorithm could be computerized and an optimization method used to find the value of a that makes

$$d - \{ma \times P(\text{death})\}\qquad(9.13)$$

close to zero for a range of values of f. Alternatively, the value of f that yields the result in equation (9.13) for a range of values of a can be found. A two-way table giving the required number of patients for different combinations of values of a and f will be particularly useful in planning a study.

The following section gives details underlying the derivation of the result in equation (9.12) and can again be omitted without loss of continuity.

9.3.1 * Derivation of the required number of patients

We begin with the general result from distribution theory that the marginal probability of a patient dying during the course of a study can be obtained from the joint probability of death and entry to the study at time t using

$$P(\text{death}) = \int_0^a P(\text{death and entry at time } t)\,dt. \qquad (9.14)$$

The joint probability can in turn be found from the result

$$P(\text{death and entry at time } t) = P(\text{death} \mid \text{entry at } t) \times P(\text{entry at } t),$$
$$(9.15)$$

which is simply a version of the result that $P(A \mid B) = P(AB)/P(B)$.

We now assume a uniform recruitment rate over the accrual period. The distribution of entry times to the study can then be taken to be uniform over the time interval $(0, a)$. Therefore, the probability of an individual being recruited to the study at time t is a^{-1}, for any value of t in the interval $(0, a)$. From equations (9.14) and (9.15), we have

$$P(\text{death}) = \int_0^a P(\text{death} \mid \text{entry at } t)a^{-1}\,dt,$$

so that

$$P(\text{death}) = 1 - \frac{1}{a} \int_0^a P(\text{survival} \mid \text{entry at } t)\,dt.$$

A patient entering the study at time t who survives for the duration of the study, that is, to time $a + f$, must have been alive for a period of length $a + f - t$ after entry. The conditional probability P(survival | entry at t) is therefore the probability of survival beyond $a + f - t$. This probability is the value of the survivor function for that individual at $a + f - t$, that is, $S(a + f - t)$. Consequently,

$$P(\text{death}) = 1 - \frac{1}{a} \int_0^a S(a + f - t)\,dt,$$

and on writing $u = a + f - t$, this result becomes

$$P(\text{death}) = 1 - \frac{1}{a} \int_f^{a+f} S(u)\,du. \qquad (9.16)$$

The integral of the survivor function is now approximated using numerical integration. According to Simpson's rule,

$$\int_u^v f(x)\,dx \approx \frac{v - u}{6} \left\{ f(u) + 4f\left(\frac{u + v}{2}\right) + f(v) \right\},$$

so that

$$\int_f^{a+f} S(u)\,du \approx \frac{a}{6} \left\{ S(f) + 4S(0.5a + f) + S(a + f) \right\},$$

and hence, using equation (9.16), the probability of death during the study is given by

$$P(\text{death}) = 1 - \frac{1}{6} \left\{ S(f) + 4S(0.5a + f) + S(a + f) \right\}.$$

From this result, the approximate probability of death for an individual in Group I, for whom the survivor function is $S_S(t)$, is

$$P(\text{death; Group I}) = 1 - \frac{1}{6}\{S_S(f) + 4S_S(0.5a + f) + S_S(a + f)\}$$

and similarly that for an individual in Group II is

$$P(\text{death; Group II}) = 1 - \frac{1}{6}\{S_N(f) + 4S_N(0.5a + f) + S_N(a + f)\}.$$

On the assumption that there is an equal probability of an individual being assigned to either of the two treatment groups, the overall probability of death is the average of these two probabilities, so that

$$P(\text{death}) = \frac{P(\text{death; Group I}) + P(\text{death; Group II})}{2}.$$

On substituting for the probabilities of death in the two treatment groups, and writing $\bar{S}(t) = \{S_S(t) + S_N(t)\}/2$, we get

$$P(\text{death}) = 1 - \frac{1}{6}\{\bar{S}(f) + 4\bar{S}(0.5a + f) + \bar{S}(a + f)\}.$$

as in equation (9.12).

If the proportion of individuals to be allocated to Groups I and II is π_1 and π_2, respectively, the overall probability of death becomes

$$\pi_1 P(\text{death; Group I}) + \pi_2 P(\text{death; Group II}),$$

and the result for the overall probability of death is modified accordingly.

Example 9.2 Survival from chronic active hepatitis
In Example 9.1, it was shown that 140 deaths needed to be observed for the study on chronic hepatitis to have sufficient power to detect a hazard ratio of 0.57 as significant. Suppose that patients are to be recruited to the study over an 18-month accrual period and that there is to be a subsequent follow-up period of 24 months. From equation (9.12), the probability of death in the 42 months of the study will then be given by

$$P(\text{death}) = 1 - \frac{1}{6}\{\bar{S}(24) + 4\bar{S}(33) + \bar{S}(42)\}.$$

Now, using the estimated survivor functions shown in Figure 9.2,

$$\bar{S}(24) = \frac{S_S(24) + S_N(24)}{2} = \frac{0.70 + 0.82}{2} = 0.76,$$

$$\bar{S}(33) = \frac{S_S(33) + S_N(33)}{2} = \frac{0.57 + 0.73}{2} = 0.65,$$

$$\bar{S}(42) = \frac{S_S(42) + S_N(42)}{2} = \frac{0.45 + 0.63}{2} = 0.54,$$

and so the probability of death is

$$1 - \frac{1}{6}\{0.76 + (4 \times 0.65) + 0.54\} = 0.35.$$

From equation (9.11), the required number of patients is

$$n = \frac{140}{0.35} = 400,$$

and so 400 patients will need to be recruited to the study over the accrual period of 18 months. This demands a recruitment rate of about 22 patients per month.

If it is only expected that 18 patients can be found each month, the accrual period will need to be extended to ensure that there is a sufficient number of individuals to give the required number of deaths. The number of individuals that could be recruited in a period of a months would be $18a$. Various values of a can then be tried in order to make this approximately equal to the value obtained from equation (9.11). For example, if we take $a = 24$ and continue with $f = 24$, the probability of death over the four years of the study is

$$P(\text{death}) = 1 - \frac{1}{6}\{\bar{S}(24) + 4\bar{S}(36) + \bar{S}(48)\}.$$

From Figure 9.2, the survivor functions for patients on each treatment at 24, 36 and 48 months can be estimated, and we find that $\bar{S}(24) = 0.76$, $\bar{S}(36) = 0.68$ and $\bar{S}(48) = 0.54$. The probability of death then turns out to be 0.33 and the required number of patients to give 140 deaths is now 424. This would be consistent with an estimated recruitment rate of 18 per month.

Now suppose that it is decided that the study will not have a follow-up period, so that the accrual period is now equal to the duration of the study. If the accrual period is taken to be 20 months, so that $a = 20$ and $f = 0$, the probability of death is given by

$$P(\text{death}) = 1 - \frac{1}{6}\{\bar{S}(0) + 4\bar{S}(10) + \bar{S}(20)\}.$$

Now, $\bar{S}(0) = 1.00$, $\bar{S}(10) = 0.84$ and $\bar{S}(20) = 0.81$, and the probability of death is 0.14. The required number of patients is now $140/0.14 = 1000$, and this would just about be met by a recruitment rate of fifty patients per month. This shows that the absence of a follow-up period leads to an increase in the number of patients that must be entered into the study.

9.4 Further reading

Full details on the issues to be considered when designing a clinical trial are given in Pocock (1983) and Altman (1991). Whitehead (1992) gives an authoritative account of sequential methodology for survival trials. The procedures described are readily implemented using the computer package PEST (Planning and Evaluation of Sequential Trials) distributed by the author.

The formula for the required number of deaths in equation (9.1) appears in many papers, including Bernstein and Lagakos (1978), Schoenfeld (1981), Schoenfeld and Richter (1982) and Schoenfeld (1983), although the assumptions on which the result is based are different. Bernstein and Lagakos (1978) obtain equation (9.1) on the assumption that the survival times in each group have exponential distributions. Lachin (1981), Rubinstein, Gail and Santner (1981) and Lachin and Foulkes (1986) also discuss sample size requirements in trials where the survival times are assumed to be exponentially distributed. See also the earlier work of George and Desu (1974).

Schoenfeld (1981) obtains the same result as Bernstein and Lagakos (1978) and others when the log-rank test is used to compare treatments, without making the assumption of exponentiality. Schoenfeld (1983) shows that equation (9.1) holds when information on the values of explanatory variables is allowed for.

Schoenfeld and Richter (1982) give nomograms that enable the required numbers of patients to be determined on the assumption of exponential survival times. Dupont and Plummer (1990), in their review of sample size formulae, only give results that assume exponential survival times, although they do describe a public domain computer program that can be used to determine sample size requirements. This is easier to use than Schoenfeld and Richter's nomograms!

The formulae for the required number of patients in Section 9.3 are based on Schoenfeld (1983). When the assumption of exponential survival times is made, these formulae do simplify to the results of Schoenfeld and Richter (1982). Although the resulting formulae are easier to use, it is dangerous to conduct sample size calculations on the basis of restrictive assumptions about survival time distributions.

A variant on the formula for the required number of deaths is given by Freedman (1982). Freedman's result has $\{(1 + \psi)/(1 - \psi)\}^2$ in place of $4/(\log \psi)^2$ in equation (9.1). However, for small values of $\log \psi$,

$$\{(1 + \psi)/(1 - \psi)\}^2 \approx 4/(\log \psi)^2,$$

and so the two expressions will tend to give similar results. Freedman's expression provides the basis for the extensive tables of sample size requirements in Machin and Campbell (1987). Freedman's subsequent result for the required number of patients is based on stronger assumptions than those made in Section 9.3 of this chapter, and will not therefore be as reliable.

More recently, Lakatos (1988) presented a method for estimating the required number of patients to compare two treatments which can accommodate matters such as staggered entry, non-compliance, loss to follow-up and non-proportional hazards. This method is based on a Markov model, and requires a computer program for its implementation. Lakatos and Lan (1992) show that the Lakatos procedure performs well in a variety of circumstances.

10

Some additional topics

In this chapter, a number of topics that arise in the practical application of the methods of survival analysis described in this book are briefly discussed. Many analyses are based on a model that assumes proportional hazards, and so some methods for processing survival data when this assumption is not satisfied are described in Section 10.1. When the assumption of proportional hazards is tenable, a Cox regression model is usually the preferred model. However, there are occasions when the Weibull proportional hazards model might be more appropriate and so the issue of model choice is discussed in Section 10.2. In Section 10.3, the implications of informative censoring are outlined, and methods for detecting this feature are suggested. Finally, multistate models used in modelling the natural progression of a disease are introduced in Section 10.4. For each of these topics, references to the literature are given when appropriate to enable the interested reader to obtain further information on them.

10.1 Non-proportional hazards

Some models that do not require the assumption of proportional hazards have been considered in previous chapters. These include the accelerated failure time model and the proportional odds model introduced in Chapter 6, and the Cox regression model that includes a time-dependent variable, described in Chapter 7. But often we are faced with a situation where the assumption of proportional hazards is untenable, and yet none of the above models is satisfactory.

As an illustration, consider a study to compare a surgical procedure with chemotherapy in the treatment of a particular form of cancer. Suppose that the survivor functions under the two treatments are as shown in Figure 10.1. Clearly the hazards are non-proportional. Death at an early stage may be experienced by patients on the surgical treatment, as a result of patients not being able to withstand the surgery or complications arising from it. In the longer term, patients who have recovered from the surgery have a better prognosis.

A similar situation arises when an aggressive form of chemotherapy is com-

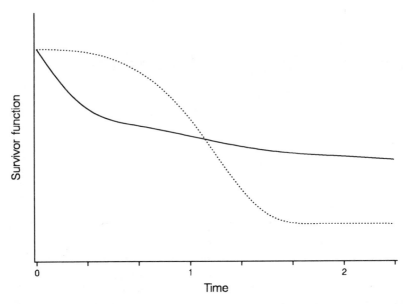

Figure 10.1 Long-term advantage of surgery (—) over chemotherapy (···).

pared to a standard. Here also, a long-term advantage to the aggressive treatment
may be at the expense of short-term excess mortality.

One approach, which is useful in the analysis of data arising from situations such
as these, is to define the end-point of the study to be survival beyond some particular time. For example, in the study leading to the survivor functions illustrated
in Figure 10.1, the treatment difference is roughly constant after two years. The
dependence of the probability of survival beyond two years on prognostic variables and treatment might therefore be modelled. This approach was discussed in
connection with the analysis of interval-censored survival data in Section 8.2. As
shown in that section, there are advantages in using a linear model for the complementary log-log transformation of the survival probability. In particular, the
coefficients of the explanatory variables in the linear component of the model can
be interpreted as logarithms of hazard ratios. The disadvantages of this approach
are that all patients must be followed until the point in time when the survival rates
are to be analysed, and that the death data cannot be used until this time. Moreover, faith in the long-term benefits of one or other of the two treatments will be
needed to ensure that the trial is not stopped early because of excess mortality in
one treatment group.

Strictly speaking, an analysis based on the survival probability at a particular
time is only valid when that time is specified at the outset of the study. In other
words, this end-point needs to be defined in the study protocol. This may be dif-

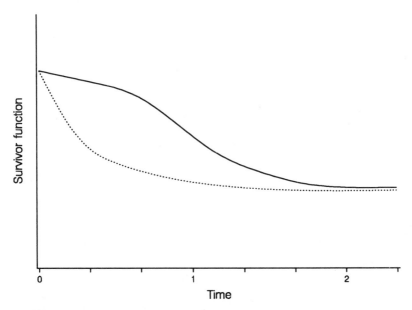

Figure 10.2 Short-term advantage of chemotherapy and surgery (—) over chemotherapy alone (···).

ficult to do. If the data are used to suggest end-points such as the probability of survival beyond two years, some caution will be needed in interpreting the results of a significance test.

In the study that leads to the survivor functions shown in Figure 10.1, it is clear that an analysis of the two-year survival rate will be appropriate. Now consider a study to compare the use of chemotherapy in addition to surgery with surgery alone, in which the survivor functions are as shown in Figure 10.2. Here, the short-term benefit of the chemotherapy may certainly be worthwhile, but an analysis of the two-year survival rates will fail to establish a treatment difference. The fact that the difference between the two survival rates is not constant makes it difficult to use an analysis based on survival rates at a given time. However, it might be reasonable to assume that the hazards are proportional over the first year of the study, and to carry out a survival analysis at that time.

10.1.1 Stratified proportional hazards models

A situation that sometimes occurs is that hazards are not proportional on an overall basis, but that they are proportional in different subgroups of the data. For example, consider a situation in which a new drug is being compared with a standard in the treatment of a particular disease. If the study involves two participating centres, it is possible that in each centre the new treatment halves the hazard of death, but

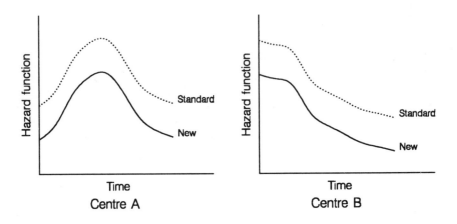

Figure 10.3 Hazard functions for individuals on a new drug (—) and a standard drug ($\cdots$) in two centres.

that the hazard functions for the standard drug are different between the subjects recruited by each centre. Then, the hazards between centres for individuals on a given drug are not proportional. This situation is illustrated in Figure 10.3.

In problems of this kind, it may be assumed that patients in each of the sub-groups or *strata* have a different baseline hazard function, but that all other explanatory variables satisfy the proportional hazards assumption within each stratum. Suppose that the patients in the j'th stratum have a baseline hazard function $h_{0j}(t)$, for $j = 1, 2, \ldots, g$, where g is the number of strata. The effect of explanatory variables on the hazard function can then be represented by a proportional hazards model for $h_{ij}(t)$, the hazard function for the i'th individual in the j'th stratum, where $i = 1, 2, \ldots, n_j$, say, and n_j is the number of individuals in the j'th stratum. We then have the *stratified proportional hazards model*, according to which

$$h_{ij}(t) = \exp(\beta' \mathbf{x}_{ij}) h_{0j}(t),$$

where $\mathbf{x}_{ij}$ is the vector of values of p explanatory variables, $X_1, X_2, \ldots, X_p$, recorded on the i'th individual in the j'th stratum.

As an example of this model, consider the particular case where there are two treatments being compared in each of g strata, and no other explanatory variables. Let x_{ij} be the value of an indicator X which is zero if the i'th subject in the j'th stratum is on the standard treatment and unity if on the new treatment. The hazard

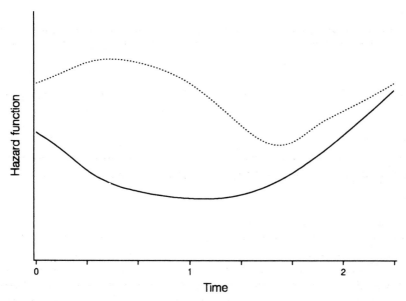

Figure 10.4 Non-proportional hazards for individuals on a new treatment (—) and a standard treatment ($\cdots$).

function for this individual is then

$$h_{ij}(t) = e^{\beta x_{ij}} h_{0j}(t).$$

On fitting this model, the estimated value of β is the log-hazard ratio for an individual on the new treatment, relative to one on the standard, in each stratum.

This model for stratified proportional hazards is easily fitted using major software packages for survival analysis. Models can be compared using the $-2 \log \hat{L}$ statistic and no new principles are involved. When two or more groups of survival data are being compared, the stratified proportional hazards model is in fact equivalent to the stratified log-rank test described in Section 2.7 of Chapter 2. Further details on the stratified proportional hazards model can be found in Kalbfleisch and Prentice (1980).

10.1.2 Non-proportional hazards between treatments

If there are non-proportional hazards between two treatments, misleading inferences can result from ignoring this phenomenon. To illustrate this point, suppose that the hazard function for two groups of individuals, on a new and standard treatment, are as shown in Figure 10.4. If a proportional hazards model were fitted, the resulting fitted hazard functions are likely to be as shown in Figure 10.5. Incorrect conclusions would then be drawn about the relative merit of the two treatments.

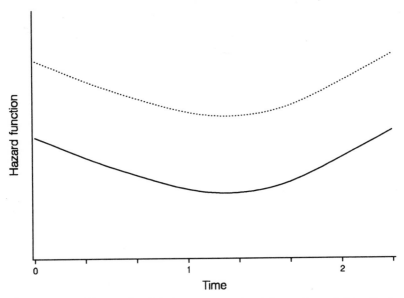

Figure 10.5 The result of fitting a proportional hazards model to the hazard functions in Figure 10.4.

Non-proportional hazards between treatments can be modelled assuming proportional hazards in a series of consecutive time intervals. This is achieved using a piecewise Cox model, which is analogous to the piecewise exponential model introduced in Chapter 6. To illustrate the use of the model, suppose that the time period over which the hazard functions in Figure 10.4 are given is divided into three intervals, namely $(0, t_1)$, (t_1, t_2) and (t_3, ∞). Within each of these intervals, hazards might be assumed to be proportional.

Now let X be an indicator variable associated with the two treatments, where $X = 0$ if an individual is on the standard treatment and $X = 1$ if an individual is on the new treatment. The piecewise Cox regression model can then be fitted by defining two time-dependent variables, $Z_2(t)$ and $Z_3(t)$, say, which are as follows:

$$Z_2(t) = \begin{cases} 1 & \text{if } t \in (t_1, t_2) \text{ and } X = 1, \\ 0 & \text{otherwise;} \end{cases}$$

$$Z_3(t) = \begin{cases} 1 & \text{if } t \in (t_2, t_3) \text{ and } X = 1, \\ 0 & \text{otherwise.} \end{cases}$$

In the absence of other explanatory variables, the model for the hazard function for the i'th individual at t can be written

$$h_i(t) = \exp\{\beta_1 x_i + \beta_2 z_{2i}(t) + \beta_3 z_{3i}(t)\} h_0(t)$$

where x_i is the value of X for the i'th individual, and $z_{2i}(t)$ and $z_{3i}(t)$ are the values of the two time-dependent variables for the i'th individual at t. Under this model, the log-hazard ratio for an individual on the new treatment, relative to one on the standard, is then β_1 for $t \in (0, t_1)$, $\beta_1 + \beta_2$ for $t \in (t_1, t_2)$ and $\beta_1 + \beta_3$ for $t \in (t_2, t_3)$. This model can be fitted in the manner described in Chapter 7.

10.1.3 Further reading

Examples of survival analyses in situations where the proportional hazards model is not applicable have been given by Stablein, Carter and Novak (1981) and Gore, Pocock and Kerr (1984). A review of methods for dealing with non-proportional hazards in the Cox regression model is included in Schemper (1992).

10.2 Model choice

One attraction of the proportional hazards model for survival data is that it is not necessary to adopt a specific probability distribution for the survival times. However, when a Weibull distribution is appropriate for the observed survival data, the parametric version of the proportional hazards model provides a more suitable basis for modelling the data.

In order to help choose between the Cox and Weibull proportional hazards models, it will often be useful to fit the Cox model and examine the shape of the baseline hazard function. If this turns out to be a monotonically increasing or decreasing function of time, the Weibull model may well be suitable. The Weibull model would then be preferred if the estimated parameters under this model were more precisely estimated, that is, if they had smaller standard errors.

Diagnostic plots based on the log-cumulative hazard function, described in Section 4.2, may throw light on whether or not the assumption of Weibull survival times is plausible, but as has already been pointed out, this technique is often not informative in the presence of explanatory variables that affect survival times.

A suitable analysis of residuals can be used to investigate whether one model fits better than the other. However, it will only be in exceptional circumstances that model-checking diagnostics provide convincing evidence that one or other of the two models is more acceptable.

In general, discrimination between a Cox and a Weibull proportional hazards model will be difficult unless the sample data contain a large number of death times. In cases where there is little to choose between the two models in terms of goodness of fit, the standard errors of the estimated β-parameters in the linear component of the two models can be compared. If those for the Weibull model are substantially smaller than those for the Cox model, the Weibull model would be preferred on grounds of efficiency. On the other hand, if these standard errors are similar, the Cox model is likely to be the model of choice in view of its less restrictive assumptions.

10.3 Informative censoring

The methods described in this book for the analysis of censored survival data are only valid if the censoring is non-informative. Essentially, this means that the censoring is not related to any factors associated with the actual survival time, as pointed out in Section 1.1 of Chapter 1.

As an example of a situation where censoring is informative, suppose that individuals are withdrawn from a survival study to compare two treatments because they experience life-threatening side-effects from one particular treatment. The effect of this will be that the survival rates observed on this particular treatment will appear greater than they should be, leading to incorrect conclusions about the extent of the treatment difference.

The assumption of uninformative censoring can be examined in a number of ways. One possibility is to plot observed survival times against the values of explanatory variables, where the censored observations are distinguished from the uncensored. If a pattern is exhibited in the censoring, such as there being more censored observations at an earlier time on one treatment than on the other, or if there is a greater proportion of censored survival times in patients with a particular range of values of explanatory variables, informative censoring is suggested.

More formally, a model could be used to examine whether the probability of censoring was related to the explanatory variables in the model. In particular, a linear logistic model could be used in modelling a binary response variable which takes the value unity if an observed survival time is censored and zero otherwise. If particular explanatory variables in the data set lead to significant changes in the deviance when they are included in the model, the assumption of non-informative censoring may have been violated.

In situations where the reasons for censoring are available, this information may be used to throw light on differences between alternative treatments. For example, if there was a greater proportion of patients with censored survival times on one treatment, and the censoring was due to the occurrence of a particular side-effect, this would give information about the merits of the two treatments. One method of analysis, useful when censoring does not occur too early in patient time, is to analyse the survival data at a time before any censoring has occurred. Alternatively, the probability of survival beyond such a time could be modelled. However, this approach is unlikely to be useful when informative censoring occurs early in the study.

In summary there is no satisfactory way to compare survival times of two or more groups of patients in the presence of informative censoring. The earlier warning that great care should be taken to ensure that informative censoring does not occur is therefore repeated.

Figure 10.6   A two-state model for survival analysis.

10.4 Multistate models

The experience of a patient in a survival study can be thought of as a process that involves two *states*. At the point of entry to the study, the patient is in a state that corresponds to their being alive. Patients then transfer from this 'live' state to the 'dead' state at some *transition rate* $h(t)$, which is the hazard of death at a given time t. The situation is expressed diagrammatically in Figure 10.6. The dependence of the rate of transition from one state to the other on explanatory variables is then modelled.

In some studies, the state representing those alive can be partitioned into two or more states, each of which corresponds to a particular stage in the natural progression of the disease. To fix ideas, consider a study concerned with a type of cancer, in which the survival times of patients are recorded from the surgical removal of a primary tumour. Following removal of the tumour, a patient is at risk of death, but there is also the competing risk of a recurrence of the cancer. If the single event of death is used as the end-point of the study, no distinction is drawn between those patients who have experienced a recurrence and those who have not.

It is unlikely that the hazard of death in patients who have had a recurrence will be the same as that in patients who have not. Moreover, the prognostic factors associated with the hazard of death may be different in these two groups of patients. This suggests that the hazard of a recurrence should be modelled as a function of explanatory variables, in much the same way as the time to death. However, this analysis cannot by itself shed light on factors affecting the time from surgery to death. We therefore adopt a three-state model for the study. The model is illustrated in Figure 10.7.

This model can be specified in terms of three hazard functions. The hazard of death at t without a recurrence of the disease is denoted $h_D(t)$, the hazard of a recurrence at t is denoted $h_R(t)$, and the hazard of death at t after a recurrence is $h_{RD}(t)$. Notice that although $h_D(t)$ and $h_{RD}(t)$ both denote the hazard of death at time t, the latter is conditional on the cancer having recurred.

Hsieh *et al.* (1983) have shown that each transition in the multistate model can be analysed separately, using a Cox proportional hazards model. It is straightforward to model the hazard of death without recurrence $h_D(t)$ and the hazard of a re-

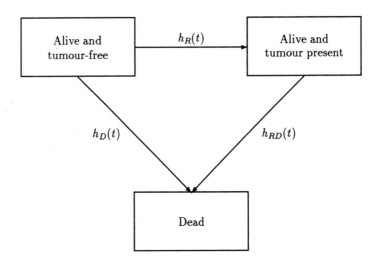

Figure 10.7   A three-state model for survival analysis.

currence $h_R(t)$. In modelling $h_D(t)$, the survival times of those patients who have suffered a recurrence are taken to be censored at the recurrence time. Patients who are alive and without a recurrence also contribute censored survival times. When modelling $h_R(t)$, the end-point is the recurrence time. The survival times of those patients who have not suffered a recurrence are regarded as censored, irrespective of whether they are still alive, or have died without experiencing a recurrence.

It is not so straightforward to model the hazard of death in those who have experienced a recurrence, $h_{RD}(t)$. This is because the set of patients at risk of death at any time consists of those who have had a recurrence of the disease and are still alive at that time. Patients who have not yet had a recurrence cannot be in this risk set.

For example, consider the survival experience of seven patients who have had a recurrence, shown schematically in Figure 10.8. In this figure, the recurrence times are denoted by a '•', censored survival times are denoted by a '×' and death is denoted by a '□'. At the time of death of patient number 6, patients 1, 2, 3, 6 and 7 are in the risk set.

The β-parameters in the hazard function $h_{RD}(t)$ can be estimated after restructuring the data set. In the new data set, the record for each individual who has died after a recurrence is followed by the records for all patients who are at risk at the death time of the individual who has died. When the data are expressed in this manner, the data can be envisaged as arising from a *matched case-control study*. The 'cases' are those who die following a recurrence. For each case there is a set

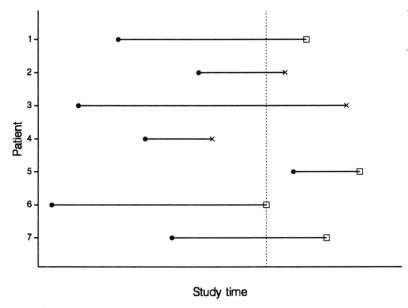

Figure 10.8   Risk set for seven patients who have had a recurrence.

of 'matched controls', which are the patients in the risk set at the death time of the case.

In the example shown in Figure 10.8, there are four cases, namely patients 1, 5, 6 and 7. The sets of matched controls for patients 1, 6 and 7 are $\{3, 5, 7\}$, $\{1, 2, 3, 7\}$ and $\{3, 5\}$, respectively. There is no control corresponding to patient 5.

After this data set has been constructed the data can be analysed using the technique of *conditional logistic regression*. This technique is described in Collett (1991) and provides a method for estimating the effects of different explanatory variables on the hazard function $h_{RD}(t)$. Fuller justification for this approach is given in the section below.

10.4.1 * The likelihood function used in modelling $h_{RD}(t)$

Suppose that there are r individuals who die following a recurrence and that t_j is the death time of the j'th such individual, $j = 1, 2, \ldots, r$. Let x_j be the vector of values of p explanatory variables, $X_1, X_2, \ldots, X_p$, for the j'th individual, so that x_j is the set of values of these explanatory variables for the case in the restructured data set. Also, let x_l be the vector of values of the explanatory variables for the l'th individual in the risk set at t_j. This is the set of individuals who have had a recurrence and are still alive at t_j, and is denoted $R(t_j)$. From equation (3.4), the partial likelihood function is

$$\prod_{j=1}^{r} \frac{\exp(\beta'\mathbf{x}_j)}{\sum_{l \in R(t_j)} \exp(\beta'\mathbf{x}_l)}.$$

This function can be written in the form

$$\prod_{j=1}^{r} \frac{\exp(\beta'\mathbf{x}_j)}{\exp(\beta'\mathbf{x}_j) + \sum_{l^* \in R^*(t_j)} \exp(\beta'\mathbf{x}_{l^*})}, \qquad (10.1)$$

where $R^*(t_j)$ is the risk set at t_j, excluding the j'th individual. This risk set is therefore formed from the matched controls corresponding to the case in the revised data set. If we now write expression (10.1) as

$$\prod_{j=1}^{r} \left[1 + \sum_{l^* \in R^*(t_j)} \exp\{\beta'(\mathbf{x}_{l^*} - \mathbf{x}_j)\} \right]^{-1},$$

it can be seen that the likelihood is expressed in terms of differences between the explanatory variables of the case and those of the controls. This form of the likelihood function is equivalent to that used in connection with conditional logistic regression. It therefore follows that computer software for analysing matched case-control studies using conditional logistic regression can be used to estimate the β- parameters in the hazard function $h_{RD}(t)$. For example, the packages SAS (proc phreg), BMDP (modules 2L and LE) and EGRET all have the necessary facilities.

10.4.2 Further reading

Another form of multistate model is the *competing risks model*. This model is relevant when there are several types of failure, corresponding to different causes of death, for example. The multistate model then has more than one death state. See Gail (1975), Prentice *et al.* (1978) and Kalbfleisch and Prentice (1980) for discussion on this type of model. Further examples of the use of multistate models in medical research have been given by Weiden *et al.* (1981), Hougaard and Madsen (1985) and Andersen (1988).

In order to apply the method presented in this section, the recurrence times must be known. Multistate models that do not rely on the recurrence time being known have been considered by many authors in connection with animal tumourigenicity experiments. In particular, see Dinse (1991), Kodell and Nelson (1980), and McKnight and Crowley (1984). A useful review of this literature is included in Lindsey and Ryan (1993).

11

Computer software for survival analysis

Most of the techniques for analysing survival data that have been presented in this book require suitable computer software for their implementation. Accordingly, the facilities for survival analysis that are available in some of the more widely used statistical packages are reviewed in this chapter.

The packages to be discussed in detail are SAS (version 6.04), BMDP (version PC90) and SPSS (release 5.0). Some information on using GLIM and Genstat for survival analysis will also be provided. There are many other packages that can be used in modelling survival data, including EGRET, STATA, SYSTAT and SPIDA, but it is not practicable to describe each of these in detail.

In addition to summarizing the capabilities of SAS, BMDP and SPSS for survival analysis, the use of these packages in the analysis of an example data set is described and illustrated. Listings of the programs are given together with the resulting output. Comments are also provided on both the input and output from each of the three packages.

Some of the diagnostics used in checking the adequacy of models fitted to survival data, described in Chapter 5, are not directly available in SAS and other packages. In order to obtain these diagnostics, suitable macros are required. SAS macros that use the SAS matrix language IML have been written to compute the values of score residuals and certain influence diagnostics for both the Cox and Weibull proportional hazards models. These macros are described in Section 11.7 and listed in Appendix C. Machine-readable versions of these macros are available on request from the author. The chapter concludes with a summary of the relative merits of the different packages.

11.1 Computational methods used in packages for survival analysis

There are a number of similarities between the packages SAS, BMDP and SPSS in terms of the computational methods that have been implemented for analysing survival data. These are summarized in this section, before each package is discussed in greater detail.

11.1.1 Estimating the survivor function

For grouped survival data, each of SAS, BMDP and SPSS provides the life-table estimate of the survivor function described in Section 2.1.1, and an estimate of the corresponding hazard function. For data in the form of actual survival times, the Kaplan–Meier estimate of the survivor function can be obtained using each package. Standard errors of the estimated survivor function at death times are obtained using Greenwood's formula described in Section 2.1.3.

11.1.2 Non-parametric tests for comparing groups

Each of the three packages can be used to provide the results of the log-rank and Wilcoxon tests for comparing two or more sets of survival data. These were described in detail in Section 2.5.

The tests are based on the statistic

$$U = \sum_{j=1}^{r} w_j(d_{1j} - e_{1j}),$$

where, as in Section 2.5, d_{1j} is the number of deaths at $t_{(j)}$, the j'th ordered death time, in the first treatment group, $j = 1, 2, \ldots, r$, and e_{1j} is the corresponding expected number of deaths, given in equation (2.19). For the log-rank test, w_j in the above expression is equal to unity, while for the Wilcoxon test, $w_j = n_j$, the number at risk at $t_{(j)}$.

The results of two other test procedures are available in BMDP and SPSS. The *Tarone–Ware test* takes $w_j = \sqrt{(n_j)}$, while for the *Peto–Prentice test*, w_j is an estimate of the survivor function at $t_{(j)}$. This estimate is such that

$$w_j = \prod_{i=1}^{j} \frac{n_i - d_i + 1}{n_i + 1},$$

in which n_i is the number at risk at the i'th ordered death time across the two groups and d_i is the total number of deaths at that time. Note that the estimated survivor function used in this formula is similar, but not identical, to the Kaplan–Meier estimate of the survivor function defined in equation (2.4).

11.1.3 Fitting the Cox regression model

The Cox proportional hazards model for the hazard of death at time t for the i'th of n individuals is given by

$$h_i(t) = \exp(\beta' \mathbf{x}_i) h_0(t),$$

in standard notation. All three packages fit this model by maximizing the partial likelihood using the Newton–Raphson procedure, as outlined in Section 3.3. The estimated β-parameters and their standard errors are provided. The packages do differ in their handling of factors, in the statistic used to summarize the fit of a model, and in the method used to estimate the cumulative hazard function. Details are given in later sections.

11.1.4 Fitting the Weibull model

The two packages that allow a fully parametric model to be fitted use the log-linear representation of the proportional hazards and accelerated failure time models. This was described in Sections 4.7 and 6.4.

From equation (4.38) in Section 4.7 of Chapter 4, the model for T_i, the random variable associated with the survival time of the i'th individual, can be expressed as

$$\log T_i = \mu + \alpha_1 x_{1i} + \alpha_2 x_{2i} + \cdots + \alpha_p x_{pi} + \sigma \epsilon_i,$$

in which μ, σ and the α's are unknown parameters and ϵ_i has a distribution that implies the Weibull, log-logistic or other distributions for T_i. The parameters μ and σ in this log-linear model are referred to as the intercept and scale parameters, respectively, and the model can be fitted using the method of maximum likelihood.

The parametrization on which this model is based is not that which has generally been adopted in this book. However, it is possible to transform the parameters in the log-linear representation to those used in the models discussed in Chapters 4 and 6. In particular, as in Section 4.7, if the Weibull proportional hazards model is expressed as

$$h_i(t) = \exp(\beta_1 x_{1i} + \beta_2 x_{2i} + \cdots + \beta_p x_{pi}) h_0(t),$$

with $h_0(t) = \lambda \gamma t^{\gamma - 1}$, the parameters λ, γ, and the β's are such that

$$\lambda = \exp(-\mu/\sigma), \qquad \gamma = \sigma^{-1}, \qquad \beta_j = -\alpha_j/\sigma,$$

for $j = 1, 2, \ldots, p$.

If the Weibull accelerated failure time model is to be used, so that

$$h_i(t) = (e^{\eta_i})^{\gamma} \lambda \gamma t^{\gamma - 1},$$

where $\eta_i = \beta_1 x_{1p} + \beta_2 x_{2p} + \cdots + \beta_p x_{pi}$, as in Section 6.3.4, the parameters μ, σ, and α_j are such that

$$\lambda = \exp(-\mu/\sigma), \qquad \gamma = \sigma^{-1}, \qquad \beta_j = -\alpha_j.$$

Now consider the log-logistic accelerated failure time model, presented in Section 6.3.5, where

$$h_i(t) = e^{\eta_i} h_0(e^{\eta_i} t),$$

and

$$h_0(t) = \frac{e^{\theta} \kappa t^{\kappa - 1}}{1 + e^{\theta} t^{\kappa}},$$

with $\eta_i = \beta_1 x_{1p} + \beta_2 x_{2p} + \cdots + \beta_p x_{pi}$. The relationship between the parameters θ, κ and β_j, and the parameters in this log-linear model, μ, σ and α_j, is such that

$$\theta = -\mu/\sigma, \qquad \kappa = \sigma^{-1}, \qquad \beta_j = -\alpha_j.$$

The log-linear parametrization makes it difficult to obtain confidence intervals for a log-hazard ratio, β, in a proportional hazards model, since only the standard error of the estimate of α is given in the output. In particular, for the Weibull proportional hazards model, the output provides the estimated value of $\alpha = -\sigma\beta$, $\hat{\alpha}$, and the standard error of $\hat{\alpha}$. The corresponding estimate of β is easily found from $-\hat{\alpha}/\hat{\sigma}$, but the standard error of $\hat{\beta}$ is more complicated to calculate.

To obtain the standard error of $\hat{\beta}$, we can use the result that the approximate variance of a function of two parameter estimates, $\hat{\theta}_1$, $\hat{\theta}_2$, say, is

$$\left(\frac{\partial g}{\partial \hat{\theta}_1}\right)^2 \text{var}\,(\hat{\theta}_1) + \left(\frac{\partial g}{\partial \hat{\theta}_2}\right)^2 \text{var}\,(\hat{\theta}_2) + 2\left(\frac{\partial g}{\partial \hat{\theta}_1}\frac{\partial g}{\partial \hat{\theta}_2}\right)\text{cov}\,(\hat{\theta}_1, \hat{\theta}_2). \quad (11.1)$$

This is an extension of the result given in equation (2.8) of Chapter 2 for the approximate variance of a function of a single random variable. To obtain the approximate variance of the function

$$g(\hat{\alpha}, \hat{\sigma}) = -\frac{\hat{\alpha}}{\hat{\sigma}},$$

the derivatives of $g(\hat{\alpha}, \hat{\sigma})$ are required. We have that

$$\frac{\partial g}{\partial \hat{\alpha}} = -\frac{1}{\hat{\sigma}}, \qquad \frac{\partial g}{\partial \hat{\sigma}} = \frac{\hat{\alpha}}{\hat{\sigma}^2},$$

and so using expression (11.1),

$$\text{var}\left(-\frac{\hat{\alpha}}{\hat{\sigma}}\right) \approx \left(-\frac{1}{\hat{\sigma}}\right)^2 \text{var}\,(\hat{\alpha}) + \left(\frac{\hat{\alpha}}{\hat{\sigma}^2}\right)^2 \text{var}\,(\hat{\sigma}) + 2\left(-\frac{1}{\hat{\sigma}}\right)\left(\frac{\hat{\alpha}}{\hat{\sigma}^2}\right)\text{cov}\,(\hat{\alpha}, \hat{\sigma}).$$

After some algebra, the approximate variance becomes

$$\frac{1}{\hat{\sigma}^4}\left\{\hat{\sigma}^2\,\text{var}\,(\hat{\alpha}) + \hat{\alpha}^2\,\text{var}\,(\hat{\sigma}) - 2\hat{\alpha}\hat{\sigma}\,\text{cov}\,(\hat{\alpha}, \hat{\sigma})\right\}, \quad (11.2)$$

and the square root of this is the standard error of $\hat{\beta}$.

Example 11.1 Chemotherapy in ovarian cancer patients
The computation of a standard error of a log-hazard ratio is now illustrated using the data on the survival times following one of two chemotherapy treatments for ovarian cancer. This data set was introduced in Example 4.9 of Chapter 4 and will be used later in this chapter to illustrate the use of SAS, BMDP and SPSS in survival analysis. In this example, output from fitting a Weibull proportional hazards model is used to illustrate the calculation of the standard error.

We will consider the model that contains the variable associated with the age of the woman, AGE, and the variable corresponding to treatment group, $TREAT$, which is unity if the treatment is the single compound and 2 if it is the mixture. In the log-linear parametrization of the model used by the major software packages, the estimated coefficients of AGE and $TREAT$ are -0.0790 and 0.5615, respectively. Also, the estimates of μ and σ are given by $\hat{\mu} = 10.4254$ and $\hat{\sigma} = 0.5489$, respectively. The estimated log-hazard ratio for the mixture treatment

$(TREAT = 2)$ relative to the single $(TREAT = 1)$ is

$$\hat{\beta} = -\frac{0.5615(2-1)}{0.5489} = -1.023,$$

as in Example 4.9. The corresponding hazard ratio is 0.36.

In this example, we are only concerned with the coefficient of $TREAT$, and the estimated coefficient from the computer output will simply be denoted $\hat{\alpha}$. The standard errors of $\hat{\alpha}$ and $\hat{\sigma}$ are given in the standard computer output, and are 0.3399 and 0.1291, respectively. The corresponding estimated variances of $\hat{\alpha}$ and $\hat{\sigma}$ are therefore 0.1155 and 0.0167. The covariance between $\hat{\alpha}$ and $\hat{\sigma}$ can be found from the software, although it is not usually part of the default output. It is found to be 0.002938.

Substituting for $\hat{\alpha}$, $\hat{\sigma}$, and their variances and covariance in expression (11.2), we get

$$\text{var}\,(\hat{\beta}) \approx 0.4217,$$

and so the standard error of $\hat{\beta}$ is given by s.e. $(\hat{\beta}) = 0.6494$. This can be used in the construction of confidence intervals for the corresponding true hazard ratio.

Incidentally, the estimated coefficient of $TREAT$ and its standard error, in terms of the parametrization of the model described in Chapter 4, is quite similar to the values obtained for the Cox proportional hazards model of -0.7959 for the estimated coefficient of $TREAT$ and 0.6329 for its standard error.

The above example has shown how expression (11.1) can be used to obtain the standard error of a log-hazard ratio from software used to fit a Weibull proportional hazards model to survival data. Much the same method can be used to obtain such a standard error in other parametric models. In principle, a generalization of the result in expression (11.1), given in Appendix B, can be used to find the standard error of median survival times, or other summary statistics, from the log-linear representation of a parametric model.

In the following sections, the facilities available for survival analysis in the packages SAS, BMDP, SPSS, GLIM and Genstat are summarized. These sections are followed by illustrations of the use of SAS, BMDP and SPSS.

11.2 SAS

There are three SAS procedures that can be used in processing survival data. The procedure proc lifetest is used for certain non-parametric procedures, such as estimation of the survivor function and the log-rank test. The procedure proc phreg can be used to fit the Cox proportional hazards model, and proc lifereg is used in fitting parametric models such as the Weibull proportional hazards model and the log-logistic accelerated failure time model. These three procedures are described below.

11.2.1 *SAS procedure* proc lifetest

The procedure proc lifetest is used to obtain the Kaplan–Meier estimate of the survivor function, the standard error of the estimates at the observed survival times, and corresponding confidence intervals for the true survival rates. The output from the procedure includes the median, 25th and 75th percentiles of estimated survivor functions.

Plots of the survivor function, and the log-cumulative hazard function can be requested. Estimates of the survivor function for different groups of individuals can be found, such as for patients in two or more treatment groups, or individuals classified according to categorical variables such as grade of tumour and sex.

The log-rank and Wilcoxon test statistics can be found using either the strata statement or the test statement. If the test statement is used alone, separate estimates of the survivor functions for the levels of the factor specified in the test statement are not given, and for this reason strata is to be preferred.

When either the strata statement or the test statement is included, the output gives the results of the log-rank and Wilcoxon tests for comparing sets of survival times defined by the combinations of levels of categorical variables specified in the statement. If a strata statement and a test statement are used together, the results of a stratified log-rank or Wilcoxon test are given. These procedures were described in Section 2.7 of Chapter 2, and are equivalent to using the stratified Cox model described in Section 10.1.1 of Chapter 10, in which a different baseline hazard function is assumed for each combination of levels of the stratifying variables.

There can be a difference between the results of the log-rank test obtained using a strata statement alone and that obtained using the test statement alone. This is due to the different ways in which tied survival times are handled by the two statements. In Section 3.9, the log-rank test was shown to be equivalent to a score test of no difference between the groups in a Cox proportional hazards model. In fact, the SAS output for the log-rank test using strata is based on Cox's method for dealing with ties in the Cox model (Section 3.3.2), and is the result of using equation (2.23) in Section 2.5. On the other hand, the result of the log-rank test from using the test statement is equivalent to the score test in the Cox proportional hazards model where Breslow's method for dealing with ties is used. This is the default method used by proc phreg for dealing with tied observations, and so there is a direct correspondence between the output from test and that from proc phreg. If Cox's method for handling ties is used in proc phreg, the score test of no group differences will give the same result as the log-rank test obtained using the strata statement in proc lifetest.

For the Wilcoxon statistic, the formula in equation (2.24) of Section 2.5.3 is used by the strata statement when there are two groups, but a slightly different expression is used by the test statement. This means that differences in the value of the Wilcoxon statistic can occur whether or not there are tied survival times.

In addition to the log-rank and Wilcoxon tests, use of strata also leads to a statistic denoted in the output by -2Log(LR). This is the result of a likelihood ratio test of the hypothesis that the data have exponential distributions with the same mean. The restrictive assumption of exponential survival times is not required for the log-rank and Wilcoxon tests. As a result, the value of the chi-squared statistic for the test of common exponential distributions will often be very different from that of the other two tests.

11.2.2 SAS procedure proc phreg

SAS proc phreg can be used to fit the Cox proportional hazards model. The model is specified in a model statement and the output includes the estimates of the β-parameters, their standard errors, and the value of $-2 \log \hat{L}$ for the model fitted. Also given is the value of $-2 \log \hat{L}$ for a model that contains no β-parameters, known as the null model. Under the null model, each individual has the same hazard function, namely $h_0(t)$. However, this model cannot be fitted explicitly using a model statement. If it is desired to do this, a single variable that takes the value zero for each individual in the data set can be fitted.

The procedure is unable to deal with factors or *class variables* in SAS terminology. This is a major limitation of proc phreg and means that indicator variables must be defined in a data step for each factor to be included in the model. A further disadvantage of the procedure is that it provides facilities for automatic variable selection! See Section 3.6.1 for remarks on the disadvantages of these procedures.

The procedure proc phreg incorporates four methods for handling tied observations, namely Breslow's method, Cox's method based on a discrete model, a method due to Efron, and what is referred to as an exact method. The first three methods were described in Section 3.2. Breslow's method is the default.

Time-dependent variables can be incorporated in the Cox model through the use of programming statements within proc phreg. These statements are designed to modify the values of the explanatory variables used in the model statement within the fitting procedure. The procedure allows the stratified version of the Cox model described in Section 10.1.1 to be fitted by using the strata statement in proc phreg.

A number of quantities can be accessed after a fit using an output statement. These include estimated values of the risk score, $\hat{\eta}_i$, and their standard errors, and the estimate of the survivor function for each individual at their failure time, calculated using equation (3.16) in Section 3.8 of Chapter 3. Martingale and deviance residuals can also be saved. The martingale residuals can subsequently be manipulated in a data step to give Cox–Snell residuals, and proc lifetest can then be employed in producing a log-cumulative hazard plot of the residuals.

A baseline statement is available which can be used to provide estimates of the survivor function for individuals with a particular set of values of the explanatory variables in the model. If desired, an estimate of the baseline survivor function, $\hat{S}_0(t)$, can be computed by setting the explanatory variables to zero. The equation $\hat{H}_0(t) = -\log \hat{S}_0(t)$ can be used to estimate the baseline cumulative hazard function.

11.2.3 SAS procedure `proc lifereg`

This is the SAS procedure used to fit parametric models to survival data. Options in the procedure allow the Weibull, lognormal, log-logistic and gamma distributions to be adopted for the survival times. The log-linear representation of the parametric model is utilized, and the parameters μ and σ in this model are referred to in SAS output as INTERCPT and SCALE, respectively.

The procedure `proc lifereg` does allow factors to be included in a model statement. When a factor is declared as such in a class statement, the effect due to the last level is set equal to zero. See Section 3.2 for a full discussion on this. However, it is not possible to include interaction terms between two class variables or between a class variable and a variate in the model statement. This means that interactions can only be fitted by including appropriate sets of indicator variables in the model.

There is a facility for saving the parametrically estimated survivor function, evaluated at the survival time of each individual. This can be manipulated within a data step to give the Cox–Snell residuals, using the result $r_{Ci} = -\log \hat{S}_i(t)$. The martingale and deviance residuals can then be calculated from r_{Ci}. Model-checking diagnostics are not provided automatically by the procedure. The output statement can also be used to save pre-specified percentiles of the survival time distribution for all individuals or a subset of them. The standard errors can also be obtained and confidence limits for corresponding true percentiles constructed.

11.3 BMDP

There are two programs for survival analysis in BMDP, labelled 1L and 2L. Program 1L is used to summarize survival data in terms of the survivor function and also gives output from non-parametric tests for comparing survival data in two or more groups. Program 2L is used to fit both the Cox proportional hazards model and parametric models.

11.3.1 BMDP program 1L

Program 1L is used for non-parametric analyses, and can be used to give the Kaplan–Meier estimate of the survivor function, and the results of non-parametric tests for the equality of survivor functions. The survivor functions for two or more groups can be obtained by using the grouping subcommand in the /estimate paragraph. The output includes the median and percentiles of the survival times in each group defined by a grouping subcommand, and a confidence interval for the true median survival time. Plots of survivor functions and cumulative hazard functions can be produced.

Four tests are available for comparing the groups. The log-rank test is referred to by BMDP as the *Generalized Savage (Mantel–Cox)* test and the Wilcoxon test is called the *Generalized Wilcoxon (Breslow)* test. In addition, the results of the Tarone–Ware test and the Peto–Prentice modification of the Wilcoxon test are available. The values of the statistics are calculated using the formulae in Section 11.1.1, and so in the presence of ties agree with the output from SAS proc lifetest with the strata statement.

The test for trend across the levels of an ordered categorical variable, described in Section 2.8 of Chapter 2, can be carried out by including the trend subcommand in the /estimate paragraph. Stratified tests can be performed by combining a grouping subcommand with a strata subcommand in the /estimate paragraph. Unlike SAS, the strata statement cannot be used without a grouping statement, but use of the grouping statement does lead to separate estimates of the survivor function.

11.3.2 BMDP program 2L

This program allows the Cox proportional hazards model, and parametric models based on the Weibull, log-logistic and lognormal distributions, to be fitted to survival data. The parametric models are based on the log-linear formulation of the accelerated failure time model described in Section 11.1.4.

In using program 2L for Cox regression, tied survival times are handled using Breslow's method. The output includes estimates of the β-parameters in the model

$$h_i(t) = \exp(\beta_1 x_{1i} + \beta_2 x_{2i} + \cdots + \beta_p x_{pi}) h_0(t),$$

and their standard errors. Also provided is the value of the maximized partial likelihood function, $\log \hat{L}$, referred to in the output as LOG LIKELIHOOD, and the result of the score test of the hypothesis that all the β-parameters in the model are equal to zero. The latter is referred to in the output as GLOBAL CHI-SQUARE. The value of $\log \hat{L}$ can be multiplied by -2 to give the statistic $-2 \log \hat{L}$ used in comparing survival models.

A stratified cox model can be fitted by including a stratification statement in the /regression paragraph. Time-dependent variables can be included in the model using either a /function paragraph, or in more complicated problems by providing a suitable FORTRAN subroutine.

When program 2L is used to fit models to survival data, factors cannot be fitted directly. Indicator variables therefore need to be defined using the /transform paragraph. Facilities for automatic variable selection are available.

The estimated survivor function can be printed for individuals with specified values of explanatory variables, but note that the estimated survivor function and allied quantities are computed in a different manner to that used in the packages SAS and SPSS. Further details are given below in the discussion of output from BMDP 2L. It is not possible to store such quantities for future manipulation. Plots of the Cox–Snell residuals or the standardized residuals, defined in Section 5.6, against specified quantities can be produced, but it is not straightforward to obtain a log-cumulative hazard plot of the residuals.

When the statement `accel` is included in the `/regress` paragraph, a Weibull model is fitted. The parameter μ in the log-linear representation of the model in equation (4.38) of Chapter 4 is referred to in BMDP output as `CONST`, and the scale parameter σ is denoted `SCALE`. Estimates of the α-parameters in the BMDP output can then be converted to the parameters used in specifying the models in this book, following the lines of Section 11.1.4.

Use of `accel` makes a further difference to the form of the output from program 2L. The value of the statistic $\log \hat{L}$ is given from which $-2 \log \hat{L}$ can be computed. However, the quantity in the output referred to as `GLOBAL CHI-SQUARE` is in fact the change in the value of $-2 \log \hat{L}$ when the variables specified in the `/regress` paragraph are added to the null model. This is different from the quantity labelled `GLOBAL CHI-SQUARE` in the output when `accel` is not used, but there is a message about this inconsistency in the output.

One final point to note is that a model with no explanatory variables can only be fitted explicitly when the `accel` statement is in use.

11.4 SPSS

SPSS provides three procedures for processing survival data, namely `survival`, `km` and `coxreg`. The procedure `survival` is used to produce a life-table estimate of the survivor function from grouped survival data, the hazard function and various plots. Because of its limited application in medical research, this procedure will not be studied in detail here. The procedure `km` is used to provide the Kaplan–Meier estimate of the survivor function and the results of certain non-parametric test procedures. The Cox proportional hazards model can be fitted using the procedure `coxreg`. At the time of writing, SPSS cannot be used to fit parametric models to survival data.

11.4.1 SPSS procedure km

The facilities that are available in `km` are similar to those of SAS `proc lifetest` and BMDP program 1L. The Kaplan–Meier estimate of the survivor function, standard errors of the estimated survival rates, and median survival times, are given for each level of a factor named in the by command in the call of `km`. Plots of the survivor function and log-cumulative hazard function are easily obtained using the `plot` subcommand. If in addition a `strata` subcommand is included, a stratified analysis can be performed. It is not possible to use the `test` subcommand without naming a factor in the by command of `km`. If the `test` subcommand is used in conjunction with `strata`, a stratified analysis is performed.

Estimates of the survivor function for individuals with specified characteristics, the standard error of the estimate and cumulative hazard function, can all be saved for subsequent manipulation using the `save` subcommand. The `trend` subcommand can be used to implement a test for trend across factor levels.

Three test statistics to compare two or more groups of survival data are available, namely the log-rank test, the Wilcoxon test, which SPSS refers to as the *Breslow test*, and the Tarone–Ware test. The test statistics are calculated using the formulae shown in Section 11.1.2 and Chapter 2, and so agree with the output from using the `strata` statement in SAS `proc lifetest` and with the BMDP output.

11.4.2 SPSS procedure `coxreg`

The procedure `coxreg` is used to fit the Cox proportional hazards model. Breslow's method appears to be used to cope with tied observations, but the manual is not specific on this point. The output includes estimates of the β-parameters and their standard errors, and the value of the statistic $-2 \log \hat{L}$. The change in the value of $-2 \log \hat{L}$ from a previously fitted model, or the null model, is given, along with the score statistic for testing the hypothesis that all the β-parameters in the model are zero.

The procedure is designed for automatic variable selection and so the output is a little strange when a single model is being fitted. This is well illustrated in the example in Section 11.6.3.

Factors can be fitted directly and a number of parametrizations for the terms corresponding to a factor are available. The default is for the parameter estimates to sum to zero, with the estimate for the last level not being printed. This can be changed by using the `contrast` subcommand. Interactions between factors can also be fitted.

After fitting a model, a number of quantities can be saved using the `save` subcommand. These include the estimated survivor function at an individual's survival time, the cumulative baseline hazard function, and the Cox–Snell residuals. Score residuals, referred to in SPSS as partial residuals, for each explanatory variable are available, as well as the unstandardized delta-beta's.

The stratified Cox model can be fitted using the `strata` subcommand and the procedure `time program` can be used to include time-dependent variables. This must precede the use of `coxreg`.

11.5 GLIM and Genstat

Neither GLIM nor Genstat can be used to fit models to survival data in a direct manner. However, both packages have extensive facilities for statistical modelling and these can be exploited by using GLIM *macros* or Genstat *procedures* to analyse survival data.

The use of GLIM in fitting the Cox proportional hazards model has been considered by a number of authors. Whitehead (1980) shows how a generalized linear model for Poisson data can be used to fit the model, Clayton and Cuzick (1985) describe the use of the EM algorithm, and an approach based on the piecewise exponential distribution is presented in Aitkin *et al.* (1989). Articles by Thompson (1981) and Whitehead (1983) are also relevant.

GLIM may also be used to fit parametric models based on the Weibull, log-logistic and other distributions. The method is described by Aitkin and Clayton (1980), but fuller detail is included in Aitkin *et al.* (1989). In addition, see Aitkin, Laird and Francis (1983). Aitkin *et al.* (1989) include a listing of GLIM macros for survival analysis which are based on version 3.12 of the GLIM package. The enhanced facilities available in version 4 of GLIM enable considerable improvements to be made to these macros, and suitable code is included in the library of macros distributed with GLIM4.

Genstat procedures for modelling survival data can be written along similar lines to the GLIM macros, but library procedures are not yet available in Genstat.

11.6 Illustrations of the use of SAS, BMDP and SPSS

The use of the packages SAS, BMDP and SPSS is now illustrated using data on the survival times of women with ovarian cancer. These data were first described in Example 4.9, but were also used in Example 11.1 of this chapter.

In this section, three analyses will be illustrated.

1. Comparison of two samples of survival times using the Kaplan–Meier estimate of the survivor functions, and non-parametric procedures such as the log-rank test.

2. Fitting a Cox proportional hazards model.

3. Fitting a Weibull proportional hazards model and its interpretation in terms of an accelerated failure time model.

The variables that will be used in these illustrations will be $TIME$, $CENS$, $TREAT$ and AGE. The corresponding identifiers to be used in the computer-based analyses will be denoted survtime, cens, treat and age. The data are assumed to be in a file named OVCANCER.DAT, which contains five columns. The first is the patient number from 1 to 26, and the remaining four contain the values of the variables $TIME$, $CENS$, $TREAT$ and AGE, respectively. For reference, the file is listed below.

1	156	1	1	66
2	1040	0	1	38
3	59	1	1	72
4	421	0	2	53
5	329	1	1	43
6	769	0	2	59
7	365	1	2	64
8	770	0	2	57
9	1227	0	2	59
10	268	1	1	74
11	475	1	2	59
12	1129	0	2	53
13	464	1	2	56

14	1206	0	2	44
15	638	1	1	56
16	563	1	2	55
17	1106	0	1	44
18	431	1	1	50
19	855	0	1	43
20	803	0	1	39
21	115	1	1	74
22	744	0	2	50
23	477	0	1	64
24	448	0	1	56
25	353	1	2	63
26	377	0	2	58

In this file, the identifier cens is zero when a survival time is censored and unity otherwise. The identifier treat is equal to 1 if the treatment is the single compound, cyclophosphamide, and 2 if it is the mixture of cyclophosphamide and adriamycin. The values for treat in this file are used for the variable $TREAT$ when models are fitted using packages that do not have facilities for defining factors directly.

In each case, optional commands designed to list the data and produce plots have not been included, nor has graphical output. This should not be taken to imply that these features of the software are not useful. On the contrary, graphical summaries of survival data are extremely valuable in model selection and model checking.

11.6.1 Use of SAS

SAS input to read in data

The following SAS code can be used to read the data into a SAS data set. The procedure proc format has been used to attach the names 'single' and 'mixture' to the two treatments.

```
proc format;
value treatfmt 1='single' 2='mixture';
run;

data survival;
infile 'ovcancer.dat';
input patient survtime cens treat age;
format treat treatfmt.;
run;
```

SAS input for `proc lifetest`

The SAS procedure `proc lifetest` is used to produce the Kaplan–Meier estimate of the survivor functions for the patients in the two treatment groups, and the
results of two non-parametric tests for comparing the two groups. The variable
that contains the survival times is indicated in the `time` statement. This statement
also defines the censoring indicator variable to be `cens`, and that a zero value for
this variable corresponds to a censored survival time. The `strata` statement is
used to provide the estimated survivor functions for the two treatment groups and
tests of the equality of the two survival distributions.

```
proc lifetest;
time survtime*cens(0);
strata treat;
run;
```

SAS input for `proc phreg`

The SAS statements required for the procedure `proc phreg` are shown below in
connection with fitting a model that contains the variables AGE and $TREAT$.

```
proc phreg;
model survtime*cens(0)=age treat;
run;
```

SAS input for `proc lifereg`

The following statements are used to fit a Weibull model to the data. Note that the
`class` statement is used to fit $TREAT$ as a factor with two levels. Equivalent
results would be obtained by fitting $TREAT$ as a variate.

```
proc lifereg;
class treat;
model survtime*cens(0)=age treat;
run;
```

Output generated by the SAS procedures `lifetest`, `phreg` and `lifereg` is
given below.

SAS output from `proc lifetest`

Product-Limit Survival Estimates
TREAT = mixture

SURVTIME	Survival	Failure	Survival Standard Error	Number Failed	Number Left
0.00	1.0000	0	0	0	13
353.00	0.9231	0.0769	0.0739	1	12
365.00	0.8462	0.1538	0.1001	2	11
377.00*	.	.	.	2	10
421.00*	.	.	.	2	9
464.00	0.7521	0.2479	0.1256	3	8
475.00	0.6581	0.3419	0.1407	4	7
563.00	0.5641	0.4359	0.1488	5	6
744.00*	.	.	.	5	5
769.00*	.	.	.	5	4
770.00*	.	.	.	5	3
1129.00*	.	.	.	5	2
1206.00*	.	.	.	5	1
1227.00*	.	.	.	5	0

* Censored Observation

Quantiles	75%	.	Mean	514.03
	50%	.	Standard Error	23.91
	25%	475.00		

NOTE: The last observation was censored so the estimate of the mean
is biased.

Product-Limit Survival Estimates
TREAT = single

SURVTIME	Survival	Failure	Survival Standard Error	Number Failed	Number Left
0.00	1.0000	0	0	0	13
59.00	0.9231	0.0769	0.0739	1	12
115.00	0.8462	0.1538	0.1001	2	11
156.00	0.7692	0.2308	0.1169	3	10
268.00	0.6923	0.3077	0.1280	4	9
329.00	0.6154	0.3846	0.1349	5	8
431.00	0.5385	0.4615	0.1383	6	7
448.00*	.	.	.	6	6

477.00*	.	.	.	6	5
638.00	0.4308	0.5692	0.1467	7	4
803.00*	.	.	.	7	3
855.00*	.	.	.	7	2
1040.00*	.	.	.	7	1
1106.00*	.	.	.	7	0

* Censored Observation

Quantiles	75%	.	Mean	448.00
	50%	638.00	Standard Error	66.83
	25%	268.00		

NOTE: The last observation was censored so the estimate of the mean
is biased.

Summary of the Number of Censored and Uncensored Values

TREAT	Total	Failed	Censored	%Censored
mixture	13	5	8	61.5385
single	13	7	6	46.1538
Total	26	12	14	53.8462

Testing Homogeneity of Survival Curves over Strata

Rank Statistics

TREAT	Log-Rank	Wilcoxon
mixture	-1.76647	-47
single	1.766469	47

Covariance Matrix for the Log-Rank Statistics

TREAT	mixture	single
mixture	2.93620	-2.93620
single	-2.93620	2.93620

Covariance Matrix for the Wilcoxon Statistics

TREAT	mixture	single
mixture	1154.00	-1154.00
single	-1154.00	1154.00

Test of Equality over Strata

Test	Chi-Square	DF	Pr > Chi-Square
Log-Rank	1.0627	1	0.3026
Wilcoxon	1.9142	1	0.1665
-2Log(LR)	1.1149	1	0.2910

Comments on the output

Much of the output is self-explanatory. Note that the estimated survivor function for the mixture treatment does not go below 0.5, which is why the 75% and 50% quantiles (percentiles) cannot be estimated. The mean is not simply the average of the observed survival times, since it must take account of censored observations. In SAS, the mean is computed from the formula

$$\sum_{j=1}^{r} \hat{S}(t_{(j)})(t_{(j)} - t_{(j-1)}),$$

in which the summation is over the r ordered death times, $\hat{S}(t_{(j)})$ is the Kaplan–Meier estimate of the survivor function at the j'th death time, $t_{(j)}$, and $t_{(0)}$ is defined to be zero. This expression will lead to an underestimate of the mean if the largest observed survival time is censored, and to the appearance of a warning message in the SAS output. The standard error of this estimate is obtained from

$$\left(\sum_{j=1}^{r-1} A_j^2 / \{n_j \hat{S}(t_{(j)})\} \right)^{1/2},$$

where

$$A_j = \sum_{i=j}^{r-1} \hat{S}(t_{(i)})(t_{(i+1)} - t_{(i)}).$$

In general, the location of survival data is better estimated using the median survival time, and so this part of the output can be ignored.

The rank statistics given are the values of U_L and U_W in the notation of Section 2.5 of Chapter 2. The two values of the log-rank and Wilcoxon statistics that appear in the output are identical, apart from a difference in sign. They arise from basing the statistic on differences between the observed and expected numbers of deaths at each time in each of the two treatment groups.

Again using the notation of Section 2.5, the two covariance matrices contain the values of V_L and V_W on the diagonals. The covariance term is simply the covariance between the two values of the rank statistic. Since these values only differ by a change in sign, this is simply $-V_L$ or $-V_W$.

The results of the log-rank and Wilcoxon tests are given in the final part of the output. The log-rank test statistic has the value 1.063 (P = 0.303) and the Wilcoxon test statistic is 1.914 (P = 0.166). The test statistic against the label −2Log(LR) is a test of the equality of the two survival distributions assuming exponential survival times. The result of this test is not therefore comparable to the results of the log-rank or Wilcoxon tests.

If the statement test treat; is used in place of the strata statement, the log-rank statistic is still 1.063, but the Wilcoxon statistic is now 1.700 (P = 0.192). The difference in the value of the Wilcoxon statistic is due to the fact that different formulae are used by these two statements in computing the statistic.

SAS output from proc phreg

The PHREG Procedure

Data Set: WORK.SURVIVAL
Dependent Variable: SURVTIME
Censoring Variable: CENS
Censoring Value(s): 0
Ties Handling: BRESLOW

Summary of the Number of
Event and Censored Values

Total	Event	Censored	Percent Censored
26	12	14	53.85

Testing Global Null Hypothesis: BETA=0

Criterion	Without Covariates	With Covariates	Model Chi-Square
−2 LOG L	69.970	54.148	15.822 with 2 DF (p=0.0004)
Score	.	.	18.609 with 2 DF (p=0.0001)
Wald	.	.	13.545 with 2 DF (p=0.0011)

Analysis of Maximum Likelihood Estimates

Variable	DF	Parameter Estimate	Standard Error	Wald Chi-Square	Pr > Chi-Square	Risk Ratio
AGE	1	0.146570	0.04585	10.21738	0.0014	1.158
TREAT	1	−0.795932	0.63294	1.58134	0.2086	0.451

Comments on the output

From this output, the value of the statistic $-2 \log \hat{L}$ for the model that contains the variables AGE and $TREAT$ is 54.15. For the model that contains no explanatory variables, that is, a model in which the same hazard function is assumed for each individual in the study, the $-2 \log \hat{L}$ statistic takes the value 69.97. The difference between these two values of 15.82 on 2 d.f. can be compared with percentage points of the chi-squared distribution to test whether both AGE and $TREAT$ are simultaneously needed in the model. This test statistic is labelled Model Chi-Square in the output. Since interest will usually centre on whether particular variables are needed in the model, this part of the output is of limited use. The score and Wald tests given in the output are asymptotically equivalent tests to that based on the statistic $-2 \log \hat{L}$.

The output gives the estimated coefficients of the two variables and their standard errors. The value labelled Wald Chi-Square is the value of $\{\hat{\beta}/\text{s.e.}(\hat{\beta})\}^2$, for any parameter estimate $\hat{\beta}$. This is a Wald test of the hypothesis that the parameter is zero, in the presence of other variables included in the model. Again, a test based on $-2 \log \hat{L}$ is the preferred way of testing such hypotheses, as discussed in Section 3.5. The risk ratio is simply $\exp(\hat{\beta})$, and is the relative hazard corresponding to a unit change in the corresponding variate. For example, 1.158 is the hazard of death in a woman of age $a + 1$ years relative to that of a woman of age a. Similarly, 0.451 is the hazard for a woman on the combined chemotherapy treatment ($TREAT = 2$) relative to one on the single treatment ($TREAT = 1$).

SAS output from proc lifereg

```
             L I F E R E G   P R O C E D U R E
                  Class Level Information


             Class    Levels    Values

             TREAT       2       mixture single

             Number of observations used = 26


Data Set           =WORK.SURVIVAL
Dependent Variable=Log(SURVTIME)
Censoring Variable=CENS
Censoring Value(s)=    0
Noncensored Values=   12  Right Censored Values=    14
Left Censored Values=  0  Interval Censored Values=   0


Log Likelihood for WEIBULL -20.56313339
```

Variable	DF	Estimate	Std Err	ChiSquare	Pr>Chi	Label/Value
INTERCPT	1	10.9868387	1.276331	74.10008	0.0001	Intercept
AGE	1	-0.0789772	0.019763	15.97014	0.0001	
TREAT	1			2.727816	0.0986	
	1	0.56145659	0.339945	2.727816	0.0986	mixture
	0	0	0	.	.	single
SCALE	1	0.54892011	0.129148			scale parameter

Comments on the output

The quantity labelled Log Likelihood for WEIBULL in the output from the SAS procedure proc lifereg is the value of $\log \hat{L}$ for the fitted model. This value therefore needs to be multiplied by -2 before comparing alternative models.

In the SAS code, $TREAT$ was specified as a factor. When interpreting output after fitting a factor, care must be taken in identifying which is the first level of the factor. In this output, the first level is that corresponding to the mixture treatment, in spite of the value of $TREAT$ being equal to 2 for this treatment. The reason for this is that SAS has reordered the levels of $TREAT$ in accord with the labels defined in the format statement. The option order=data could be appended to the proc lifereg statement to override this.

Fitting $TREAT$ as a factor is therefore equivalent to fitting an indicator variable, C, say, which takes the values shown in the table below.

Treatment	Value of $TREAT$	Value of C
Single	1	0
Mixture	2	1

If instead $TREAT$ had been fitted as a variate, its estimated coefficient would also have been 0.5615. However, since $C = TREAT - 1$, the value of the quantity labelled INTERCPT would then have been $10.987 - 0.562 = 10.425$.

As indicated in Sections 4.7 and 11.1.4, the parametrization of the Weibull model for the hazard function of the i'th individual is such that

$$h_i(t) = \frac{1}{\sigma t} \exp\left(\frac{\log t - \mu - \alpha_1 x_{1i} - \cdots - \alpha_p x_{pi}}{\sigma} \right),$$

where p is the number of explanatory variables in the model. The estimates of the parameters μ and σ are 10.987 and 0.549 respectively, and the estimated coefficients of AGE and C are -0.079 and 0.561 respectively.

In terms of the parametrization of the Weibull proportional hazards model presented in this book, $\hat{\lambda} = \exp(-\hat{\mu}/\hat{\sigma}) = 2.03 \times 10^{-9}$, and $\hat{\gamma} = \hat{\sigma}^{-1} = 1.822$. The estimated coefficients in this model are $\hat{\beta} = -\hat{\alpha}/\hat{\sigma}$, and so the estimated coefficients of AGE and C are 0.144 and -1.023 respectively. These can be interpreted as log-hazard ratios.

The parameters can also be interpreted in terms of a Weibull accelerated failure time model. Again, estimates of γ and λ in this formulation of the model are 1.822 and 2.03×10^{-9}, respectively, but the estimated coefficients of AGE and C are now 0.079 and -0.561, respectively. The acceleration factor corresponding to the treatment effect is therefore $\exp(-0.561) = 0.57$. An interpretation of this is that the single chemotherapy treatment accelerates the survival time by a factor of about 2, relative to the mixture.

11.6.2 Use of BMDP

BMDP input to read in data

The code shown below is used to read the data on the five variables from the data file named OVCANCER.DAT. In this program, the print level, which controls the amount of output provided, is set to `brief`. The default when using BMDP interactively is `min`, but when this is in force, estimates of the survivor function are not shown.

```
/input      file='ovcancer.dat'.
            variables=5.
            format=free.

/print      level=brief.

/variable   names=patient,survtime,cens,treat,age.
```

BMDP input for program 1L

Program 1L of BMDP is used to produce the Kaplan–Meier estimate of the survivor function, and the results of a number of non-parametric tests of the equality of the two survival time distributions. In the code which follows, the /form paragraph is used to specify that survtime is the identifier that contains the survival times and that cens is the censoring indicator which is unity for an uncensored observation.

The /group paragraph has been used to attach the labels 'single' and 'mixture' to the two levels of the factor associated with the treatment group, and this links up with the grouping statement in the /estimate paragraph.

The method statement in the /estimate paragraph specifies that the Kaplan–Meier estimate of the survivor function is required. The statistics statement in this paragraph has been used to provide output from the log-rank and other non-parametric tests described in Section 11.1.2. In most applications, only the log-rank or Wilcoxon test will be required.

The code below must follow that used to input the data.

```
/form      time=survtime.
           status=cens.
           response=1.

/group     codes(treat)=1,2.
           names(treat)=single,mixture.

/estimate  method=product.
           grouping=treat.
           statistics=breslow,mantel,tarone,peto.
```

BMDP input for program 2L

To use BMDP 2L a /form paragraph and a /regress paragraph are required. The /form paragraph is as for program 1L, while the /regress paragraph is used to specify which variables are to be included in the Cox proportional hazards model.

```
/form      time=survtime.
           status=cens.
           response=1.

/regress   covariates=age,treat.
```

If the statement accel=weibull. is included in the /regress paragraph, the Weibull model is fitted. No other changes need to be made to the code shown above, and so the input is as follows.

```
/regress   covariates=age,treat.
           accel=weibull.
```

Output from BMDP programs 1L and 2L follows.

BMDP output from program 1L

```
NUMBER OF CASES READ. . . . . . . . . . . . . .     26

                 RESPONSE CODES           1    DEAD
                 CENSORED CODES           0    CENSORED
```

PRODUCT-LIMIT SURVIVAL ANALYSIS   GROUPING VARIABLE IS treat
LEVEL IS single

TIME VARIABLE IS survtime

CASE NUMBER	TIME DAYS	STATUS	CUMULATIVE SURVIVAL	STANDARD ERROR	CUM DEAD	CUM LOST	REMAIN AT RISK
3	59.00	DEAD	0.9231	0.0739	1	0	12
21	115.00	DEAD	0.8462	0.1001	2	0	11
1	156.00	DEAD	0.7692	0.1169	3	0	10
10	268.00	DEAD	0.6923	0.1280	4	0	9
5	329.00	DEAD	0.6154	0.1349	5	0	8
18	431.00	DEAD	0.5385	0.1383	6	0	7
24	448.00	CENSORED			6	0	6
23	477.00	CENSORED			6	0	5
15	638.00	DEAD	0.4308	0.1467	7	0	4
20	803.00	CENSORED			7	0	3
19	855.00	CENSORED			7	0	2
2	1040.00	CENSORED			7	0	1
17	1106.00	CENSORED			7	0	0

MEAN SURVIVAL TIME = 649.60 LIMITED TO 1106.00   S.E. = 129.765

QUANTILE	ESTIMATE	ASYMPTOTIC STANDARD ERROR
75TH	268.00	127.30
MEDIAN (50TH)	638.00	245.49

LOWER ONE-SIDED BROOKMEYER-CROWLEY 95% CONFIDENCE LIMIT FOR MEDIAN
SURVIVAL TIME =   268.00

PRODUCT-LIMIT SURVIVAL ANALYSIS   GROUPING VARIABLE IS treat
LEVEL IS mixture

TIME VARIABLE IS survtime

CASE NUMBER	TIME DAYS	STATUS	CUMULATIVE SURVIVAL	STANDARD ERROR	CUM DEAD	CUM LOST	REMAIN AT RISK
25	353.00	DEAD	0.9231	0.0739	1	0	12
7	365.00	DEAD	0.8462	0.1001	2	0	11
26	377.00	CENSORED			2	0	10
4	421.00	CENSORED			2	0	9
13	464.00	DEAD	0.7521	0.1256	3	0	8
11	475.00	DEAD	0.6581	0.1407	4	0	7

16	563.00	DEAD	0.5641	0.1488	5	0	6
22	744.00	CENSORED			5	0	5
6	769.00	CENSORED			5	0	4
8	770.00	CENSORED			5	0	3
12	1129.00	CENSORED			5	0	2
14	1206.00	CENSORED			5	0	1
9	1227.00	CENSORED			5	0	0

MEAN SURVIVAL TIME = 888.60 LIMITED TO 1227.00 S.E. = 128.946

QUANTILE	ESTIMATE	ASYMPTOTIC STANDARD ERROR
75TH	475.00	82.33

LOWER ONE-SIDED BROOKMEYER-CROWLEY 95% CONFIDENCE LIMIT FOR MEDIAN
SURVIVAL TIME = 475.00

SUMMARY TABLE

	TOTAL	DEAD	CENSORED	PROPORTION CENSORED
single	13	7	6	0.4615
mixture	13	5	8	0.6154
	----	----	----	
TOTALS	26	12	14	

TEST STATISTICS

	STATISTIC	D.F.	P-VALUE
GENERALIZED SAVAGE (MANTEL-COX)	1.063	1	0.3026
TARONE-WARE	1.485	1	0.2230
GENERALIZED WILCOXON (BRESLOW)	1.914	1	0.1665
GENERALIZED WILCOXON (PETO-PRENTICE)	1.699	1	0.1924

Comments on the output

Following the estimates of the two survivor functions, quantiles that can be estimated are given along with their standard errors. The mean survival times are

calculated from the formula

$$t_{(1)} + \sum_{j=1}^{r-1} \hat{S}(t_{(j)})(t_{(j+1)} - t_{(j)}),$$

in which the summation is over the first $r - 1$ ordered death times and $\hat{S}(t_{(j)})$ is the Kaplan–Meier estimate of the survivor function at the j'th death time, $t_{(j)}$. The standard error of this estimate is obtained from

$$\left(\frac{r}{r-1} \sum_{j=1}^{r-1} \frac{A_j^2}{(n-j)(n-j+1)} \right)^{1/2},$$

where

$$A_j = \sum_{i=j}^{r-1} \hat{S}(t_{(i)})(t_{(i+1)} - t_{(i)}).$$

Both of these formulae differ from the expressions used by SAS. When the last observation is censored, the phrase LIMITED TO, followed by the largest survival time, is included in the output. This signifies that the mean is then underestimated.

Standard errors for the median and other percentiles are calculated from equation (2.15) of Chapter 2, while confidence limits for the median are computed using a method described by Brookmeyer and Crowley (1982).

The log-rank statistic, here called the Generalized Savage (Mantel–Cox) statistic, is the same as that given using SAS proc lifetest with the strata statement. The Wilcoxon test, referred to in BMDP output as the Generalized Wilcoxon (Breslow) test is also in agreement with the value given earlier in the SAS output.

BMDP output from fitting a Cox model using program 2L

```
BMDP2L--SURVIVAL ANALYSIS WITH COVARIATES

NUMBER OF CASES READ. . . . . . . . . . . . . .      26

         COX PROPORTIONAL HAZARDS MODEL
         -------------------------------
         RISK TYPE IS  LOGLIN

   LOG LIKELIHOOD =     -27.0738
GLOBAL CHI-SQUARE =      18.61  D.F.=   2  P-VALUE =0.0001
NORM OF THE SCORE VECTOR=    0.318E-05
```

VARIABLE	COEFFICIENT	STANDARD ERROR	COEFF./S.E.	EXP(COEFF.)
5 age	0.1466	0.0459	3.1965	1.1579
4 treat	-0.7959	0.6329	-1.2575	0.4512

TIME VARIABLE IS survtime

SURVIVAL TIME	STATUS	KAPLAN MEIER SURVIVAL	--- PROPORTIONAL HAZARDS MODEL ---		COX-SNELL RESIDUAL
			BASELINE SURVIVAL	BASELINE CUM HAZARD	
59.0000	DEAD	0.9615	0.9887	0.0114	0.1842
115.0000	DEAD	0.9231	0.9749	0.0254	0.5497
156.0000	DEAD	0.8846	0.9556	0.0454	0.3046
268.0000	DEAD	0.8462	0.9337	0.0686	1.4854
329.0000	DEAD	0.8077	0.8914	0.1150	0.0265
353.0000	DEAD	0.7692	0.8505	0.1620	0.3157
365.0000	DEAD	0.7308	0.8076	0.2137	0.4822
377.0000	CENSORED		0.7983	0.2252	0.2110
421.0000	CENSORED		0.7652	0.2677	0.1205
431.0000	DEAD	0.6878	0.7578	0.2773	0.1783
448.0000	CENSORED		0.7295	0.3155	0.4885
464.0000	DEAD	0.6419	0.7037	0.3513	0.2455
475.0000	DEAD	0.5961	0.6509	0.4294	0.4657
477.0000	CENSORED		0.6487	0.4328	2.1649
563.0000	DEAD	0.5464	0.5610	0.5780	0.3488
638.0000	DEAD	0.4967	0.4765	0.7413	1.1481
744.0000	CENSORED				
769.0000	CENSORED				
770.0000	CENSORED				
803.0000	CENSORED				
855.0000	CENSORED				
1040.0000	CENSORED				
1106.0000	CENSORED				
1129.0000	CENSORED				
1206.0000	CENSORED				
1227.0000	CENSORED				

Comments on the output

The quantity labelled LOG LIKELIHOOD is the value of $\log \hat{L}$ for the fitted model, and so has to be multiplied by -2 to obtain the statistic $-2 \log \hat{L}$ for comparing models. The quantity GLOBAL CHI-SQUARE is a score test of the null hypothesis that all the β-parameters in the fitted model are zero. In general, this statistic will not have the same value as the difference between the values of $-2 \log \hat{L}$ for the fitted model and the null model.

The quantity termed NORM OF THE SCORE VECTOR is the square root of the sum of squares of the elements of the vector of efficient scores, evaluated at the parameter estimates. It is therefore the *length* of this vector. Since this quantity will be zero at the maximum (partial) likelihood estimates of the parameters, the norm provides an assessment of how close the estimates in the output are to those that make all the elements of the score vector equal to zero. It is not particularly useful.

The parameter estimates and their standard errors are preceded by a number which signifies the order of the variable in the input data set. The ratio of the co-efficient to its standard error can be compared to percentage points of the standard normal distribution to test the null hypothesis that the true value of the corresponding parameter is zero. This is the square root of the Wald chi-squared value in the SAS output. The values in the column headed EXP (COEFF.) are the hazard ratios corresponding to a unit change in the value of the associated variable. The use of such quantities was discussed in the section on the SAS output from `proc phreg`.

The final part of the output includes the Kaplan–Meier estimate of the survivor function for the data, ignoring the explanatory variables. Also given are quantities referred to as the estimated baseline survivor function at the survival time of each individual, the baseline cumulative hazard function, and the Cox–Snell residuals. In fact, the so-called baseline estimates are evaluated at the means of the explanatory variables in the data set, rather than at values of zero. They are therefore estimates of

$$\exp(\beta_1 \bar{x}_1 + \beta_2 \bar{x}_2 + \cdots + \beta_p \bar{x}_p) H_0(t)$$

where $\bar{x}_j$ is the mean of the values of the j'th explanatory variable, $j = 1, 2, \ldots, p$, taken over the individuals in the data set and $H_0(t)$ is the cumulative baseline hazard function.

In BMDP, the estimated value of $H_0(t)$ is obtained from an estimate of the survivor function, which is in turn computed using a result given in Link (1984). This is different from the estimate given in equation (3.16) in Section 3.8, which is used by both SAS and SPSS. This means that estimates of the survivor function for individuals with particular explanatory variables, derived from BMDP output, will be different from the corresponding values calculated using SAS or SPSS. For the same reason, the Cox–Snell residuals listed in the BMDP output will tend to differ from those found using other packages.

BMDP output from fitting a Weibull model using program 2L

BMDP2L--SURVIVAL ANALYSIS WITH COVARIATES

NUMBER OF CASES READ.   26

```
              ACCELERATED FAILURE TIME MODEL
              ------------------------------
              DISTRIBUTION IS  WEIBULL

THE NATURAL LOGARITHM OF SURVIVAL TIME IS USED IN THE ANALYSIS

              INITIAL VALUES OF PARAMETERS
              CALCULATED BY LEAST SQUARES

        CONST          age          treat        SCALE
        8.374433     -0.052942     0.511464     0.364589


    LOG LIKELIHOOD =     -20.5631
  GLOBAL CHI-SQUARE =     18.41  D.F.=   2  P-VALUE =0.0001
  FOR ACCELERATED FAILURE TIME MODEL, GLOBAL CHI-SQUARE IS BASED ON
  LIKELIHOOD RATIO TEST
  NORM OF THE SCORE VECTOR=    0.818E-06

                                    STANDARD
        VARIABLE     COEFFICIENT     ERROR      COEFF./S.E.    EXP(COEFF.)
        --------     -----------     --------   -----------    -----------
  -1  CONST            10.4254       1.4340        7.2703      33704.3603
   5  age              -0.0790       0.0198       -3.9963          0.9241
   4  treat             0.5615       0.3399        1.6516          1.7532
  -2  SCALE             0.5489       0.1291        4.2503          1.7314

                    TIME VARIABLE IS survtime


                                    --ACCELERATED FAILURE TIME MODEL --
        SURVIVAL                     COX-SNELL    STANDARDIZED
          TIME      STATUS           RESIDUAL      RESIDUAL

         59.0000   DEAD               0.1078       -2.2279
        115.0000   DEAD               0.4847       -0.7243
        156.0000   DEAD               0.2672       -1.3198
        268.0001   DEAD               2.2637        0.8170
        329.0000   DEAD               0.0380       -3.2696
        353.0000   DEAD               0.2762       -1.2866
        365.0001   DEAD               0.3390       -1.0819
        377.0000   CENSORED           0.1516       -1.8862
        421.0001   CENSORED           0.0903       -2.4045
        431.0001   DEAD               0.1702       -1.7705
        447.9999   CENSORED           0.4331       -0.8368
        464.0000   DEAD               0.1660       -1.7957
        475.0000   DEAD               0.2668       -1.3214
        477.0001   CENSORED           1.5350        0.4285
```

563.0000	DEAD	0.2045	-1.5872
638.0000	DEAD	0.8247	-0.1927
744.0000	CENSORED	0.1655	-1.7988
769.0001	CENSORED	0.6417	-0.4437
769.9999	CENSORED	0.4824	-0.7291
802.9998	CENSORED	0.1087	-2.2196
855.0000	CENSORED	0.2166	-1.5298
1040.0000	CENSORED	0.1507	-1.8923
1106.0000	CENSORED	0.3997	-0.9170
1129.0000	CENSORED	0.5448	-0.6074
1206.0000	CENSORED	0.1683	-1.7821
1227.0000	CENSORED	1.5031	0.4075

Comments on the output

In the first part of the output, an indication is given that the analysis is based on a linear model for the logarithm of the survival time. After the initial estimates of the parameters, the maximized log-likelihood for the Weibull model is given. Again, this needs to be multiplied by -2 to use it in comparing alternative models. The quantity GLOBAL CHI-SQUARE is now the difference between the values of the statistic $-2 \log \hat{L}$ for the fitted model and the null model.

In the listing of the parameter estimates, those labelled CONST and SCALE are $\hat{\mu}$ and $\hat{\sigma}$, respectively, in the notation of Section 11.1.4. The estimates of σ and the parameters associated with AGE and $TREAT$ are the same as those given by SAS. The parameter μ is now estimated by 10.425. In the discussion of the output from SAS proc lifereg, it was shown that taking $TREAT$ to be a variate in the model, rather than a factor, would lead to an estimate of μ of 10.425, as given in this BMDP output.

The statistics in the column headed COEFF./S.E. can be used to test the hypothesis that the corresponding parameter is zero. The values of this quantity alongside CONST and SCALE therefore correspond to the hypotheses $\mu = 0$ and $\sigma = 0$, respectively. The former corresponds to $\lambda = 1$ in the notation of Section 11.1.4 and elsewhere, which may be of interest in exceptional circumstances. However, testing the hypothesis that $\sigma = 0$ corresponds to testing $\gamma = \infty$, and this will certainly not be of any relevance.

Furthermore, the parameter estimates associated with the explanatory variables in the model are not estimates of the β-parameters in the proportional hazards model, but of the α's in the log-linear representation of the model, with the relation $\beta = -\alpha/\sigma$. The estimates of the α's are not log-hazard ratios, and so the quantities under the heading EXP(COEFF.) in the output are not estimates of hazard ratios. They should also be ignored.

In the final part of the output, the Cox–Snell residuals are given alongside standardized residuals. The latter were described in Section 5.6, where it was shown that they can simply be obtained by exponentiating the Cox–Snell residuals.

11.6.3 Use of SPSS

SPSS input to read in data

The data are read from the file OVCANCER.DAT using the code shown below. The value labels statement has been used to attach the names 'single' and 'mixture' to the two treatment groups.

```
data list file = "ovcancer.dat" free
    / patient survtime cens treat age.
value labels treat 1 "single" 2 "mixture".
```

SPSS input for procedure km

The procedure km is the SPSS analogue of SAS proc lifetest or BMDP program 1L. That the variable cens has a value of unity for a death time is specified using the /status subcommand. In the /test subcommand, the results of the three available tests for comparing the two survivor functions have been requested.

```
km survtime by treat
    /status = cens event(1)
    /test = logrank breslow tarone.
```

SPSS input for procedure coxreg

The response variable and explanatory variables to be included in the model are specified in the coxreg command, and as for the km procedure, the /status subcommand is used to define which observations are actual death times. The variable *TREAT* is declared to be a factor using the /categorical subcommand. The default parametrization of factors will be used, in which the estimates sum to zero. The subcommand /print = all is used to produce all the output, but /print = summary will usually be sufficient.

```
coxreg survtime with treat age
    /status = cens event(1)
    /categorical = treat
    /print = all.
```

Output from the SPSS procedures km and coxreg is given below.

SPSS output from procedure km

Survival Analysis for SURVTIME

Factor TREAT = single

Time	Status	Cumulative Survival	Standard Error	Cumulative Events	Number Remaining
59.00	1.00	.9231	.0739	1	12
115.00	1.00	.8462	.1001	2	11
156.00	1.00	.7692	.1169	3	10
268.00	1.00	.6923	.1280	4	9
329.00	1.00	.6154	.1349	5	8
431.00	1.00	.5385	.1383	6	7
448.00	.00			6	6
477.00	.00			6	5
638.00	1.00	.4308	.1467	7	4
803.00	.00			7	3
855.00	.00			7	2
1040.00	.00			7	1
1106.00	.00			7	0

Number of Cases: 13   Censored: 6   (46.15%)   Events: 7

	Survival Time	Standard Error	95% Confidence Interval
Mean:	649.60	120.14	(414.13, 885.07)
(Limited to	1106.0)		
Median:	638.00	245.49	(156.83, 1119.17)

Survival Analysis for SURVTIME

Factor TREAT = mixture

Time	Status	Cumulative Survival	Standard Error	Cumulative Events	Number Remaining
353.00	1.00	.9231	.0739	1	12
365.00	1.00	.8462	.1001	2	11
377.00	.00			2	10
421.00	.00			2	9
464.00	1.00	.7521	.1256	3	8
475.00	1.00	.6581	.1407	4	7
563.00	1.00	.5641	.1488	5	6
744.00	.00			5	5
769.00	.00			5	4

770.00	.00	5	3
1129.00	.00	5	2
1206.00	.00	5	1
1227.00	.00	5	0

Number of Cases: 13 Censored: 8 (61.54%) Events: 5

	Survival Time	Standard Error	95% Confidence Interval	
Mean:	888.60	115.33	(662.55,	1114.65)
(Limited to	1227.0)			
Median:	.	.	(.,	.)

Survival Analysis for SURVTIME

		Total	Number Events	Number Censored	Percent Censored
TREAT	single	13	7	6	46.15
TREAT	mixture	13	5	8	61.54
Overall		26	12	14	53.85

Test Statistics for Equality of Survival Distributions for TREAT

	Statistic	df	Significance
Log Rank	1.06	1	.3026
Breslow	1.91	1	.1665
Tarone-Ware	1.49	1	.2230

Comments on the output

Following the estimated survivor function for each treatment group is an estimate of the mean survival time, and a standard error. The manual does not explain how these quantities are calculated. The estimate of the mean survival time is the same as that given by BMDP, but the standard error does not agree with the BMDP output. When the survivor function goes below 0.5, the estimated median survival time is also given, with an accompanying standard error. These values are in agreement with the BMDP output.

The three test statistics for comparing the two survivor functions are the logrank test, the Wilcoxon test, referred to in the output as the Breslow test, and the Tarone–Ware test. The values of all three agree with the results obtained from SAS proc lifetest with a strata statement and with the output from BMDP 1L.

SPSS output from procedure coxreg

```
          Deviation Parameter Coding
          Value    Freq     (1)
TREAT
          single    13   1.000
          mixture   13  -1.000

     26  Total cases read
      0  Cases with missing values
      0  Valid cases with non-positive times
      0  Censored cases before the earliest event in a stratum
      0  Total cases dropped
     26  Cases available for the analysis
```

Dependent Variable: SURVTIME

```
          Events  Censored

           12       14 (53.8%)
```

Beginning Block Number 0.   Initial Log Likelihood Function

-2 Log Likelihood   69.970

Beginning Block Number 1. Method: Enter

Variable(s) Entered at Step Number 1..
    TREAT
    AGE

Log likelihood converged after 4 iterations.

-2 Log Likelihood   54.148

```
                    Chi-Square   df    Sig
Overall (score)       18.609      2   .0001
Change (-2LL) from
 Previous Block        15.822      2   .0004
 Previous Step         15.822      2   .0004
```

--------------- Variables in the Equation ----------------

Variable	B	S.E.	Wald	df	Sig	R
TREAT	.3980	.3165	1.5813	1	.2086	.0000
AGE	.1466	.0459	10.2174	1	.0014	.3427

| | 95% CI for Exp(B) | |
Variable	Exp(B)	Lower	Upper
TREAT	1.4888	.8007	2.7683
AGE	1.1579	1.0583	1.2667

Correlation Matrix of Regression Coefficients (B)

                  TREAT

AGE   -.16908

| | Baseline | ---- At mean of covariates --- | | |
Time	Cum Hazard	Survival	SE	Cum Hazard
59.00	3.571E-06	.9875	.0133	.0126
115.00	8.289E-06	.9712	.0231	.0292
156.00	1.439E-05	.9505	.0351	.0508
268.00	2.350E-05	.9204	.0469	.0829
329.00	3.674E-05	.8784	.0636	.1296
353.00	5.070E-05	.8362	.0751	.1788
365.00	6.627E-05	.7915	.0846	.2338
431.00	8.470E-05	.7417	.0946	.2988
464.00	1.062E-04	.6874	.1043	.3748
475.00	1.294E-04	.6336	.1111	.4563
563.00	1.735E-04	.5422	.1236	.6121
638.00	2.269E-04	.4492	.1254	.8004

Covariate Means

Variable	Mean
TREAT	.0000
AGE	55.7308

Comments on the output

After preliminary information about the number of observations and the proportion censored, the value of $-2 \log \hat{L}$ for the null model is given. This is followed by that for the model that contains terms corresponding to AGE and $TREAT$. Also given is the result of a score test for the null hypothesis that neither AGE nor $TREAT$ is needed in the model, and the likelihood ratio test of this same hypothesis. The latter is the change in $-2 \log \hat{L}$ on adding both AGE and $TREAT$ to the null model.

The output then gives the parameter estimates in the fitted model. Note that

$TREAT$ was declared to be a factor with two levels. The coding used for this factor is equivalent to using an indicator variable which is equal to 1 for the single chemotherapy treatment and -1 for the mixture. This is in fact detailed in the first part of the SPSS output. This coding is equivalent to fitting the variable $3 - 2 \times TREAT$, where $TREAT$ is equal to 1 for the single treatment and 2 for the mixture. Thus, multiplying the SPSS estimate of 0.398 by -2 gives -0.796, the estimated coefficient of $TREAT$ given by SAS proc phreg and program 2L of BMDP.

This part of the output also includes the standard errors of the estimates, the values of the Wald chi-squared statistic, calculated as $\{\hat{\beta}/\text{s.e.}(\hat{\beta})\}^2$, and the P-value associated with this test. In the column headed R is a quantity designed to measure the *partial correlation* of each variable in the model. This is calculated from

$$R = \text{sgn}(\hat{\beta}) \left(\frac{\text{Wald statistic} - (2 \times \text{df})}{-2 \log \hat{L}_0} \right)^{1/2},$$

where df is the number of degrees of freedom associated with the variable and $\log \hat{L}_0$ is the maximized log-likelihood for the null model. If the Wald statistic is less than $2 \times \text{df}$, R is set equal to zero. The idea of this quantity is that it can be used to assess the correlation between a variable in the model and the hazard function, after allowing for the effects of other variables in the model. Since this is better examined by comparing the values of $-2 \log \hat{L}$ for alternative models, this statistic is not considered further.

The next part of the output gives the values of $\exp(\hat{\beta})$ for each variable. These quantities can be interpreted as hazard ratios corresponding to a change of one unit in any variate in the model. This quantity is not so easy to interpret when factors are included in the model. For example, in the above output, the default parametrization of the factor associated with treatment has been used. This means that the logarithm of the hazard ratio for the mixture treatment relative to the single is then -2 multiplied by the estimate alongside TREAT in the SPSS output. We find that $-2 \times 0.398 = -0.796$ which is the estimate given in the output from both SAS and BMDP. The default parametrization for factors used by SPSS is not the most convenient for estimating hazard ratios. In fact, such estimates are more readily obtained using parametrizations which set the effect due to one level of a factor equal to zero.

After the correlation matrix of the parameter estimates in the model, an estimate of the baseline cumulative hazard at the survival time of each individual is given. This is the quantity $\hat{H}_0(t_i)$, and appears to be calculated using equation (3.16) in Section 3.8, although the manual does not say so. Also given in this part of the output are the estimated values of the survivor function, the standard error of the estimate and the estimated cumulative hazard function for an individual for whom all the explanatory variables are at their mean. The means used in these computations are given at the end of the output.

11.7 SAS macros for model checking

In Chapter 5 a number of diagnostics were introduced which cannot be obtained directly from most statistical packages. In that chapter, the values of the diagnostics given in the illustrative examples were computed using SAS macros written for the purpose. These macros are described in this section and listed in Appendix C.

11.7.1 The SAS macro coxdiag

This macro gives the values of certain diagnostics associated with the Cox proportional hazards model. It is used after fitting a Cox model using proc phreg, and yields the score residuals introduced in Section 5.1.5, and the unstandardized delta-beta's described in Section 5.4.1. The code for this macro is easily amended to give standardized rather than unstandardized delta-beta's. The macro also gives the values of the likelihood displacement diagnostic, LD_i, and the quantity l_{max}, described in Section 5.4.2.

The macro has the following parameters:

SURVDATA   name of the SAS data set that contains the survival data.

TIME   name of the variable in SURVDATA that contains the survival times.

CENS   name of the variable in SURVDATA that contains the values of the censoring indicator.

VCENS   value of the variable CENS that indicates right-censored observations.

XVARS   names of the explanatory variables in SURVDATA that are contained in the fitted model. These are given as a list of names, each separated by a single space.

NXVARS   number of explanatory variables fitted in the model. This is equal to the number of variables defined in XVARS.

RSCORE   name of the variable that contains the values of the risk score, $\sum_{j=1}^{p} \hat{\beta}_j x_{ji}$. This variable can be obtained using the output statement in proc phreg with the keyword xbeta, and it must be defined in SURVDATA.

OUTVALS   name of the SAS data set that will contain the values of the score residuals, delta-beta's, and the values of the influence diagnostics LD_i, and l_{max}.

After using the macro, the variables sres_x1, sres_x2, ... contain the score residuals, while db_x1, db_x2, ... contain the delta-beta's. The variables LD and Lmax contain the values of LD_i and l_{max}, respectively.

Use of the macro `coxdiag`

To illustrate the use of this macro, consider again the data on the survival times of women with ovarian cancer that has been used throughout this chapter. The SAS code needed to fit a Cox proportional hazards model that contains the variables *AGE* and *TREAT*, and which then uses the macro `coxdiag`, is as follows:

```
proc phreg;
model survtime*cens(0)=age treat;
output out=ovdata xbeta=eta;
run;

%coxdiag(SURVDATA=ovdata, TIME=survtime, CENS=cens,
   VCENS=0, XVARS=age treat, NXVARS=2, RSCORE=eta,
   OUTVALS=results);

proc print data=results;
run;
```

Plots based on the diagnostics can be obtained using `proc plot`.

11.7.2 The SAS macro `weibdiag`

This macro is used after `proc lifereg` has been used to fit a Weibull model. It gives the score residuals defined in Section 5.6 and unstandardized delta-beta's defined in Section 5.7.1. The code is easily amended to give standardized delta-beta's. Also given are the two influence diagnostics described in Section 5.7.2, and denoted there by F_i and C_i.

The macro has the following parameters:

SURVDATA	name of the SAS data set that contains the survival data.
TIME	name of the variable in SURVDATA that contains the survival times.
CENS	name of the variable in SURVDATA that contains the values of the censoring indicator.
VCENS	value of the variable CENS that indicates right-censored observations.
XVARS	names of the explanatory variables in SURVDATA that are contained in the fitted model. These are given as a list of names, each separated by a single space.
NXVARS	number of explanatory variables fitted in the model. This is equal to the number of variables defined in XVARS.
RSCORE	name of the variable that contains the values of the risk score. The log-linear representation of the model is used for this,

and so the values of the risk score are $\hat{\mu} + \sum_{j=1}^{p} \hat{\alpha}_j x_{ji}$, where $\hat{\mu}$ is the estimate labelled INTERCPT in the output from proc lifereg, and the $\hat{\alpha}_j$ are the estimated coefficients of the explanatory variables given in the output. This variable is obtained using the output statement in proc lifereg with the keyword xbeta, and must be defined in SURVDATA.

SIGMA   value of the estimated scale parameter, $\hat{\sigma}$, in the fitted model. This will be the estimate labelled SCALE in the output from proc lifereg.

OUTVALS   name of the SAS data set that will contain the values of the score residuals, delta-beta's and the values of the influence diagnostics F_i and C_i.

In the version of this macro given in Appendix C, factors cannot be used directly. Instead, appropriate indicator variables are used after having defined them in a data step.

After using the macro, the variables sres_x1, sres_x2, . . . contain the score residuals, while db_x1, db_x2, . . . contain the delta-beta's. The variables F and C contain the values of F_i and C_i, respectively.

Use of the macro weibdiag

The SAS code required to fit a Weibull proportional hazards model that contains the variables AGE and $TREAT$ to the data on the survival times of ovarian cancer patients, and to then call the macro weibdiag, is shown below:

```
proc lifereg;
model survtime*cens(0)=age treat;
output out=ovdata xbeta=eta;
run;

%weibdiag(SURVDATA=ovdata, TIME=survtime, CENS=cens,
    VCENS=0,XVARS=age treat, NXVARS=2, RSCORE=eta,
    SIGMA=0.5489, OUTVALS=results);

proc print data=results;
run;
```

Again, plots constructed from the diagnostics produced by this macro can easily be obtained.

11.8 Relative merits of SAS, BMDP and SPSS for survival analysis

While much can be deduced about the relative merits of the three packages for survival analysis considered in detail in this chapter, some of the main points are summarized in this final section.

No single package is ideal for survival analysis in terms of scope. Most of the analyses described in this book can be carried out using either SAS or BMDP, but both packages are limited in terms of their facilities for model checking. SPSS does offer diagnostics not available in the other two packages, such as the score residuals and delta-beta's, but this is counterbalanced by the lack of SPSS software for fully parametric modelling.

Data management is readily accomplished using all three packages, but the facilities for this in SAS are the most extensive. High-quality graphical output can be obtained from each package.

There is little to choose between the packages in terms of their use for non-parametric analyses. All of them can be used to produce Kaplan–Meier estimates of survivor functions and the results of a log-rank and Wilcoxon test. However, it is unfortunate that there is inconsistency between the methods used in the strata and test statements of SAS proc lifetest to calculate the values of these statistics.

Of the three packages, SPSS seems to provide the most extensive facilities for use in fitting the Cox proportional hazards model. This package allows factors to be defined and included in the model and provides a wider range of diagnostics than the other two. However, the scope of SAS can be extended by writing SAS macros for non-standard analyses, such as those described in the previous section.

In terms of the output provided, SAS proc phreg and the SPSS procedure coxreg give the value of the statistic $-2 \log \hat{L}$ on fitting a model, whereas program 2L of BMDP merely give the value of $\log \hat{L}$. All packages give parameter estimates and their standard errors, together with the ratio, or the square of the ratio, of the coefficient to its standard error, and the exponential of the parameter estimate. For a variate, the latter can be interpreted as a hazard ratio, but this interpretation is more difficult when factors are fitted. Unless care is taken, these quantities can be misinterpreted. When fitting factors, the user must certainly be aware of the manner in which indicator variables have been defined within the package. Each of the three packages can be used to provide an estimate of the cumulative hazard function or survivor function. It is important to note that the method used by BMDP to compute these functions is different from that used by SAS and SPSS, but the difference in the results will not usually be very great.

Only SAS and BMDP can be used to fit a parametric model to survival data, and of these, SAS is the most versatile. This is the only package that allows factors to be incorporated in a model, but interactions between factors, or between a variate and a factor, cannot be fitted directly. Both packages suffer from the disadvantage that it is the value of the maximized log-likelihood function, $\log \hat{L}$, that is given in the output rather than $-2 \log \hat{L}$.

Each of the three packages can be run interactively or in batch mode. Those parts of the manuals concerned with the software for survival analysis are comprehensive, and all include a number of illustrations of the use of the software. However, the SPSS manual does not include much material on the computational methods used by the package.

Appendix A

Maximum likelihood estimation

This appendix gives a summary of results on maximum likelihood estimation that are relevant to survival analysis. The results presented apply equally to inferences based on a partial likelihood function, and so can be used in conjunction with the Cox regression model described in Chapter 3 and the fully parametric models introduced in Chapters 4 and 6. A full treatment of the theory of maximum likelihood estimation and likelihood ratio testing is given by Cox and Hinkley (1974).

A.1 Inference about a single unknown parameter

Suppose that the likelihood function of n observed survival times, $t_1, t_2, \ldots, t_n$, is a function of a single unknown parameter β, and denoted $L(\beta)$. The *maximum likelihood estimate* of β is then the value $\hat{\beta}$ for which this function is a maximum. In almost all applications, it is more convenient to work with the natural logarithm of the likelihood function, $\log L(\beta)$. The value $\hat{\beta}$ which maximizes the log-likelihood is the same value which maximizes the likelihood function itself, and is generally found using differential calculus.

Specifically, $\hat{\beta}$ is the value of β for which the derivative of $\log L(\beta)$, with respect to β, is equal to zero. In other words, $\hat{\beta}$ is such that

$$\frac{\mathrm{d} \log L(\beta)}{\mathrm{d}\beta} \bigg|_{\hat{\beta}} = 0.$$

The first derivative of $\log L(\beta)$ with respect to β is known as the *efficient score* for β, and is denoted $u(\beta)$. Therefore,

$$u(\beta) = \frac{\mathrm{d} \log L(\beta)}{\mathrm{d}\beta},$$

and so the maximum likelihood estimate of β, $\hat{\beta}$, satisfies the equation

$$u(\hat{\beta}) = 0.$$

The asymptotic variance of the maximum likelihood estimate of β can be found from

$$\left(-\mathrm{E}\left\{\frac{\mathrm{d}^2 \log L(\beta)}{\mathrm{d}\beta^2}\right\}\right)^{-1}, \tag{A.1}$$

or from the equivalent formula

$$\left(\mathrm{E}\left\{\frac{\mathrm{d} \log L(\beta)}{\mathrm{d}\beta}\right\}^2\right)^{-1}.$$

The variance calculated from either of these expressions can be regarded as the approximate variance of $\hat{\beta}$, although it is usually more straightforward to use expression (A.1). When the expected value of the derivative in expression (A.1) is difficult to obtain, a further approximation to the variance of $\hat{\beta}$ is found by evaluating the derivative at $\hat{\beta}$. The approximate variance of $\hat{\beta}$ is then given by

$$\mathrm{var}\,(\hat{\beta}) \approx -\left(\frac{\mathrm{d}^2 \log L(\beta)}{\mathrm{d}\beta^2}\right)^{-1}\Bigg|_{\hat{\beta}}. \tag{A.2}$$

The second derivative of the log-likelihood function is sometimes known as the *Hessian*, and the quantity

$$-\mathrm{E}\left\{\frac{\mathrm{d}^2 \log L(\beta)}{\mathrm{d}\beta^2}\right\}$$

is called the *information function*. Since the information function is formed from the expected value of the second derivative of $\log L(\beta)$, it is sometimes called the *expected information function*. In contrast, the negative second derivative of the log-likelihood function itself is called the *observed information function*. This latter quantity will be denoted $i(\beta)$, so that

$$i(\beta) = -\left\{\frac{\mathrm{d}^2 \log L(\beta)}{\mathrm{d}\beta^2}\right\}.$$

The reciprocal of this function, evaluated at $\hat{\beta}$, is then the approximate variance of $\hat{\beta}$ given in equation (A.2), that is

$$\mathrm{var}\,(\hat{\beta}) \approx \frac{1}{i(\hat{\beta})}.$$

The standard error of $\hat{\beta}$, that is, the square root of the estimated variance of $\hat{\beta}$, is then found from

$$\mathrm{s.e.}\,(\hat{\beta}) = \frac{1}{\sqrt{\{i(\hat{\beta})\}}}.$$

This standard error can be used to construct confidence intervals for β.

In order to test the null hypothesis that $\beta = 0$, three alternative test statistics can be used. The *likelihood ratio test statistic* is the difference between the values of $-2\log L(\hat{\beta})$ and $-2\log L(0)$. The *Wald test* is based on the statistic

$$\hat{\beta}^2 i(\hat{\beta}),$$

and the *score test statistic* is

$$\frac{\{u(0)\}^2}{i(0)}.$$

Each of these statistics has an asymptotic chi-squared distribution on 1 d.f., under the null hypothesis that $\beta = 0$. Note that the Wald statistic is equivalent to the statistic

$$\frac{\hat{\beta}}{\text{s.e.}\,(\hat{\beta})},$$

which has an asymptotic standard normal distribution.

A.2 Inference about a vector of unknown parameters

The results in Section A.1 can be extended to the situation where n observations are used to estimate the values of p unknown parameters, $\beta_1, \beta_2, \ldots, \beta_p$. These parameters can be assembled into a p-component vector, β, and the corresponding likelihood function is $L(\beta)$. The maximum likelihood estimates of the p unknown parameters are the values $\hat{\beta}_1, \hat{\beta}_2, \ldots, \hat{\beta}_p$ which maximize $L(\beta)$. They are therefore found by solving the p equations

$$\frac{\mathrm{d}\log L(\beta)}{\mathrm{d}\beta_j}\bigg|_{\hat{\beta}} = 0,$$

$j = 1, 2, \ldots, p$, simultaneously.

The vector formed from $\hat{\beta}_1, \hat{\beta}_2, \ldots, \hat{\beta}_p$ is denoted $\hat{\beta}$, and so the maximized likelihood is $L(\hat{\beta})$. The efficient score for β_j, $j = 1, 2, \ldots, p$, is

$$u(\beta_j) = \frac{\mathrm{d}\log L(\beta)}{\mathrm{d}\beta_j},$$

and these quantities can be assembled to give a p-component vector of efficient scores, denoted $\mathbf{u}(\beta)$. The vector of maximum likelihood estimates is therefore such that

$$\mathbf{u}(\hat{\beta}) = \mathbf{0},$$

where $\mathbf{0}$ is the $p \times 1$ vector of zeroes.

Now let the matrix $\mathbf{H}(\beta)$ be the $p \times p$ matrix of second partial derivatives of the log-likelihood function, $\log L(\hat{\beta})$. The (j, k)th element of $\mathbf{H}(\beta)$ is then

$$\frac{\partial^2 \log L(\hat{\beta})}{\partial \beta_j \partial \beta_k},$$

for $j = 1, 2, \ldots, p$, $k = 1, 2, \ldots, p$, and $\mathbf{H}(\beta)$ is called the *Hessian matrix*. The matrix

$$\mathbf{I}(\beta) = -\mathbf{H}(\beta)$$

is called the *observed information matrix*.

The (j, k)th element of the corresponding *expected information matrix* is

$$- \mathrm{E} \left(\frac{\partial^2 \log L(\beta)}{\partial \beta_j \partial \beta_k} \right).$$

The variance–covariance matrix of the p maximum likelihood estimates, $\hat{\beta}_1, \hat{\beta}_2,$ $\ldots, \hat{\beta}_p$, written var $(\hat{\beta})$, can then be approximated by the inverse of the observed information matrix, evaluated at $\hat{\beta}$, so that

$$\mathrm{var}\,(\hat{\beta}) \approx \mathbf{I}^{-1}(\hat{\beta}).$$

The square root of the (j, j)th element of this matrix can be taken to be the standard error of $\hat{\beta}_j$, for $j = 1, 2, \ldots, p$.

The test statistics given in Section A.1 can be generalized to the multiparameter situation. Consider the test of the hypothesis that all the β-parameters in a fitted model are equal to zero. The likelihood ratio test statistic is the value of

$$2 \left\{ \log L(\hat{\beta}) - \log L(\mathbf{0}) \right\},$$

the Wald test is based on

$$\hat{\beta}' \mathbf{I}(\hat{\beta}) \hat{\beta},$$

and the score test statistic is

$$\mathbf{u}'(\mathbf{0}) \mathbf{I}^{-1}(\mathbf{0}) \mathbf{u}(\mathbf{0}).$$

Each of these statistics has a chi-squared distribution on p d.f. under the null hypothesis that $\beta = \mathbf{0}$.

The values of these statistics are given under the heading Model Chi-Square in the output from SAS proc phreg. The value of the score statistic is referred to as GLOBAL CHI-SQUARE in BMDP output from program 2L, and as Overall (score) in SPSS output from the procedure coxreg.

In comparing alternative models, interest centres on the hypothesis that some of the β-parameters in a model are equal to zero. To test this hypothesis, the likelihood ratio test is the most suitable, and so we only consider this procedure here. Suppose that a model that contains $p + q$ parameters, $\beta_1, \beta_2, \ldots, \beta_p, \beta_{p+1}, \ldots,$ β_{p+q}, is to be compared with a model that only contains the p parameters $\beta_1, \beta_2,$ $\ldots, \beta_p$. This amounts to testing the null hypothesis that the q parameters $\beta_{p+1},$ $\beta_{p+2}, \ldots, \beta_{p+q}$ in the model with $p + q$ unknown parameters are all equal to zero. Let $\hat{\beta}_1$ denote the vector of estimates under the model with $p + q$ parameters and $\hat{\beta}_2$ that for the model with just p parameters. The likelihood ratio test of the null hypothesis that $\beta_{p+1} = \beta_{p+2} = \cdots = \beta_{p+q} = 0$ in the model with $p + q$ parameters is then based on the statistic

$$2 \left\{ \log L(\hat{\beta}_1) - \log L(\hat{\beta}_2) \right\}.$$

This has a chi-squared distribution on q d.f. under the null hypothesis. This test forms the basis for comparing alternative models, and was described in greater detail in Section 3.5 of Chapter 3.

Appendix B

Standard error of Weibull percentiles

In this appendix, an expression for the standard error for the percentiles of the Weibull distribution is derived. A more general result for the standard error of the pth percentile in the Weibull proportional hazards model is also given.

B.1 Standard error of a percentile of the Weibull distribution

In equation (4.23) of Chapter 4, the estimated pth percentile of the Weibull distribution with scale parameter λ and shape parameter γ was shown to be given by

$$\hat{t}(p) = \left[\frac{1}{\hat{\lambda}} \log \left(\frac{100}{100 - p} \right) \right]^{1/\hat{\gamma}}$$

The variance of $\hat{t}(p)$ is most easily found by first obtaining the variance of $\log \hat{t}(p)$. Now,

$$\log \hat{t}(p) = \frac{1}{\hat{\gamma}} \log \left\{ \hat{\lambda}^{-1} \log \left(\frac{100}{100 - p} \right) \right\},$$

and so

$$\log \hat{t}(p) = \frac{1}{\hat{\gamma}} \left\{ c_p - \log \hat{\lambda} \right\},$$

where

$$c_p = \log \log \left(\frac{100}{100 - p} \right).$$

This is a function of two parameter estimates, $\hat{\lambda}$ and $\hat{\gamma}$, and so we can use the result in equation (??) of Chapter 11 to obtain the approximate variance of $\log \hat{t}(p)$. From equation (??),

$$\text{var} \left\{ \log \hat{t}(p) \right\} \approx \left(\frac{\partial \log \hat{t}(p)}{\partial \hat{\lambda}} \right)^2 \text{var} (\hat{\lambda}) + \left(\frac{\partial \log \hat{t}(p)}{\partial \hat{\gamma}} \right)^2 \text{var} (\hat{\gamma})$$

$$+ 2 \frac{\partial \log \hat{t}(p)}{\partial \hat{\lambda}} \frac{\partial \log \hat{t}(p)}{\partial \hat{\gamma}} \text{cov} (\hat{\lambda}, \hat{\gamma}).$$

Now, the derivatives of $\log \hat{t}(p)$ with respect to $\hat{\lambda}$ and $\hat{\gamma}$ are given by

$$\frac{\partial \log \hat{t}(p)}{\partial \hat{\lambda}} = -\frac{1}{\hat{\lambda}\hat{\gamma}},$$

$$\frac{\partial \log \hat{t}(p)}{\partial \hat{\gamma}} = -\frac{c_p - \log \hat{\lambda}}{\hat{\gamma}^2},$$

and so the approximate variance of $\log \hat{t}(p)$ is

$$\frac{1}{\hat{\lambda}^2\hat{\gamma}^2} \text{var}(\hat{\lambda}) + \frac{\left(c_p - \log \hat{\lambda}\right)^2}{\hat{\gamma}^4} \text{var}(\hat{\gamma}) + \frac{2\left(c_p - \log \hat{\lambda}\right)}{\hat{\lambda}\hat{\gamma}^3} \text{cov}(\hat{\lambda}, \hat{\gamma}). \quad \text{(B.1)}$$

The variance of $\hat{t}(p)$ itself is found from the result in equation (2.8) of Chapter 2, from which

$$\text{var}\{\hat{t}(p)\} \approx \hat{t}(p)^2 \, \text{var}\{\log \hat{t}(p)\}.$$

Therefore, from expression (B.1),

$$\text{var}\{\hat{t}(p)\} \approx \frac{\hat{t}(p)^2}{\hat{\lambda}^2\hat{\gamma}^4} \left\{ \hat{\gamma}^2 \, \text{var}(\hat{\lambda}) + \hat{\lambda}^2 \left(c_p - \log \hat{\lambda}\right)^2 \text{var}(\hat{\gamma}) \right. $$
$$\left. + 2\hat{\lambda}\hat{\gamma} \left(c_p - \log \hat{\lambda}\right) \text{cov}(\hat{\lambda}, \hat{\gamma}) \right\},$$

and so the standard error of $\hat{t}(p)$ is the square root of this expression, that is

$$\text{s.e.}\{\hat{t}(p)\} = \frac{\hat{t}(p)}{\hat{\lambda}\hat{\gamma}^2} \left\{ \hat{\gamma}^2 \, \text{var}(\hat{\lambda}) + \hat{\lambda}^2 \left(c_p - \log \hat{\lambda}\right)^2 \text{var}(\hat{\gamma}) \right. $$
$$\left. + 2\hat{\lambda}\hat{\gamma} \left(c_p - \log \hat{\lambda}\right) \text{cov}(\hat{\lambda}, \hat{\gamma}) \right\}^{1/2}. \quad \text{(B.2)}$$

A $100(1 - \alpha)\%$ confidence interval for the true pth percentile is found from exponentiating the corresponding confidence limits for $\log t(p)$. These limits are

$$\log \hat{t}(p) \pm z_{\alpha/2} \, \text{s.e.}\{\log \hat{t}(p)\},$$

where s.e. $\{\log \hat{t}(p)\}$ is the square root of expression (B.1), and $z_{\alpha/2}$ is the upper $\alpha/2$-point of the standard normal distribution.

Similar methods can be used to obtain expressions for s.e. $\{\hat{t}(p)\}$ in terms of the log-linear parametrization of the Weibull model.

Note that for the special case of the exponential distribution, where the shape parameter, γ, is equal to unity, the standard error of the estimated pth percentile is

$$\frac{\hat{t}(p)}{\hat{\lambda}} \, \text{s.e.}(\hat{\lambda}).$$

From equation (4.13) of Chapter 4,

$$\text{s.e.}(\hat{\lambda}) = \hat{\lambda}/\sqrt{r},$$

where r is the number of death times in the data set, and so

$$\text{s.e.}\,\{\hat{t}(p)\} = \hat{t}(p)/\sqrt{r},$$

as in equation (4.17).

B.2 Standard error of a percentile in the Weibull model

Using the parametrization of the general Weibull proportional hazards model adopted in Section 4.5 and elsewhere, the estimated pth percentile for an individual for whom the values of the explanatory variables form the vector $\mathbf{x}$ is

$$\hat{t}(p) = \left\{ \frac{1}{\hat{\lambda}\exp(\hat{\beta}'\mathbf{x})} \log\left(\frac{100}{100-p}\right) \right\}^{1/\hat{\gamma}}$$

Again, the variance of $\log \hat{t}(p)$ is found first, and writing

$$c_p = \log\log\left(\frac{100}{100-p}\right),$$

we get

$$\log \hat{t}(p) = \frac{1}{\hat{\gamma}}\left(c_p - \log\hat{\lambda} - \hat{\beta}'\mathbf{x}\right). \tag{B.3}$$

This is a function of the $p+2$ parameter estimates $\hat{\lambda}$, $\hat{\gamma}$ and $\hat{\beta}_1, \hat{\beta}_2, \ldots, \hat{\beta}_p$, and the approximate variance of this function can be found using a further generalization of the result in equation (11.1). This generalization is best expressed in matrix form.

Suppose that $\hat{\theta}$ is a vector formed from k parameter estimates, $\hat{\theta}_1, \hat{\theta}_2, \ldots, \hat{\theta}_k$, and that $g(\hat{\theta})$ is a function of these k estimates. The approximate variance of $g(\hat{\theta})$ is then given by

$$\text{var}\,\{g(\hat{\theta})\} \approx \mathbf{d}(\hat{\theta})'\,\text{var}\,(\hat{\theta})\mathbf{d}(\hat{\theta}), \tag{B.4}$$

where $\text{var}\,(\hat{\theta})$ is the $k \times k$ variance–covariance matrix of the estimates in $\hat{\theta}$, and $\mathbf{d}(\hat{\theta})$ is the k-component vector whose ith element is

$$\left.\frac{\partial g(\hat{\theta})}{\partial \hat{\theta}_i}\right|_{\hat{\theta}},$$

for $i = 1, 2, \ldots, k$.

We now write $\mathbf{V}$ for the $(p + 2) \times (p + 2)$ variance–covariance matrix of $\hat{\lambda}$, $\hat{\gamma}$ and $\hat{\beta}_1, \hat{\beta}_2, \ldots, \hat{\beta}_p$. Next, from equation (B.3), the derivatives of $\log \hat{t}(p)$ with respect to these parameter estimates are

$$\frac{\partial \log \hat{t}(p)}{\partial \hat{\lambda}} = -\frac{1}{\hat{\lambda}\hat{\gamma}}$$

$$\frac{\partial \log \hat{t}(p)}{\partial \hat{\gamma}} = -\frac{c_p - \log\hat{\lambda} - \hat{\beta}'\mathbf{x}}{\hat{\gamma}^2}$$

$$\frac{\partial \log \hat{t}(p)}{\partial \hat{\beta}_j} = -\frac{x_j}{\hat{\gamma}},$$

for $j = 1, 2, \ldots, p$, where x_j is the jth component of $\mathbf{x}$. The vector $\mathbf{d}$ in equation (B.4) can be written as

$$\mathbf{d} = -\hat{\gamma}\mathbf{d}_0,$$

where $\mathbf{d}_0$ is a vector with components $\hat{\lambda}^{-1}$, $\hat{\gamma}^{-1}\left\{c_p - \log \hat{\lambda} - \hat{\beta}'\mathbf{x}\right\}$, and $x_1, x_2,$ $\ldots, x_p$. Then, the standard error of $\log \hat{t}(p)$ is given by

$$\text{s.e.} \left\{\log \hat{t}(p)\right\} = \hat{\gamma}^{-1}\sqrt{(\mathbf{d}_0'\mathbf{V}\mathbf{d}_0)}, \tag{B.5}$$

whence

$$\text{s.e.} \{\hat{t}(p)\} = \frac{\hat{t}(p)}{\hat{\gamma}} \text{ s.e.} \left\{\log \hat{t}(p)\right\}.$$

Notice that for the null model, which contains no explanatory variables, the standard error of $\log \hat{t}(p)$ in equation (B.5) reduces to the result in equation (B.2).

Appendix C

Two SAS macros

In Chapter 11, two SAS macros for obtaining the values of certain diagnostics used in model checking were described. These macros, named coxdiag and weibdiag, are listed in this Appendix. They may be reproduced and used with no implied warranty from the author or publisher. Machine readable versions of these macros are available from the author on request.

C.1 The SAS macro coxdiag

```
%macro coxdiag(SURVDATA,TIME,CENS,VCENS,XVARS,NXVARS,
                RSCORE,OUTVALS);
data zzzsetup;
length &TIME &CENS &RSCORE 8;
set &SURVDATA;
%do i=1 %to &NXVARS;
    x&i=%scan(&XVARS,&i,%str( ));
%end;
index=_n_;
keep &TIME &CENS &RSCORE x1 -- x&NXVARS index;
proc sort; by &TIME;
data zzztemp1;
set zzzsetup;
drop index;
proc iml wrksize=55;
reset noprint;
use zzztemp1;
read all into xx;
nr=nrow(xx);
nc=ncol(xx);
np=nc-3;
t=xx[,1];
cens=xx[,2];
```

```
eta=xx[,3];
s_xe=j(nr,1);
s_e=j(nr,1);
x=j(nr,1);
ld=j(nr,1);
s_res=j(nr,np);
a_mat=j(nr,np);
delta=j(nr,np);
s_xxe=j(nr,1);
varcov=j(np,np);
s_e[1]=sum(exp(eta));
do i=2 to nr;
   s_e[i]=s_e[i-1]-exp(eta[i-1]);
end;
do i=2 to nr;
   if t[i]=t[i-1] then s_e[i]=s_e[i-1];
end;
do h=4 to nc;
   x=xx[,h];
   s_xe[1]=sum(x#exp(eta));
   do i=2 to nr;
      s_xe[i]=s_xe[i-1]-x[i-1]#exp(eta[i-1]);
   end;
   do i=2 to nr;
      if t[i]=t[i-1] then s_xe[i]=s_xe[i-1];
   end;
   a_mat[,h-3]=s_xe/s_e;
   s_res[,h-3]=cens#(x-a_mat[,h-3]);
   do k=1 to nr;
      sum1=0;
      sum2=0;
      do i=1 to nr while (t[i]<=t[k]);
         sum1=sum1+cens[i]#a_mat[i,h-3]/s_e[i];
         sum2=sum2+cens[i]/s_e[i];
      end;
      delta[k,h-3]=sum1#exp(eta[k])-sum2#exp(eta[k])#x[k];
   end;
   delta[,h-3]=s_res[,h-3]+delta[,h-3];
end;
do i=4 to nc;
  do j=4 to nc;
    s_xxe[1]=sum(xx[,i]#xx[,j]#exp(eta));
    do k=2 to nr;
       s_xxe[k]=s_xxe[k-1]-(xx[k-1,i]#xx[k-1,j]#exp(eta[k-1]));
    end;
    do k=2 to nr;
       if t[k]=t[k-1] then s_xxe[k]=s_xxe[k-1];
```

```
      end;
      varcov[i-3,j-3]=sum(cens#((s_xxe/s_e)-(a_mat[,i-3]#a_mat[,j-3])));
      end;
   end;
free a_mat xx s_e s_xe s_xxe;
varcov=inv(varcov);
do i=1 to nr;
   ld[i]=delta[i,]*varcov*delta[i,]';
end;
dbeta=delta*varcov;
s=root(varcov);
v=s*delta'*delta*s';
evec=eigvec(v);
lmax=delta*s'*evec[,1];
lmax=lmax/sqrt(lmax'*lmax);
evec=eigvec(delta*varcov*delta');
lmax=evec[,1];
clabel1={sres_x1,sres_x2,sres_x3,sres_x4,sres_x5,sres_x6,
         sres_x7,sres_x8};
create zzztemp2 from s_res [colname=clabel1];
append from s_res;
clabel2={db_x1,db_x2,db_x3,db_x4,db_x5,db_x6,db_x7,db_x8};
create zzztemp3 from dbeta [colname=clabel2];
append from dbeta;
create zzztemp4 var{ld lmax};
append var{ld lmax};
close zzztemp2 zzztemp3 zzztemp4;
quit;
data zzztemp5;
merge zzzsetup zzztemp2 zzztemp3 zzztemp4;
proc sort; by index;
data &OUTVALS;
set zzztemp5;
%do i=1 %to &NXVARS;
   rename x&i=%scan(&XVARS,&i,%str( ));
%end;
drop index;
%mend;
```

C.2 The SAS macro weibdiag

```
%macro weibdiag(SURVDATA,TIME,CENS,VCENS,XVARS,NXVARS,
                RSCORE,SIGMA,OUTVALS);
data zzzsetup;
length &TIME &CENS &RSCORE 8;
set &SURVDATA;
```

```
%do i=1 %to &NXVARS;
    x&i=%scan(&XVARS,&i,%str( ));
%end;
keep &TIME &CENS &RSCORE x1 -- x&NXVARS;
proc iml wrksize=55;
sigma=&SIGMA;
reset noprint;
use zzzsetup;
read all into xx;
nr=nrow(xx);
nc=ncol(xx);
np=nc-3;
t=xx[,1];
cens=xx[,2];
eta=xx[,3];
x=j(nr,1);
z=(log(t)-eta)/sigma;
score=j(nr,np+2);
sres=j(nr,np);
I=j(np+2,np+2);
f=j(nr,1);
c=j(nr,1);
dbeta=j(nr,np+2);
udbeta=j(nr,np);
xx[,3]=j(nr,1,1);
do j=1 to np+1;
    score[,j]=(1/sigma)#(-cens#xx[,j+2]+xx[,j+2]#exp(z));
    score[,np+2]=-(1/sigma)#(cens+cens#z-z#exp(z));
end;
do j=1 to np+1;
    do k=1 to np+1;
        I[j,k]=(1/(sigma#sigma))#sum(xx[,j+2]#xx[,k+2]#exp(z));
    end;
    I[j,np+2]=-sum(cens#xx[,j+2]-xx[,j+2]#(1+z)#exp(z));
    I[j,np+2]=I[j,np+2]/(sigma#sigma);
    I[np+2,j]=I[j,np+2];
end;
I[np+2,np+2]=-sum(cens+2#cens#z-z#(2+z)#exp(z));
I[np+2,np+2]=I[np+2,np+2]/(sigma#sigma);
varcov=inv(I);
r1=score'*score;
do i=1 to nr;
    g1=score[i,]*inv(r1)*score[i,]';
    g2=score[i,]*varcov*score[i,]';
    f[i]=g1/((np+2)#(1-g1));
    c[i]=g2/((1-g2)#(1-g2));
end;
```

```
do i=1 to nr;
   dbeta[i,]=score[i,]*varcov;
end;
sres[,1:np]=score[,2:np+1];
udbeta[,1:np]=dbeta[,2:np+1];
clabel1={sres_x1,sres_x2,sres_x3,sres_x4,sres_x5,sres_x6,
         sres_x7,sres_x8};
create zzzdata1 from sres [colname=clabel1];
append from sres;
clabel2={db_x1,db_x2,db_x3,db_x4,db_x5,db_x6,db_x7,db_x8};
create zzzdata2 from udbeta [colname=clabel2];
append from udbeta;
create zzzdata3 var{f c};
append var{f c};
close  zzzsetup zzzdata1 zzzdata2 zzzdata3;
quit;
data &OUTVALS;
merge zzzsetup zzzdata1 zzzdata2 zzzdata3;
%do i=1 %to &NXVARS;
   rename x&i=%scan(&XVARS,&i,%str( ));
%end;
%mend;
```

References

Abramowitz, M. and Stegun, I.A. (1972) *Handbook of Mathematical Functions with Formulas, Graphs and Mathematical Tables*, U.S. Government Printing Office, Washington.

Aitkin, M., Anderson, D.A., Francis, B. and Hinde, J.P. (1989) *Statistical Modelling in GLIM*, Clarendon Press, Oxford.

Aitkin, M. and Clayton, D. (1980) The fitting of exponential, Weibull and extreme value distributions to censored survival data using GLIM. *Applied Statistics*, **29**, 156–63.

Aitkin, M., Laird, N. and Francis, B. (1983) A reanalysis of the Stanford heart transplant data (with comments). *Journal of the American Statistical Association*, **78**, 264–92.

Akaike, H. (1974) A new look at the statistical model identification. *IEEE Transactions on Automatic Control*, **19**, 716–23.

Altman, D.G. (1991) *Practical Statistics for Medical Research*, Chapman and Hall, London.

Andersen, P.K. (1982) Testing goodness of fit of Cox's regression and life model. *Biometrics*, **38**, 67–77.

Andersen, P.K. (1988) Multistate models in survival analysis: a study of nephropathy and mortality in diabetes. *Statistics in Medicine*, **7**, 661–70.

Andersen, P.K. (1992) Repeated assessment of risk factors in survival analysis. *Statistical Methods in Medical Research*, **1**, 297–315.

Andrews, D.F. and Herzberg, A.M. (1985) *Data*, Springer-Verlag, New York.

Arjas, E. (1988) A graphical method for assessing goodness of fit in Cox's proportional hazards model. *Journal of the American Statistical Association*, **83**, 204–12.

Armitage, P. and Berry, G. (1987) *Statistical Methods in Medical Research*, 2nd edn, Blackwell Scientific Publications, Oxford.

Atkinson, A.C. (1985) *Plots, Transformations and Regression*, Clarendon Press, Oxford.

Barlow, R.E. and Proschan, F. (1975) *Statistical Theory of Reliability and Life Testing*, Holt, Rinehart and Winston, New York.

Barlow, W.E. and Prentice, R.L. (1988) Residuals for relative risk regression.

Biometrika, **75**, 65–74.

Barnett, V. (1982) *Comparative Statistical Inference,* 2nd edn, Wiley, London.

Becker, N.G. and Melbye, M. (1991) Use of a log-linear model to compute the empirical survival curve from interval-censored data, with application to data on tests for HIV positivity. *Australian Journal of Statistics,* **33**, 125–33.

Bennett, S. (1983a) Analysis of survival data by the proportional odds model. *Statistics in Medicine,* **2**, 273–7.

Bennett, S. (1983b) Analysis of Survival Data by the Proportional Odds Model, Ph.D. Thesis, University of Reading.

Bennett, S. (1983c) Log-logistic regression models for survival data. *Applied Statistics,* **32**, 165–71.

Bennett, S. and Whitehead, J.R. (1981) Fitting logistic and log-logistic regression models to censored survival data using GLIM. *GLIM Newsletter,* **4**, 12–9. Correction, **5**, 3.

Bernstein, D. and Lagakos, S.W. (1978) Sample size and power determination for stratified clinical trials. *Journal of Statistical Computation and Simulation,* **8**, 65–73.

Breslow, N.E. (1972) Contribution to the discussion of a paper by D.R. Cox. *Journal of the Royal Statistical Society, B,* **34**, 216–7.

Breslow, N.E. (1974) Covariance analysis of censored survival data. *Biometrics,* **30**, 89–100.

Breslow, N.E. and Crowley, J. (1974) A large sample study of the life table and product limit estimates under random censorship. *Annals of Statistics,* **2**, 437–53.

Breslow, N.E. and Day, N.E. (1987) *Statistical Methods in Cancer Research. 2: The Design and Analysis of Cohort Studies,* I.A.R.C., Lyon.

Brookmeyer, R. and Crowley, J. (1982) A confidence interval for the median survival time. *Biometrics,* **38**, 29–41.

Burdette, W.J. and Gehan, E.A. (1970) *Planning and Analysis of Clinical Studies,* Charles C. Thomas, Springfield, Illinois.

Cain, K.C. and Lange, N.T. (1984) Approximate case influence for the proportional hazards regression model with censored data. *Biometrics,* **40**, 493–9.

Chatfield, C. (1988) *Problem Solving: A Statisticians Guide,* Chapman and Hall, London.

Chatfield, C. (1989) *The Analysis of Time Series,* 4th edn, Chapman and Hall, London.

Christensen, E. (1987) Multivariate survival analysis using Cox's regression model. *Hepatology,* **7**, 1346–58.

Ciampi, A. and Etezadi-Amoli, J. (1985) A general model for testing the proportional hazards and the accelerated failure time hypotheses in the analysis of censored survival data with covariates. *Communications in Statistics, A,* **14**, 651–67.

Clayton, D. and Cuzick, J. (1985) The EM algorithm for Cox's regression model using GLIM. *Applied Statistics,* **34**, 148–56.

Collett, D. (1991) *Modelling Binary Data,* Chapman & Hall, London.

Cook, R.D. (1986) Assessment of local influence (with discussion). *Journal of the Royal Statistical Society, B,* **48**, 133–69.

Cook, R.D. and Weisberg, S. (1982) *Residuals and Influence in Regression*, Chapman and Hall, London.

Cox, D.R. (1972) Regression models and life tables (with discussion). *Journal of the Royal Statistical Society, B*, **74**, 187–220.

Cox, D.R. (1975) Partial likelihood. *Biometrika*, **62**, 269–76.

Cox, D.R. (1979) A note on the graphical analysis of survival data. *Biometrika*, **66**, 188–90.

Cox, D.R. and Hinkley, D. V. (1974) *Theoretical Statistics*, Chapman and Hall, London.

Cox, D.R. and Oakes, D. (1984) *Analysis of Survival Data*, Chapman and Hall, London.

Cox, D.R. and Snell, E.J. (1968) A general definition of residuals (with discussion). *Journal of the Royal Statistical Society, A*, **30**, 248–75.

Cox, D.R. and Snell, E.J. (1981) *Applied Statistics: Principles and examples*, Chapman and Hall, London.

Cox, D.R. and Snell, E.J. (1989) *Analysis of Binary Data*, 2nd edn, Chapman and Hall, London.

Crowder, M.J., Kimber, A.C., Smith, R.L. and Sweeting, T.J. (1991) *Statistical Analysis of Reliability Data*, Chapman & Hall, London.

Crowley, J. and Hu, M. (1977) Covariance analysis of heart transplant survival data. *Journal of the American Statistical Association*, **72**, 27–36.

Crowley, J. and Storer, B.E. (1983) Comment on a paper by M. Aitkin *et al.* *Journal of the American Statistical Association*, **78**, 277–81.

Cryer, J.D. (1986) *Time Series Analysis*, Duxbury Press, Bolton.

Day, L. (1985) Residual analysis for Cox's proportional hazards model. In *Proceedings of STATCOM-MEDSTAT '85'*, MacQuarie University, Sydney.

Dinse, G.E. (1991) Constant risk differences in the analysis of animal tumourigenicity data. *Biometrics*, **47**, 681–700.

Dobson, A.J. (1990) *An Introduction to Generalized Linear Models*, Chapman and Hall, London.

Draper, N.R. and Smith, H. (1981) *Applied Regression Analysis*, 2nd edn, Wiley, New York.

Dupont, W.D. and Plummer, W.D. (1990) Power and sample size calculations: a review and computer program. *Controlled Clinical Trials*, **11**, 116–28.

Edmunson, J.H., Fleming, T.R., Decker, D.G., Malkasian, G.D., Jorgenson, E.O., Jeffries, J.A., Webb, M.J. and Kvols, L.K. (1979) Different chemotherapeutic sensitivities and host factors affecting prognosis in advanced ovarian carcinoma versus minimal residual disease. *Cancer Treatment Reports*, **63**, 241–7.

Efron, B. (1977) The efficiency of Cox's likelihood function for censored data. *Journal of the American Statistical Association*, **72**, 557–65.

Efron, B. (1981) Censored data and the bootstrap. *Journal of the American Statistical Association*, **76**, 312–9.

Elandt-Johnson, R.C. and Johnson, N.L. (1980) *Survival Models and Data Analysis*, Wiley, New York.

Elashoff, J.D. (1983) Surviving proportional hazards. *Hepatology*, **3**, 1031–5.

Emerson, J.D. (1982) Nonparametric confidence intervals for the median in the presence of right censoring. *Biometrics*, **38**, 17–27.

Escobar, L.A. and Meeker, W.Q. (1988) Using the SAS system to assess local influence in regression analysis with censored data. *Proceedings of the Annual SAS User's Group Conference*, 1036–41.

Escobar, L.A. and Meeker, W.Q. (1992) Assessing influence in regression analysis with censored data. *Biometrics*, **48**, 507–28.

Everitt, B.S. (1987) *Introduction to Optimisation Methods*, Chapman and Hall, London.

Finkelstein, D.M. (1986) A proportional hazards model for interval-censored failure time data. *Biometrics*, **42**, 845–54.

Fisher, L.D. (1992) Discussion of a paper by L.J. Wei. *Statistics in Medicine*, **11**, 1881–5.

Fleming, T.R. and Harrington, D.P. (1991) *Counting Processes and Survival Analysis*, Wiley, New York.

Freedman, L.S. (1982) Tables of the number of patients required in clinical trials using the logrank test. *Statistics in Medicine*, **1**, 121–9.

Gail, M.H. (1975) A review and critique of some models used in competing risks analysis. *Biometrics*, **31**, 209–22.

Gaver, D.P. and Acar, M. (1979) Analytical hazard representations for use in reliability, mortality and simulation studies. *Communications in Statistics*, **8**, 91–111.

Gehan, E.A. (1969) Estimating survival functions from the life table. *Journal of Chronic Diseases*, **21**, 629–44.

George, S.L. and Desu, M.M. (1974) Planning the size and duration of a clinical trial studying the time to some critical event. *Journal of Chronic Diseases*, **27**, 15–24.

Gill, R.D. and Schumacher, M. (1987) A simple test of the proportional hazards assumption. *Biometrika*, **74**, 289–300.

Gore, S.M., Pocock, S.J. and Kerr, G.R. (1984) Regression models and nonproportional hazards in the analysis of breast cancer survival. *Applied Statistics*, **33**, 176–95.

Gray, R. (1990) Some diagnostic methods for Cox regression models through hazard smoothing. *Biometrics*, **46**, 93–102.

Greenwood, M. (1926) The errors of sampling of the survivorship tables. *Reports on Public Health and Statistical Subjects*, number 33, Appendix 1, HMSO, London.

Gross, A.J. and Clark, V.A. (1975) *Survival Distributions: Reliability Applications in the Biomedical Sciences*, Wiley, New York.

Hall, W.J., Rogers, W.H. and Pregibon, D. (1982) Outliers matter in survival analysis. *Rand Corporation Technical Report P-6761*, Santa Monica, California.

Hall, W.J. and Wellner, J.A. (1980) Confidence bands for a survival curve from censored data. *Biometrika*, **67**, 133–43.

Harris, E.K. and Albert, A. (1991) *Survivorship Analysis for Clinical Studies*, Marcel Dekker, New York.

Hastie, T. and Tibshirani, R. (1990) *Generalized Additive Models*, Chapman and Hall, London.

Henderson, R. and Milner, A. (1991) On residual plots for relative risk regression.

Biometrika, **78**, 631–6.

Hinkley, D.V., Reid, N. and Snell, E.J. (1991) *Statistical Theory and Modelling*, Chapman and Hall, London.

Hjorth, U. (1980) A reliability distribution with increasing, decreasing and bathtub-shaped failure rate. *Technometrics*, **22**, 99–107.

Hougaard, P. and Madsen, E.B. (1985) Dynamic evaluation of short-term prognosis of myocardial infarction. *Statistics in Medicine*, **4**, 29–38.

Hsieh, F.Y., Crowley, J. and Tormey, D.C. (1983) Some test statistics for use in multi-state survival analysis. *Biometrika*, **70**, 111–9.

Johnson, N.L. and Kotz, S. (1969) *Distributions in Statistics: Discrete Distributions*, Houghton Mifflin, Boston.

Johnson, N.L. and Kotz, S. (1970) *Distributions in Statistics: Continuous Univariate Distributions*, Volume 1, Houghton Mifflin, Boston.

Kahn, H.A. and Sempos, C.T. (1989) *Statistical Methods in Epidemiology*, Oxford University Press, New York.

Kalbfleisch, J.D. and Prentice, R.L. (1972) Contribution to the discussion of a paper by D.R. Cox. *Journal of the Royal Statistical Society, B*, **34**, 215–6.

Kalbfleisch, J.D. and Prentice, R.L. (1973) Marginal likelihoods based on Cox's regression and life model. *Biometrika*, **60**, 267–78.

Kalbfleisch, J.D. and Prentice, R.L. (1980) *The Statistical Analysis of Failure Time Data*, Wiley, New York.

Kaplan, E.L. and Meier, P. (1958) Nonparametric estimation from incomplete observations. *Journal of the American Statistical Association*, **53**, 457–81.

Kay, R. (1977) Proportional hazard regression models and the analysis of censored survival data. *Applied Statistics*, **26**, 227–37.

Kay, R. (1984) Goodness of fit methods for the proportional hazards model. *Revue Epidemiologie et de Santé Publique*, **32**, 185–98.

Kodell, R.L. and Nelson, C.J. (1980) An illness-death model for the study of the carcinogenic process using survival/sacrifice data. *Biometrics*, **36**, 267–77.

Krall, J.M., Uthoff, V.A. and Harley, J.B. (1975) A step-up procedure for selecting variables associated with survival. *Biometrics*, **31**, 49–57.

Lachin, J.M. (1981) Introduction to sample size determination and power analysis for clinical trials. *Controlled Clinical Trials*, **2**, 93–113.

Lachin, J.M. and Foulkes, M.A. (1986) Evaluation of sample size and power for analyses of survival with allowance for nonuniform patient entry, losses to follow-up, noncompliance, and stratification. *Biometrics*, **42**, 507–19.

Lagakos, S.W. (1980) The graphical evaluation of explanatory variables in proportional hazards models. *Biometrika*, **68**, 93–8.

Lakatos, E. (1988) Sample sizes based on the log-rank statistic in complex clinical trials. *Biometrics*, **44**, 229–41.

Lakatos, E. and Lan, K.K.G. (1992) A comparison of sample size methods for the log-rank statistic. *Statistics in Medicine*, **11**, 179–91.

Lawless, J.F. (1982) *Statistical Models and Methods for Lifetime Data*, Wiley, New York.

Leathem, A.J. and Brooks, S.A. (1987) Predictive value of lectin binding on breast-cancer recurrence and survival. *The Lancet*, **I**, 1054–6.

Lee, E.T. (1992) *Statistical Methods for Survival Data Analysis*, Wiley, New York.

Lin, D.Y. and Wei, L.J. (1991) Goodness-of-fit tests for the general Cox regression model. *Statistica Sinica*, **1**, 1–17.

Lindley, D.V. and Scott, W.F. (1984) *New Cambridge Elementary Statistical Tables*, Cambridge University Press, Cambridge.

Lindsey, J.C. and Ryan, L.M. (1993) A three state multiplicative model for rodent tumourigenicity experiments. *Applied Statistics*, **42**, 283–300.

Link, C.L. (1984) Confidence intervals for the survival function using Cox's proportional hazards model with covariates. *Biometrics*, **40**, 601–10.

McCullagh, P. and Nelder, J.A. (1989) *Generalized Linear Models*, 2nd edn, Chapman and Hall, London.

McGilchrist, C.A. and Aisbett, C.W. (1991) Regression with frailty in survival analysis. *Biometrics*, **47**, 461–6.

McKnight, B. and Crowley, J. (1984) Tests for differences in tumour incidence based on animal carcinogenesis experiments. *Journal of the American Statistical Association*, **79**, 639–48.

Machin, D. and Campbell, M.J. (1987). *Statistical Tables for the Design of Clinical Trials*, Blackwell, Oxford.

Mann, N.R., Shafer, R.E. and Singpurwalla, N.D. (1974) *Methods for Statistical Analysis of Reliability and Life Data*, Wiley, New York.

Mantel, N. (1966) Evaluation of survival data and two new rank order statistics arising in its consideration. *Cancer Chemotherapy Reports*, **50**, 163–70.

Mantel, N. and Haenszel, W. (1959) Statistical aspects of the analysis of data from retrospective studies of disease. *Journal of the National Cancer Institute*, **22**, 719–48.

Meier, P. (1975) Estimation of a distribution function from incomplete observations. In *Perspectives in Probability and Statistics* (ed. J. Gani), Academic Press, London, pp. 67–87.

Miller, R.G. (1981) *Survival Analysis*, Wiley, New York.

Miller, A.J. (1990) *Subset Selection in Regression*, Chapman and Hall, London.

Montgomery, D.C. and Peck, E.A. (1982) *Introduction to Linear Regression Analysis*, Wiley, New York.

Moreau, T., O'Quigley, J. and Mesbah, M. (1985) A global goodness-of-fit statistic for the proportional hazards model. *Applied Statistics*, **34**, 212–8.

Morgan, B.J.T. (1992) *Analysis of Quantal Response Data*, Chapman & Hall, London.

Nagelkerke, N.J.D., Oosting, J. and Hart, A.A.M. (1984) A simple test for goodness of fit of Cox's proportional hazards model. *Biometrics*, **40**, 483–6.

Nair, V.N. (1984) Confidence bands for survival functions with censored data: a comparative study. *Technometrics*, **26**, 265–75.

Nelder, J.A. (1977) A reformulation of linear models (with discussion). *Journal of the Royal Statistical Society*, A, **140**, 48–77.

Nelson, W. (1982) *Applied Life Data Analysis*, Wiley, New York.

O'Quigley, J. and Pessione, F. (1989) Score tests for homogeneity of regression effect in the proportional hazards model. *Biometrics*, **45**, 135–44.

Petersen, T. (1986) Fitting parametric survival models with time-dependent covariates. *Applied Statistics*, **35**, 281–8.

Peto, R. (1972) Contribution to the discussion of a paper by D.R. Cox. *Journal of the Royal Statistical Society, B,* **34**, 205–7.

Peto, R. and Peto, J. (1972) Asymptotically efficient rank invariant procedures. *Journal of the Royal Statistical Society, A,* **135**, 185–207.

Peto, R., Pike, M.C., Armitage, P., Breslow, N.E., Cox, D.R., Howard, S.V., Mantel, N., McPherson, K., Peto, J. and Smith, P.G. (1976) Design and analysis of randomized clinical trials requiring prolonged observation of each patient. I. Introduction and design. *British Journal of Cancer,* **34**, 585–612.

Peto, R., Pike, M.C., Armitage, P., Breslow, N.E., Cox, D.R., Howard, S.V., Mantel, N., McPherson, K., Peto, J. and Smith, P.G. (1977) Design and analysis of randomized clinical trials requiring prolonged observation of each patient. II. Analysis and examples. *British Journal of Cancer,* **35**, 1–39.

Pettitt, A.N. and Bin Daud, I. (1989) Case-weighted measures of influence for proportional hazards regression. *Applied Statistics,* **38**, 51–67.

Pettitt, A.N. and Bin Daud, I. (1990) Investigating time dependence in Cox's proportional hazards model. *Applied Statistics,* **39**, 313–29.

Pocock, S.J. (1983) *Clinical Trials: A Practical Approach,* Wiley, Chichester.

Pollard, A.H., Yusuf, F. and Pollard, G.N. (1990) *Demographic Techniques,* 3rd edn, Pergamon Press, Sydney.

Prentice, R.L. and Gloeckler, L.A. (1978) Regression analysis of grouped survival data with application to breast cancer data. *Biometrics,* **34**, 57–67.

Prentice, R.L. and Kalbfleisch, J.D. (1979) Hazard rate models with covariates. *Biometrics,* **35**, 25–39.

Prentice, R.L., Kalbfleisch, J.D., Peterson, A.V., Jr., Flournoy, N.S., Farewell, V.T. and Breslow, N.E. (1978) The analysis of failure times in the presence of competing risks. *Biometrics,* **34**, 541–54.

Reid, N. and Crépeau, H. (1985) Influence functions for proportional hazards regression. *Biometrika,* **72**, 1–9.

Rubinstein, L.V., Gail, M.H. and Santner, T.J. (1981) Planning the duration of a comparative clinical trial with loss to follow-up and a period of continued observation. *Journal of Chronic Diseases,* **34**, 469–79.

Schemper, M. (1992) Cox analysis of survival data with non-proportional hazard functions. *The Statistician,* **41**, 455–65.

Schoenfeld, D.A. (1980) Chi-squared goodness of fit tests for the proportional hazards regression model. *Biometrika,* **67**, 145–53.

Schoenfeld, D.A. (1981) The asymptotic properties of comparative tests for comparing survival distributions. *Biometrika,* **68**, 316–9.

Schoenfeld, D.A. (1982) Partial residuals for the proportional hazards regression model. *Biometrika,* **69**, 239–41.

Schoenfeld, D.A. (1983) Sample-size formula for the proportional-hazards regression model. *Biometrics,* **39**, 499–503.

Schoenfeld, D.A. and Richter, J.R. (1982) Nomograms for calculating the number of patients needed for a clinical trial with survival as an endpoint. *Biometrics,* **38**, 163–70.

Sellke, T. and Siegmund, D. (1983) Sequential analysis of the proportional hazards model. *Biometrika,* **70**, 315–26.

Simon, R. (1986) Confidence intervals for reporting results of clinical trials. *Annals of Internal Medicine,* **105**, 429–35.

Simon, R. and Lee, Y.J. (1982) Nonparametric confidence limits for survival probabilities and median survival time. *Cancer Treatment Reports*, **66**, 37–42.

Slud, E.V., Byar, D.P. and Green, S.B. (1984) A comparison of reflected versus test-based confidence intervals for the median survival time based on censored data. *Biometrics*, **40**, 587–600.

Stablein, D.M., Carter, W.H., Jr. and Novak, J.W. (1981) Analysis of survival data with nonproportional hazard functions. *Controlled Clinical Trials*, **2**, 148–59.

Storer, B.E. and Crowley, J. (1985) A diagnostic for Cox regression and general conditional likelihoods. *Journal of the American Statistical Association*, **80**, 139–47.

Therneau, T.M. (1986) The COXREGR Procedure. In *SAS SUGI Supplemental Library User's Guide*, version 5 edn, SAS Institute Inc., Cary, North Carolina.

Therneau, T.M., Grambsch, P.M. and Fleming, T.R. (1990) Martingale-based residuals for survival models. *Biometrika*, **77**, 147–60.

Thisted, R.A. (1988) *Elements of Statistical Computing*, Chapman and Hall, London.

Thompson, R. (1981) Survival data and GLIM. Letter to the editor of *Applied Statistics*, **30**, 310.

Tibshirani, R. (1982) A plain man's guide to the proportional hazards model. *Clinical and Investigative Medicine*, **5**, 63–8.

Wei, L.J. (1992) The accelerated failure time model: a useful alternative to the Cox regression model in survival analysis. *Statistics in Medicine*, **11**, 1871–9.

Weiden, P.L., Sullivan, K.M., Flournoy, N., Storb, R., Thomas, E.D. and the Seattle Marrow Transplant Team (1981) Antileukemic effect of chronic graft-versus-host disease. Contribution to improved survival after allogenic marrow transplantation. *New England Journal of Medicine*, **304**, 1529–33.

Weissfeld, L.A. (1990) Influence diagnostics for the proportional hazards model. *Statistics and Probability Letters*, **10**, 411–7.

Weissfeld, L.A. and Schneider, H. (1990) Influence diagnostics for the Weibull model fit to censored data. *Statistics and Probability Letters*, **9**, 67–73.

Whitehead, J.R. (1980) Fitting Cox's regression model to survival data using GLIM. *Applied Statistics*, **29**, 268–75.

Whitehead, J.R. (1983) Fitting survival models using GLIM. In *Computer Science and Statistics: The Interface* (ed. J. Gentle), North Holland, Amsterdam.

Whitehead, J.R. (1989) The analysis of relapse clinical trials, with application to a comparison of two ulcer treatments. *Statistics in Medicine*, **8**, 1439–54.

Whitehead, J.R. (1992) *The Design and Analysis of Sequential Clinical Trials*, 2nd edn, Ellis Horwood, Chichester.

WHO Special Programme of Research, Development and Research Training in Human Reproduction (1987) Vaginal bleeding patterns – the problem and an example data set. *Applied Stochastic Models and Data Analysis*, **3**, 27–35.

Index of examples

For each illustrative example used in this book, the number of the example is given, followed by the page number in parentheses.

Index